Environmental Technologies to Treat Rare Earth Elements Pollution: Principles and Engineering

Environmental Technologies to Treat Rare Earth Elements Pollution: Principles and Engineering

Edited by

Arindam Sinharoy and Piet N. L. Lens

IWA PUBLISHING

Published by

IWA Publishing
Unit 104–105, Export Building
1 Clove Crescent
London E14 2BA, UK
Telephone: +44 (0)20 7654 5500
Fax: +44 (0)20 7654 5555
Email: publications@iwap.co.uk
Web: www.iwapublishing.com

First published 2022

British Library Cataloguing in Publication Data
A CIP catalogue record for this book is available from the British Library

ISBN: 9781789062229 (paperback)
ISBN: 9781789062236 (eBook)
ISBN: 9781789062243 (ePub)

Cover photographs

Cover photographs – **from right to left:** Pollution by rare earth elements and environmental technologies to remove and recover them from waste streams. **Top right:** Use of gadolinium to measure diffusion and flow in methanogenic granules and biofilms using nuclear magnetic imaging. Photograph of data published in Bartacek *et al.* (2016) *Frontiers in Environmental Science* 4, 13. **Central:** Rare earth elements containing pit at the Osamu Utsumi Mine, located in Caldas (Minas Gerais, Brazil). Photograph provided from personal collection of Elis W. Nogueira, University of São Paulo (São Paulo, Brazil). **Bottom left:** Bioleaching of rare earth elements from discarded printed circuit boards and e-waste shredding dust. Photograph of data published in Marra *et al.* (2018) *Journal of Environmental Management* 210:180–190. **Bottom right:** Bioreactor set-up containing a down-flow fixed-structured bed reactor operated under sulfidogenic conditions for removal of rare earth elements from acid mine drainage. Photograph of data published in Nogueira *et al.* (2019) *Water Science and Technology* 80(8), 1485–1493.

Contents

Part III Recovery of Rare Earth Elements from Waste Resources

Part IV Technologies to Recover Rare Earth Elements

Preface

Rare earth elements (REE) are a group of seventeen elements having comparable chemical properties and tend to occur within the same mineral ore deposits. REE have applications in various modern technologies ranging from semiconductors to mobile phones, magnets, televisions and even satellites. Many countries have categorized REE as critical raw materials due to their strategic importance in economies and high-risks associated with their supply chain. Developing more sustainable practices for efficient recycling of end-of-use products containing REE is an important approach to ensure a safe supply of REE. To achieve this, a lot of research activities – including many novel REE extraction and recovery technologies - have been developed in the past decade.

The discovery and historical development in REE research are given in the first chapter of this book volume. Then, geological deposits, exploration, detection and industrial applications of REE are overviewed. Their exposure to the environment due to industrial release (such as mining) or unsafe disposal of end-of-use products can potentially cause adverse impact. This book not only focuses on the geochemical cycling of rare earths, their mining and hydrometallurgical aspects of REE extraction, but also covers biological and other green processing methods of REE. Due to the growing interest in biological methods, increased importance to the different biological methods to extract and recover REE is given in this book, such as those involving microbes (biosorption, bioaccumulation, bioreduction, biomineralization and bioleaching) and plants (rhizofiltration, phytoextraction, phytotransformation, phytostimulation and phytostabilization). In addition, the use of sulfidogenic systems for REE recovery from mine drainage is presented as well. The

doi: 10.2166/9781789062236_xiii

chapters of this book show that recovery of REE from secondary sources, such as red mud, electronic wastes, brine solutions and coal fly ash, is becoming a reality. Equally emerging are nanoparticles of rare earth elements due to their wide spread applications in the medical and energy sectors.

Each contributed chapter is presented on a stand-alone basis, so that the reader will find it helpful to consider only the theme of each chapter. There are nevertheless many connections between what may at first seem to be quite different topics. As in all the books of the *Integrated Environmental Technology* series, it was one of our purposes to draw out and emphasize these interdisciplinary linkages. For this reason, a comprehensive index is included to facilitate cross-referencing. We hope that the work described in this book will inspire those working in the field and will encourage those who are beginning to investigate this field.

We wish to thank all contributors to this book for their valuable contributions by sharing their expertise in the form of the various chapters included in the book. We also thank all past and present co-workers as well as all collaborators who joined in unravelling different areas of REE and their application in environmental technology as described in this book, especially those at Wageningen University, UNESCO-IHE and National University Ireland Galway. We would also like to thank all the reviewers who put a lot of effort in improving the quality of this book. In addition, the national and international granting agencies who supported our work on various aspects of metal removal and recovery over the years are gratefully acknowledged, in particular the Science Foundation Ireland (SFI), who financially supported the open access publication of this book volume through the SFI Research Professorship Programme *Innovative Energy Technologies for Biofuels, Bioenergy and a Sustainable Irish Bioeconomy* (IETSBIO^3; grant number 15/RP/2763). We are also grateful to the editorial team of IWA Publishing, in particular Niall Cunniffe and Mark Hammond, for their help and editorial support in realizing this book.

Arindam Sinharoy
Piet N. L. Lens
National University of Ireland Galway, Ireland

List of Contributors

Agnihotri Anupam
Jawaharlal Nehru Aluminium Research
Development and Design Centre
Amravati Road, Wadi
Nagpur – 440023
Maharashtra
India

Avishek Kirti
Deptartment of Civil and Environmental
Engineering
Birla Institute of Technology
Ranchi
Jharkhand – 835215
India

Balaram Vysetti
CSIR - National Geophysical Research
Institute
Hyderabad – 500 007
Telangana
India

Bruckard Warren
CSIRO
Normanby Rd.
Clayton
Victoria 3168
Australia

Budnyk Andriy P.
Southern Scientific Center of the Russian
Academy of Sciences
Chekhova Street 41
344006 Rostov-on-Don
Russia

Brucha Gunther
Universidade Federal de Alfenas
Rodovia José Aurélio Vilela
11999 (BR 267 Km 533) Cidade
Universitária Poços de Caldas
Minas Gerais
Brazil

Castillo Julio
Department of Microbial, Biochemical and Food Biotechnology,
University of the Free State
9301 Bloemfontein
Republic of South Africa

Cebekhulu Sanele
Department of Microbial, Biochemical and Food Biotechnology
University of the Free State
9301 Bloemfontein
Republic of South Africa

Damianovic Márcia H. R. Z.
Biological Processes Laboratory (LPB)
São Carlos School of Engineering (EESC)
University of São Paulo (USP)
Av. João Dagnone
1100 Santa Angelina
São Carlos
São Paulo 13563-120
Brazil

Gómez-Arias Alba
Department of Microbial, Biochemical and Food Biotechnology
University of the Free State
9301 Bloemfontein
Republic of South Africa

Haque Nawshad
CSIRO
Normanby Rd
Clayton
Victoria 3168
Australia

Hazra Moushumi
Department of Hydrology
Indian Institute of Technology Roorkee
Uttarakhand – 247667
India

Holanda Roseanne B.
SENAI Innovation Institute for Mineral Technologies,
Av. Com Brás de Aguiar
548 Belém, PA
66035-405
Brazil

Hughes Anthony E.
CSIRO
Normanby Rd.
Clayton
Victoria 3168
Australia

Jowitt Simon M.
Department of Geoscience
University of Nevada Las Vegas
4505 Maryland Parkway
Las Vegas
Nevada 89154-4010
USA

Lastovina Tatiana A.
Southern Scientific Center of the Russian Academy of Sciences
Chekhova Street 41
344006 Rostov-on-Don
Russia

Lens Piet N. L.
Department of Microbiology
National University of Ireland Galway
University Road
Galway H91 TK33
Ireland

Lhamo Pema
Department of Biotechnology
Karunya Institute of Technology and Sciences
Coimbatore – 641114
Tamil Nadu
India

Mahanty Biswanath
Department of Biotechnology
Karunya Institute of Technology and Sciences
Coimbatore – 641114
Tamil Nadu
India

Maleke Maleke
Department of Life Sciences
Faculty of Health and Environmental Science
Central University of Technology
9301 Bloemfontein
Republic of South Africa

Manikandan N. Arul
Department of Microbiology
National University of Ireland Galway
University Road
Galway H91 TK33
Ireland

Möller Peter
Helmholtz Centre Potsdam
German Research center for Geosciences (GFZ)
Section 3.4
Potsdam
Germany

Nogueira Elis W.
Biological Processes Laboratory (LPB)
São Carlos School of Engineering (EESC)
University of São Paulo (USP)
Av. João Dagnone
1100 Santa Angelina
São Carlos
São Paulo 13563-120
Brazil

Podlesnaia Ekaterina O.
Department of Nanobiophotonics
Leibniz Institute of Photonic Technology
Albert-Einstein-Straße 9
07745 Jena
Germany

Pramanik Biplob K.
School of Engineering
RMIT University
Melbourne VIC 3001
Australia

Sarker Shuronjit K.
School of Engineering
RMIT University
Melbourne VIC 3001
Australia

Singh Upendra
Jawaharlal Nehru Aluminium Research Development and Design Centre
Amravati Road, Wadi
Nagpur – 440023
Maharashtra
India

Sinharoy Arindam
Department of Microbiology
National University of Ireland Galway
University Road
Galway H91 TK33
Ireland

Sultana Shanjida
Institute of Mining
Mineralogy and Metallurgy
Bangladesh Council of Scientific and Industrial Research (BCSIR)
Joypurhat 5900
Bangladesh

Thawrani Sonali A.
Jawaharlal Nehru Aluminium Research
Development and Design Centre
Amravati Road, Wadi
Nagpur – 440023
Maharashtra
India

Unuofin John
Department of Microbial, Biochemical and
Food Biotechnology
University of the Free State
9301 Bloemfontein
Republic of South Africa

Part I

Environmental Technologies to Treat Rare Earth Pollution

Chapter 1

Environmental technologies to treat pollution by rare earth elements

Arindam Sinharoy and Piet N. L. Lens

1.1 INTRODUCTION

Rare earth elements (REEs) are a group of 17 transition metals that includes scandium, yttrium and 15 of the lanthanide series elements (Balaram, 2019). They are essential raw materials for emerging high-technology applications such as computers, mobile phones, rechargeable batteries, super magnets, LED lights, magnetic resonance imaging, and for renewable energy technologies such as solar cells and wind turbines (Chapter 4). These metals are being consumed for a variety of applications at an unprecedented rate and their demand is at an all-time high (Klossek *et al.*, 2016). Due to their importance in modern technologies and society in general, studies focused on REEs have garnered great interest among researchers in the past few decades. Research on rare earth elements has been increasing in recent times and many research and review papers have been published on this topic (Figures 1.1 and 1.2). This book volume synthesizes the available knowledge on relevant topics in the area of pollution, removal and recovery of rare earth elements (Figure 1.3).

1.2 BIOGEOCHEMICAL CYCLES OF RARE EARTH ELEMENTS

Part II of this book focuses on the biogeochemical cycle of REEs to understand their sources and role. Chapter 2 describes the discovery and historical context of REE

doi: 10.2166/9781789062236_0003

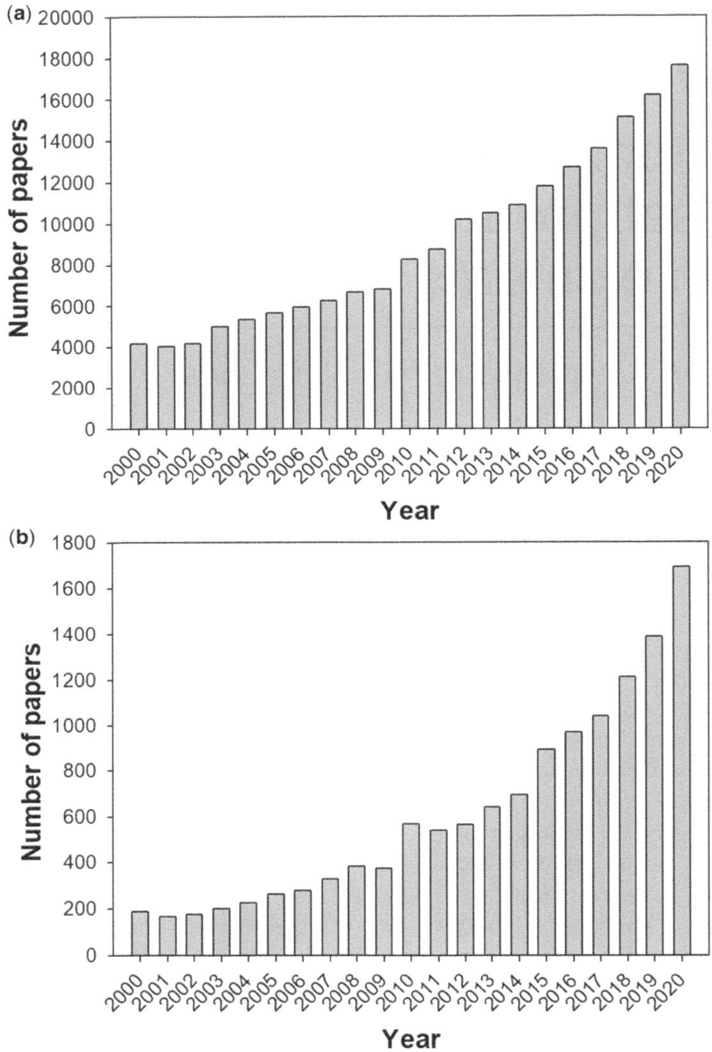

Figure 1.1. Increase in number of publication on (a) rare earth elements and (b) their extraction over the last two decades (*Source*: google scholar; data obtained by searching using keywords (a) 'rare earth elements' and (b) 'rare earth elements, extraction').

chemistry. Their occurrence and distribution on the earth's surface is described in Chapter 3. In Chapter 4, details on sources and different industrial applications of REEs are overviewed.

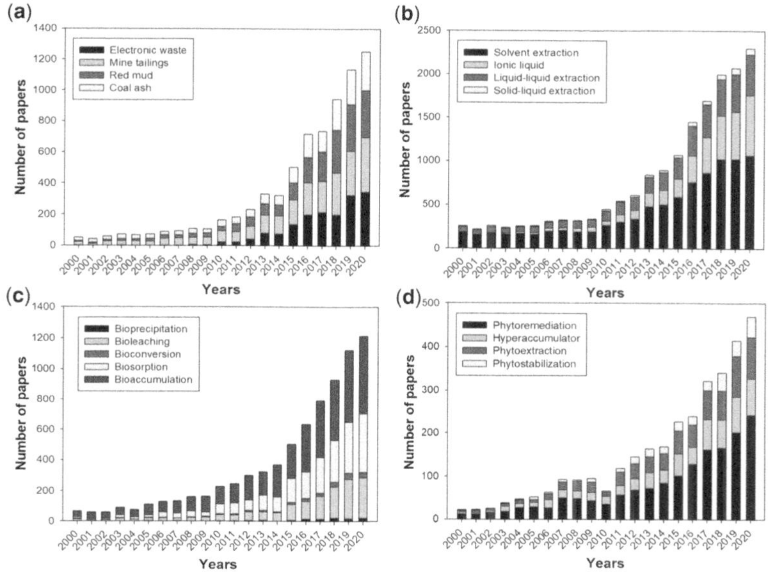

Figure 1.2. Trends in research in rare earth elements recovery from waste and secondary resources using different methods (a) source of waste materials, (b) chemical extraction methods as well as (c) microbial and (d) plant based recovery methods (*Source*: google scholar; keywords used along with 'rare earth elements' for obtaining the data are mentioned in the respective figure legends).

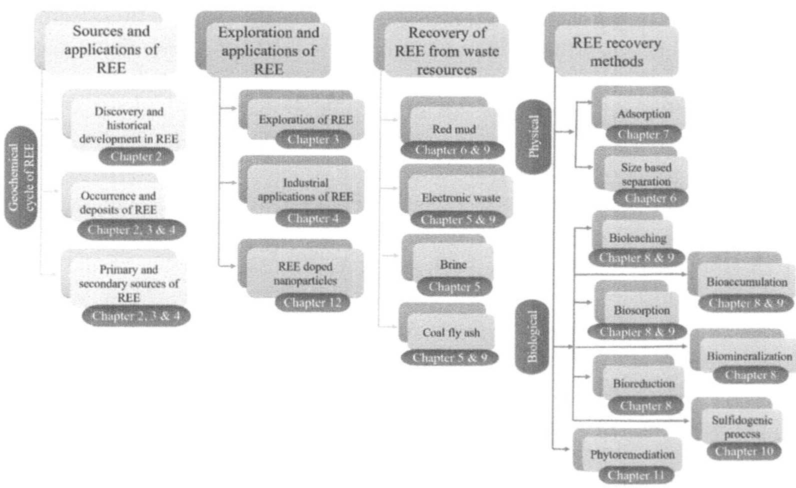

Figure 1.3. Schematic representation of the topics covered in this book.

Unlike their name suggests, rare earth elements are not rare in the earth's crust and can be found all over the world as constituents of many mineral deposits (Balaram, 2019). For example, cerium is the most abundantly available REE and the 25th most commonly found element in the earth's crust. REEs are present in different rock formations, soils and deep-sea beds, and also in the biosphere, including in plants and microorganisms (Ramos *et al.*, 2016). Primarily these elements are deposited due to igneous and hydrothermal processes on earth, and their secondary deposits are formed due to the earth's natural weathering. REEs are also present in several waste materials, that is mining waste such as red mud, coal derived fly ash, rocks associated with mining and stone quarries (Chapter 4). Natural (air, water and microbial activity) and anthropogenic activities (mining and industrial processing) further cause REEs to be released into the environment, where they may come into contact with constituents of the ecosystems and can cause hazardous impacts (Chapter 3).

Although it was assumed earlier that REEs are biologically inert and do not play any active role inside living cells, recent studies have shown contrary findings. For instance, lanthanides have been shown as co-factors of enzymes in a particular group of alcohol dehydrogenase enzymes (Huang *et al.*, 2019). REEs have the ability to change the behaviour of certain biomolecules and by doing so they can alter the biological role of these biomolecules in living organisms.

1.3 RECOVERY OF RARE EARTH ELEMENTS FROM WASTE RESOURCES

Due to the widespread applications of REEs in modern technologies, their demand is ever growing. REEs are categorized as critical raw materials by the European Union, which implies that this group of compounds is essential for economic development but there is risk associated with their supply. More than 60% of the commercially used REEs is supplied by a single country, China, imposing a higher risk for their continuous supply. As a result, many countries including the European Union are encouraging recovery of REEs from alternative sources.

Different secondary sources have been reported in the literature as potential sources for REE recovery (Figure 1.2a). Part III of this book deals with recovery of REEs from secondary sources such as brine, fly ash (Chapter 5) and red mud (Chapter 6). Most of the methods used for REE recovery from such waste resources are conventional hydrometallurgical methods involving chemical leaching followed by solvent extraction. However, size separation of REE-containing particles (particularly red mud) using physical force prior to their extraction can improve the process economics (Chapter 6). In addition, biological methods are also applied for REE recovery from electronic, mining and other waste substrates, these are based on microorganisms (Chapters 8, 9 and 10) or plants (Chapter 11).

1.4 TECHNOLOGIES TO RECOVER RARE EARTH ELEMENTS

REEs can be recovered using different physico-chemical and biological methods (Part IV; Opare *et al.*, 2021). Different methods have advantages and drawbacks that need to be studied in advance and depending on these a suitable method (or combination of methods) can be applied to ensure the economic viability of a process. Among the physical methods, size reduction, physical separation (Chapter 6) and adsorption (Chapter 7) are frequently used for recovery of REEs. These physical methods can also be used in combination with other chemical and biological methods as pre- or post-treatment to recover REEs.

Chemical processes include leaching using chemical reagents, such as acids and caustic soda, as well as solvent extraction, such as liquid-liquid extraction and ionic liquid extraction (Hidayah & Abidin, 2017). Chapters 5 and 6 address the use of these chemical methods for REE recovery from selected materials.

Biological methods involving bacteria, fungi, algae and plants are another important technique for REE recovery from environmental samples and waste resources. Biological methods have gained significant interest because these methods are low cost, sustainable and environmentally friendly. Microbial processes for REEs include bioleaching, biosorption, bioaccumulation, biomineralization and bioprecipitation (Figure 1.2c; Chapters 8 and 9). In bioleaching, acid produced by microorganisms solubilizes REEs and helps in their recovery (Rasoulnia *et al.*, 2021). Biosorption is the fastest process among the biological methods, and interaction between REEs and active groups on the surface of microbes results in their retention within the biomass (Andres *et al.*, 2003). The REEs can be transported from the microbial surface to the intracellular space, where they can be either detoxified, reduced or stored within vacuoles. This bioaccumulation of REEs occurs to a smaller extent in comparison to the other methods. Bioprecipitation can occur by either direct microbial activity (enzymatic reduction) or indirectly due to the production of reactants (which precipitate REE) and/or change in the environmental condition (such as pH or redox). In most cases involving microbial treatment of REEs, a combination of these methods contributes to the overall removal and recovery. Chapter 9 further elaborates on the use of biological methods for REE recovery from waste sources such as red mud, fly ash and electronic wastes.

Another aspect of biological REE removal and recovery is the use of sulfidic precipitation. Use of sulfide produced by biological sulfate reduction for precipitation of heavy metals is a common biological metal removal process (Lens, 2020; Sinharoy *et al.*, 2020). However, there are only a few studies on REE removal using sulfidogenic processes and most of them concluded that the REE removal in such systems is mainly due to increased pH and precipitation with phosphate, and not necessarily linked with sulfide precipitation (Chapter 10).

Plants have been routinely used for treatment of contaminated water and soil. Different methods involving plants such as rhizofiltration, phytoextraction, phytotransformation, phytostimulation and phytostabilization are used to remove a wide range of pollutants from contaminated environments (Wei *et al.*, 2021). In the case of REE removal, primarily rhizofiltration (adsorption of REEs in the plant root system), phytoextraction (uptake of REEs inside plants) and phytostabilization (immobilization and/or precipitation of REEs in the root system and surrounding soil) are utilized by plants (Chapter 11). There are hyperaccumulator plant species that have shown high tolerance towards REEs and are capable of accumulating high concentrations of REEs. Such plants are suitable for treating REE pollution and have been applied to treat REE-containing mine impacted soil with success (Chapter 11). This shows the suitability of plant based systems for onsite (*in situ*) treatment. In this regard, the role of hyperaccumulator plants and their application in wetlands for recovery of REEs is discussed in Chapter 11.

1.5 APPLICATION OF RARE EARTH ELEMENTS AS NANOPARTICLES

Apart from conventional applications of REEs in modern technology and appliances, in recent times they have also found usage in nanotechnology (Gao *et al.*, 2019; Zhu *et al.*, 2019). REEs are directly incorporated into nanoparticles (NPs) by various doping techniques available, forming oxides/hydroxides, alloys and intermetallic compounds (Chapter 12). Doping of REEs to conventional nanomaterials such as iron oxides or titanium oxides improves their properties manifold and can even change or impart new properties to the nanomaterials. REE based nanoparticles are used in photonics, sensing, quantum memory and signal processing, catalysis, diagnostics, drug delivery and other therapeutics (Chapter 12). In addition, these nanoparticles are also used in different electronic applications such as bioelectrochemical cells, supercapacitors, high conductivity materials and as bifunctional catalysts in batteries (Chapter 12). Due to such a wide range of commercial applications and renewed interest in this field of research, a chapter comprising various applications of REE based nanoparticles is provided in Part V of this book.

REFERENCES

Andres Y., Texier A. C. and Le Cloirec P. (2003). Rare earth elements removal by microbial biosorption: a review. *Environmental Technology*, **24**(11), 1367–1375.

Balaram V. (2019). Rare earth elements: a review of applications, occurrence, exploration, analysis, recycling, and environmental impact. *Geoscience Frontiers*, **10**(4), 1285–1303.

Gao W., Wen D., Ho J. C. and Qu Y. (2019). Incorporation of rare earth elements with transition metal–based materials for electrocatalysis: a review for recent progress. *Materials Today Chemistry*, **12**, 266–281.

Hidayah N. N. and Abidin S. Z. (2017). The evolution of mineral processing in extraction of rare earth elements using solid-liquid extraction over liquid-liquid extraction: a review. *Minerals Engineering*, **112**, 103–113.

Huang J., Yu Z., Groom J., Cheng J. F., Tarver A., Yoshikuni Y. and Chistoserdova L. (2019). Rare earth element alcohol dehydrogenases widely occur among globally distributed, numerically abundant and environmentally important microbes. *The ISME Journal*, **13**(8), 2005–2017.

Klossek P., Kullik J. and van den Boogaart K. G. (2016). A systemic approach to the problems of the rare earth market. *Resources Policy*, **50**, 131–140.

Lens P. N. L. (2020). Environmental Technologies to Treat Sulfur Pollution: Principles and Engineering, 2nd edn. IWA Publishing, London, United Kingdom. doi: 10.2166/9781789060966

Opare E. O., Struhs E. and Mirkouei A. (2021). A comparative state-of-technology review and future directions for rare earth element separation. *Renewable and Sustainable Energy Reviews*, **143**, 110917. doi: 10.1016/j.rser.2021.110917

Ramos S. J., Dinali G. S., Oliveira C., Martins G. C., Moreira C. G., Siqueira J. O. and Guilherme L. R. (2016). Rare earth elements in the soil environment. *Current Pollution Reports*, **2**(1), 28–50.

Rasoulnia P., Barthen R. and Lakaniemi A. M. (2021). A critical review of bioleaching of rare earth elements: the mechanisms and effect of process parameters. *Critical Reviews in Environmental Science and Technology*, **51**(4), 378–427.

Sinharoy A., Pakshirajan K. and Lens P. N. L. (2020). Biological sulfate reduction using gaseous substrates to treat acid mine drainage. *Current Pollution Reports*, **6**, 328–344.

Wei Z., Van Le Q., Peng W., Yang Y., Yang H., Gu H., Lam S. S. and Sonne C. (2021). A review on phytoremediation of contaminants in air, water and soil. *Journal of Hazardous Materials*, **403**, 123658. doi: 10.1016/j.jhazmat.2020.123658

Zhu X., Zhang J., Liu J. and Zhang Y. (2019). Recent progress of rare-earth doped upconversion nanoparticles: synthesis, optimization, and applications. *Advanced Science*, **6**(22), 1901358. doi: 10.1002/advs.201901358

Part II

Biogeochemical Cycles of Rare Earth Elements

Chapter 2

Discovery and occurrence of lanthanoids and yttrium

Peter Möller

2.1 NAMING AND STRUCTURE OF LANTHANOIDS AND YTTRIUM

2.1.1 Nomenclature

Notwithstanding their generic names, rare earth elements (REE) are neither rare nor *earths*. Some of them are more abundant in the earth's crust than gold, silver, mercury or tungsten. REE are ubiquitously present in low concentrations in virtually all minerals. Being prepared from rare and uncommon minerals, they were considered as *earths*, an old expression for oxides in use at the time of their discovery.

According to the regulations of the International Union of Pure and Applied Chemistry (IUPAC, 2005) the series of lanthanum (La) to lutetium (Lu) is collectively named 'lanthanoids', and the term 'rare earth elements' comprises scandium (Sc), yttrium (Y) and the lanthanoids. Although lanthanum cannot logically be a lanthanoid, IUPAC accepts its inclusion based on common usage. In geochemical practice the names 'rare earth elements' and 'lanthanides' are interchangeable. The term 'lanthanides' or REE comprises the series La-Lu which are commonly split up into light REE (LREE; La-Nd), medium REE (MREE; Sm-Dy) and heavy REE (HREE; Ho-Lu) (Rollinson, 1993). In the periodic table of elements Sc, Y and La are placed in the third main group and Ce-Lu in a subgroup with increasing 4*f* electrons.

doi: 10.2166/9781789062236_0013

2.1.2 Structure

Because Y^{3+} has the same charge and about the same ionic radius as Ho^{3+}, both species are not fractionated during crystallization of minerals from melts. Therefore, it was suggested to include Y in the lanthanoid suite, which was either abbreviated as YREE or REY (Bau & Dulski, 1992). In aqueous systems, however, Y^{3+} and Ho^{3+} are fractionated because Y^{3+} behaves more hydrophilically than Ho^{3+} (Nozaki *et al.*, 1997). Their most common mineral is xenotime (Table 2.1).

Y, La (no 4*f* electrons) and Ce-Lu (increasing number of 4*f* electrons) resemble each other in both elemental state and compounds, their co-occurrence in nature, and the difficulty in separating one element from another. The small but continuous differences in thermodynamic behaviour of their tervalent ions are outstanding. The coherence makes REY an invaluable tracer in geochemical, planetary and biochemical processes. Although coherent and 'inseparable', REY do fractionate

Table 2.1 Composition of REY minerals (Burt, 1989).

Aeschynite	$(REY,Ca,Fe,Th)(TiNb)_2(O,OH)_6$
Allanite	$Ca(REY,Ca)Al(Al,Fe^{3+})(Fe^{2+},Al)(SiO_4)_3OH$
Ancylite	$(LREE,Sr,Ca)(CO_3)(OH,H_2O)$
Bastnaesite	$(REY)(CO_3)F$
Beiyenite	resembles bastnaesite
Brannerite	$(U,Ca,REY)/Ti,Fe)_2O_6$
Burbankite	$(Na,Ca)_3(Sr,Ba,Ca,Ce)_3(CO_3)_5$
Cerite	$(Ce_9,Ca)(Fe^{3+},Mg)\ (SiO_4)_6[SiO_3(OH)](OH)_3$
Eudialyte	$Na_4(Ca(LREE))(Fe,Mn)_2ZrSi_6O_{17}(OH,Cl)_2$
Euxenite	$(Y,Ca,Ce,U,Th)(Nb,Ta,Ti)_2O_6$
Fergusonite	$(REY)(Nb,Ti,Ta)(O,OH)_4$
Florencite	$(LREE)Al_3(PO_4)_2(OH)_6$
Fluorobastnaesite	resembles bastnaesite
Fluoroapatite	$(Ca,REY,Na)_5(PO_4)_3(F,OH)$
Gadolinite	$(REY)_2Be_2Si_2O_8(O,OH)_2$
Loparite	$(Na,REE)(Ti,Nb)O_3$
Monazite	$(REY)PO_4$
Parasite	$Ca(REE)_2(CO_3)_3F_2$
Rinkite	$Na_2Ca_4CeTi(Si_2O_7)_2OF_3$
Sahamalite	$(Mg,Fe)Ce_2(CO_3)_4$
Samarskite	$(Y,Ce,U,Fe^{2+},Fe^{3+})_3(Nb,Ta,Ti)O_{16}$
Synchesite	$(Ca,REY)(CO_3)_2F$
Xenotime	$(Y,HREE)PO_4$

Table 2.2 Electronic configuration of atomic and tervalent REY.

Compounds	Full Name	Atomic	3 + Ions
Y	Yttrium	$[Kr]4d^1 5s^2$	[Kr]
La	Lanthanum	$[Xe]4f^0 5d^1 6ds^2$	$[Xe]4f^0$
Ce	Cerium	$[Xe]4f^1 5d^1 6ds^2$	$[Xe]4f^1$
Pr	Praseodymium	$[Xe]4f^3 6s^2$	$[Xe]4f^2$
Nd	Neodymium	$[Xe]4f^4 6s^2$	$[Xe]4f^3$
Pm	Promethium	$[Xe]4f^5 6s^2$	$[Xe]4f^4$
Sm	Samarium	$[Xe]4f^6 6s^2$	$[Xe]4f^5$
Eu	Europium	$[Xe]4f^7 6s^2$	$[Xe]4f^6$
Gd	Gadolinium	$[Xe]4f^7 5d^1 6s^2$	$[Xe]4f^7$
Tb	Terbium	$[Xe]4f^9 6s^2$	$[Xe]4f^8$
Dy	Dysprosium	$[Xe]4f^{10} 6s^2$	$[Xe]4f^9$
Ho	Holmium	$[Xe]4f^{11} 6s^2$	$[Xe]4f^{10}$
Er	Erbium	$[Xe]4f^{12} 6s^2$	$[Xe]4f^{11}$
Tm	Thulium	$[Xe]4f^{13} 6s^2$	$[Xe]4f^{12}$
Yb	Ytterbium	$[Xe]4f^{14} 6s^2$	$[Xe]4f^{13}$
Lu	Lutetium	$[Xe]4f^{14} 5d^1 6s^2$	$[Xe]4f^{14}$

in many processes and thereby provide insight into chemical and physical processes. REY fractionation depends on their environmental conditions during mineral formation from either melts or aqueous systems. Their similar chemical behaviour in reactions is the result of their uniform electron configurations in the tervalent state (Table 2.2). Their differences are given by the number of 4*f* electrons, which vary between $[Xe]4f^0$ and $[Xe]4f^{14}$. Caused by filling the 4*f* levels, the number of protons increase and thereby the attraction of electrons causes the contraction of lanthanides (Goldschmidt, 1925; Shannon, 1976). $HREE^{3+}$ show ionic radii comparable to those of Y^{3+} (Figure 2.1). Except for Ce^{4+} and Eu^{2+}, the ionic status of REE and Y is tervalency under natural conditions. Any deviating ionic charge is only obtained under either strongly oxidizing or reducing conditions. In the atomic state the electron configuration of La, Ce, Gd and Lu is the [Xe] $4f^n 5d^1 6s^2$ configuration, whereas Y shows $[Kr]4d^1 5s^2$. The remaining REE show $[Xe]4f^n 6s^2$ (Gupta, 1993).

2.1.3 Other fancy names of REE

With the exception of Pm, the two minerals cerite and gadolinite were essentially the source of both confused and complex discoveries of rare earth elements. In the beginning the possible number of rare earth elements was unknown. This kept the chemists busy for decades and more elements were announced than could be

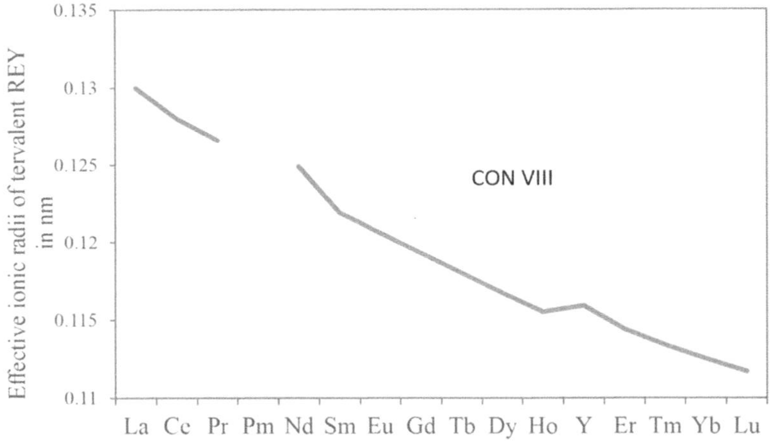

Figure 2.1 Ionic radii of rare earths and yttrium (Shannon, 1976). The crystal radii of tervalent REY decline from La to Lu (coordination number VIII).

fitted into the Periodic Table. The physicists Niels Bohr and Henry Gwyn Jeffreys Moseley came up with the number of only 15 elements in the suite of lanthanoids with 0 to 14 4*f* electrons. This number was derived from the structure of the electron shells of atoms with increasing number of protons which became the base of the Periodic Table. Spencer (1919) summarized numerous fancy names such as *jonium* in allanite (Thomsen, 1811), *austrium* in orthite/allanite (Linnemann, 1986), *russium* in monzonite (Chroustschoff, 1889), *monium*, renamed to *victorium*, being obtained by decomposition of $Y(NO_3)_3$ (Crookes, 1899), *euxenium* in euxenite (Hofmann & Prandtl, 1901) and many more. All these fanciful names soon faded because of a lack of chemical evidence (Chakhmouradian & Wall, 2012; Marshal & Marshal, 2016).

2.2 HISTORY OF Y AND REE DISCOVERY

2.2.1 Discovery of REE

In the last 90 years numerous compilations in the discovery of yttrium and lanthanides were published (Emsley, 2001; Gschneidner, 1966; Moeller, 1973; Weeks, 1932; numerous contributions on the internet under keywords such as 'lanthanides', 'lanthanoids' or 'periodic table'). Remarkable is the report of Mary Elvira Weeks (1932) because it presents some insight into the life of scientists at that time through reprints of numerous passages from letters and notes circulating between those scientists involved in the discovery of REE.

The discovery of rare earth elements started in 1751, when the chemist and assayer Axel Fredrik Cronstedt found a piece of reddish-brown mineral of Ce silicate at a mine in the Swedish community of Skinnskatteberg near Bastnäs in Vestmanland. He assumed that this was a tungsten mineral because of its high density and named it *tungstein*. Wilhelm von Hisinger sent an aliquot of this *tungstein* to the chemist Carl Wilhelm Scheele who did not find tungsten in this mineral and renamed the mineral *false tungstein*. This sample was later carefully analysed by Wilhelm von Hisinger and Jöns Jacob Berzelius in Sweden proving that it contained a previously unknown element. Its oxide was named *ceria*. The mineral was named cerite.

In 1787, the lieutenant Carl Axel Arrhenius found a piece of black mineral in a feldspar quarry in the village of Ytterby, near Stockholm (Sweden). This mineral was named *ytterite* after the village of Ytterby by C.A. Arrhenius, but was later renamed gadolinite by Anders Gustav Ekeberg in honour of Prof. Johan Gadolin at the University of Åbo.

The chemistry of *lanthanides* began with the discovery of *ceria* and *yttria*, both composite *earths*, of either dominant Ce or Y oxides, respectively. Based on either *ceria* or *yttria*, the discovery of yttrium and 13 out of 14 rare earth elements La-Lu (except Pm) occurred (Figure 2.2). The history of the discovery of the individual rare earth minerals is given below. The ordering follows their increasing atomic numbers.

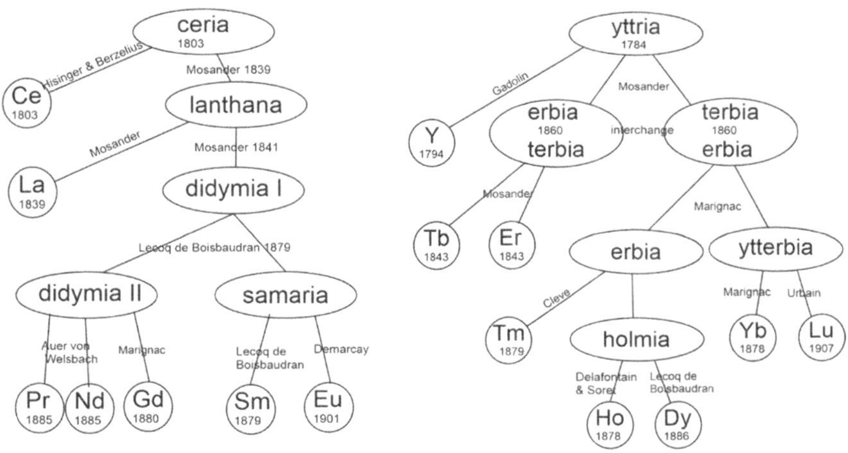

Figure 2.2 Historical outlines showing the ultimate simplifications of *ceria* and *yttria* as sources for REY discovery. Chemical elements and the year of their discovery are shown in circles. The years following names indicate the years of separation. The earths from which the elements were discovered are shown in ellipses. Berlin (1860) recorded the rose-coloured compound as *terbia* in contrast to Mosander (1843) for whom it was *erbia* (interchange) (after Moeller (1973) with corrections).

2.2.2 Yttrium, Y, 1794

Yttria was isolated from a piece of black mineral found by Karl Axel Arrhenius at a quarry near the village of Ytterby, now part of the Stockholm archipelago. Assuming it to be a tungsten ore, he handed this piece of rock to Prof. Gadolin at the University of Åbo (Turku, Finland). In 1794 he announced that the sample contained a new element. The isolated material was named *yttria* after the locality of Ytterby. This *yttria,* however, contained more than one further metal oxide. Carl Gustav Mosander (1843) analysed another sample of *yttria* and found three oxides: *terbia, erbia and ytterbia*. All these names are related to the village of Ytterby.

Impure metallic yttrium was first produced by the German chemist Friedrich Wöhler in 1828 by heating yttria with potassium.

2.2.3 Lanthanum, La, 1839

The chemist Carl Gustav Mosander converted $Ce(NO_3)_3$ into Ce-oxide and stirred the obtained oxide in diluted HNO_3, in which cerium oxide did not dissolve. Some of the oxides, however, dissolved. The filtrate was evaporated to dryness, yielding the new nitrate from which *lanthana* was derived. This compound was not chemically pure. He succeeded in extracting lanthanum and another 'new element' that he named *didymium* after the Greek word *didymos,* meaning twin (of La). Later it turned out that *didymium* again was not chemically pure. Although lanthanum was discovered by C.G. Mosander in Stockholm, Sweden, in 1839, its name was suggested by his friend Jöns Jacob Berzelius in agreement with C.G. Mosander. The name is derived from the Greek word *lanthanein* meaning 'concealed' or 'hidden'.

In 1839, the Swedish geologist Axel Erdmann described another La rich mineral from Norway, which he named in honour of C.G. Mosander, *mosandrite*. This name was later discredited and replaced by 'altered ringite' (Burt, 1989).

The metallic La was produced by reduction with alkali or alkali earth elements in an argon atmosphere by Norton and Hillebrand (1875).

2.2.4 Cerium, Ce, 1803

In 1751, the chemist and assayer Axel Cronstedt found a rock sample which he thought could be a tungsten ore because of its heavy weight (*tungstein*). The Swedish geologist Wilhelm von Hisinger sent an aliquot of this sample to the German chemist Carl Scheele, living in Sweden, asking for analysis of the assumed tungsten mineral. He did not find tungsten and named the mineral *false tungstein*. In 1804, Hisinger and Jöns Jacob Berzelius realized that this brownish sample was the ore of an unknown element. Its *earth* was named *ceria* after the newly discovered asteroid Ceres, which is categorized nowadays as a dwarf planet. Ceres was the Roman goddess of agriculture. This name was chosen by Berzelius and Hisinger. Independently, but also in 1803/4, *cererium* was found by the German chemist Martin Heinrich Klaproth in barite from Norway. A little later, another Ce rich mineral, allanite, was found in Greenland.

Metallic Ce was first produced in 1875 by the American chemists T. Norton and W. Hillebrand by passing an electric current through molten $CeCl_3$.

2.2.5 Praseodymium, Pr, 1885

In 1841, Carl Gustav Mosander assumed that *ceria* harboured two other elements: La and *didymium.* In 1882, Bohuslav Brauer (Prague) suspected, from atomic spectroscopy, that *didymium* was not a pure compound. Carl Auer von Welsbach, a student of the German chemist Robert Wilhelm Bunsen (Vienna Academy of Sciences, Austria) succeeded in splitting *didymium* into Pr and Nd oxides. The green-coloured earth was named praseodymium meaning *green didymium.* Metallic Pr was first produced in 1931.

2.2.6 Neodymium, Nd, 1885

The Austrian Carl Auer von Welsbach (Vienna Academy of Sciences) separated from *didymium*, gained from Carl Gustav Mosander's ceria, a mixture of oxides of praseodymium and a new element named *new didymium*, which was altered to neodymium. In the names of praseodymium and neodymium, part of the original name *didymium* is still maintained. Metallic Nd was first produced in 1925.

2.2.7 Promethium, Pm, 1947

In 1902, the Czech chemist Bohuslav Brauner predicted that there should be an element between Nd and Sm based on atomic spectroscopy. In 1914, Henry Gwyn Jeffreys Moseley suggested that there must be an element between Nd and Sm with the atomic number 61. In the 1920s, two independent groups of chemists in the USA and Italy claimed to have found it. Each group named this element after the place they worked: Luigi Rolla with his graduate Lorenzo Fernandes in Florence (Italy) suggested *florentium* after Florence, and B. Smith Hopkins and his co-workers Len F. Yntema and a Canadian student, J. Allen Harris, of the University of Illinois-Urbana suggested *illinium* after Illinois (Brauner, 1926). Neither claim was justified because all isotopes of Pm are radioactive with half-lives too short for them to have survived the time taken for the earth to form. Radioactivity was not considered as chemical proof of any element by IUPAC regulations.

Pm is continuously formed in uranium ores or U-bearing minerals as the result of spontaneous fission of ^{235}U by which the necessary neutrons are generated for the nuclear reaction: $^{146}Nd\ (n, \beta\ )^{147}Pm$.

M.L. Pool, J.D. Kurbatov and Lawrence L. Quill at Ohio State University bombarded Pr and Nd with neutrons, deuterons and α-particles from a cyclotron in 1938. They detected isotopes of a new element which they named cyclomium. This name was not accepted by IUPAC (2005) for reasons already mentioned above.

Applying the new method of ion-exchange chromatography (Spedding *et al.*, 1947), Charles D. Coryell engaged in the Manhattan Project at Oak Ridge (USA) and his associates Jacob A. Marinsky and Lawrence E. Gledenin succeeded in separating ^{147}Pm (Marinsky *et al.*, 1947). They named this element clintonium after the Clinton Laboratories, where this work was done. It was Coryell's wife who suggested the name prometeum after Prometheus, the Titan of Greek mythology, who stole fire from heaven for the use of mankind (Marinsky & Glendenin, 1948). The IUPAC accepted the name promethium.

At the University of Technology, Helsinki, the Finnish Professor Olavi Erämetsä (1965) and his co-workers extracted 20 tons of lanthanide oxide out of 6000 metric tons of apatite. From these 20 tons they separated a fraction of 82 g, by fractionated crystallization, which mainly contained compounds of Nd and Sm. By measurements of the radioactivity they derived a content of $4.5 \times 10^{17}\%$ of Pm in the primary lanthanide oxide. The ^{147}Pm with a half-life of 2.62 years was produced by natural neutron irradiation originating from the spontaneous fission of U and Th hosted in the apatite.

H. Meier *et al.* (1970) starting with 200 kg of gadolinite determined Pm to be $7.3 \times 10^{17}\%$ of the La oxide prepared from the sample. This result corresponds to that of O. Erämetsä (1965) albeit in both studies different sources and masses were involved.

Metallic Pm was produced by Fritz Weigel (1963).

2.2.8 Samarium, Sm, 1879

The *didymium*, discovered by Carl Gustav Mosander in 1839, turned out to be mixtures of several oxides according to atomic spectrometry in 1879. These spectra varied depending on the minerals used for isolating *didymium*. Paul-Emile Lecoq de Boisbaudran in Paris used the mineral samarskite for preparing his *didymium* nitrate. When adding NH_4OH he separated the first precipitate and measured its atomic spectrum. This hydroxide contained a previously unknown element which he named samarium after the Russian mine official, Colonel Vasilij Evgraforvic Samarskij after whom the mineral samarskite was named. In 1878, the Swiss chemist Marc Delafontain described an element named *decipium* in *didyma* but again it proved not to be a pure substance.

The pure element was produced by E.A. Demarcay (1901).

2.2.9 Europium, Eu, 1901

The *didymia* of Carl Gustav Mosander proved to be a mixture of several earths. The Austrian chemist Carl Auer von Welsbach identified Pr and Nd and Paul-Emile Lecoq de Boisbaudran observed spectral lines of a new element in the atomic spectra of Sm and Gd. E.A. Demarcay separated samarium magnesium nitrate in a painstaking sequence of crystallizations, thereby separating a new element which he named europium after the continent Europe (Demarcay, 1901). It would have been much easier to precipitate insoluble $EuSO_4$, if it had been known that Eu (III) could be reduced to Eu(II).

2.2.10 Gadolinium, Gd, 1880

The Swiss Charles Galissard de Marignac in Geneva (Switzerland) suspected that the *didymium* of Carl Gustav Mosander was not a pure oxide. Marc Delafontain and Paul-Emile Lecoq de Boisbaudran in Paris noticed that the atomic spectra of *didymium* varied depending on their sources. In 1880 Galissard de Marignac, and in 1886 Lecoq de Boisbaudran, isolated another rare earth from *didymium* for which the latter suggested the name gadolinium, namesake of gadolinite, to which Marignac agreed (Boisbaudran, 1886).

The metallic gadolinium was produced by reacting GdF_3 with Ca (Bünzli and McGill, 2018).

2.2.11 Terbium, Tb, 1843

In Sweden, the Austrian chemist Carl Gustav Mosander suspected that *yttria* may host further *earths*. He succeeded in extracting *terbia* which was, however, not a pure compound. Pure terbium was first gained after 1945 by applying the new technique of ion exchange chromatography (Spedding *et al.*, 1947). Its name is derived from the name of the village of Ytterby, Sweden.

2.2.12 Dysprosium, Dy, 1886

Paul-Emile Lecoq de Boisbaudran (1886) separated *erbia* from *yttria* in Paris (France). After 26 cycles of dissolving *holmia* in ammonia, followed by precipitation of the insoluble oxalate, he succeeded in separating dysprosium. This name is derived from the Greek term *dysprositos* which means 'difficult to get' or 'inaccessible' because it was one of the last discoveries before that of Lu and Pm. The American chemist Frank Spedding developed the technique of ion exchange chromatography by which Dy was easily separated.

Metallic Dy is obtained by reacting DyF_3 with Ca. Highly pure Dy is obtained by vacuum distillation (McGill, 2012).

2.2.13 Holmium, Ho, 1879

Marc Delafontain and Jacques Louis Soret (Geneva, Switzerland), and independently the Swedish chemist Per Teodor Cleve (Uppsala, Sweden), produced *holmia* from *yttria* that turned out not to be chemically pure. Only in 1911 did the Swedish chemist Holmberg succeed in producing pure *holmia*. It is not known whether the name holmium, suggested by Cleve is named after Holmberg or after Holmia, the latinized city of Stockholm.

Metallic Ho was first produced in 1940.

2.2.14 Erbium, Er, 1843

Carl Gustav Mosander, Stockholm, separated *yttria* into rose-coloured terbia, light yellow *erbia* and pure *yttria*. These names relate to the village of Ytterby (Sweden). In 1860, the Swedish chemist Nils Berlin extracted a rose-coloured compound which he recorded as terbia (in contrast to Mosander's nomenclature). In 1862 Marc Delafontain proved Mosander's work but turned to Berlin's names of earths. Between 1843 und 1905 many chemists produced pure Er compounds. In 1905 the American chemist Charles James and the French chemist Georges Urbain produced pure Er oxide independently.

In 1934 the chemists Wilhelm Klemm and Heinrich Bommer reduced anhydrous $ErCl_3$ with potassium vapour and obtained metallic Er.

2.2.15 Thulium, Tm, 1879

Per Teodor Cleve at the University of Uppsala (Sweden) extracted Tm oxide from *erbia* (Cleve, 1879). It was named after Thule, the ancient name of Scandinavia, the most northerly country or the mythical island on the northern edge of the world. After 15,000 recrystallizations, the American chemist Theodore William Richards was able to determine the atomic weight of Tm in 1911 (James, 1911).

2.2.16 Ytterbium, Yb, 1878

The Swiss chemist Jean Charles Galissard de Marignac at the University of Geneva separated *ytterbia* from *erbia*. He thermally decomposed $Er(NO_3)_3$ to get the Er-oxide and leached this oxide by water. This way he obtained two differently coloured oxides: the known Er oxide and an unknown oxide which he named *ytterbia* after the village of Ytterby (Sweden). In 1907, the French Georges Urbain, the Austrian Carl Auer von Welsbach and the American Charles James independently showed that Marignac's *ytterbia* was a mixture of two oxides. The elements of the two oxides were named *aldebaranium* (after the star Aldebaran) and *cassiopeium* (Welsbach, 1908), whereas Urbain (1908) suggested *neoytterbium* and lutetium. Pure Yb metal was first produced in 1953.

2.2.17 Lutetium, Lu, 1907

The French chemist Georges Urbain at the Sorbonne (Paris, France) segmented *erbia* into holmium-oxide in 1978 and thulium oxide in 1879. Finally, as the result of painstaking work, he found the element *lutecium* (Urbain, 1908), which was later changed into lutetium, named after Lutetia, the Roman name of Paris (Koppenol, 2002). Independently, the Austrian chemist Carl Auer von Welsbach working in Germany reported $YbNH_4$-oxalate contained two elements *cassiopeium* (later renamed lutetium) and *aldebaranium* (corresponds to ytterbium). Charles James at the University of New Hampshire (USA) produced pure Lu-salts but waived his claim of discovery of this element. Auer von

Welsbach's *cassiopeium* refers to the stellar constellation of Cassiopeia (Urbain, 1910; Welsbach, 1908). The metal was first produced in 1953.

2.3 PRESENTATION OF THE SUITE OF LANTHANOIDS AND Y

In plots of abundances *vs* atomic numbers of REE, the abundance of REE decreases from La towards Lu, and a 'sawtooth' line is obtained in which the even numbered elements are more abundant than the uneven ones (Figure 2.3). The Oddo-Harkins rule (Gioseppe Oddo, Italian chemist; William Draper Harkins, US American physical chemist) argues that elements with odd atomic numbers have one unpaired proton and are more likely to capture another one during the nucleosynthesis of elements in our solar system, thus increasing their atomic number. It seems that protons of elements with even atomic numbers, are paired. Each number of the pair balances the spin of the other; the resulting parity thus enhances the stability.

The 'sawtooth'-like patterns are difficult to use in comparative studies (Rollinson, 1993) (Figure 2.4a), but the measured concentration of REY in samples can easily be smoothed out by normalization to REY values in some reference material (Figure 2.4b). There are several suggestions for reference materials such as the C1-chondrite composition resembling the solar composition

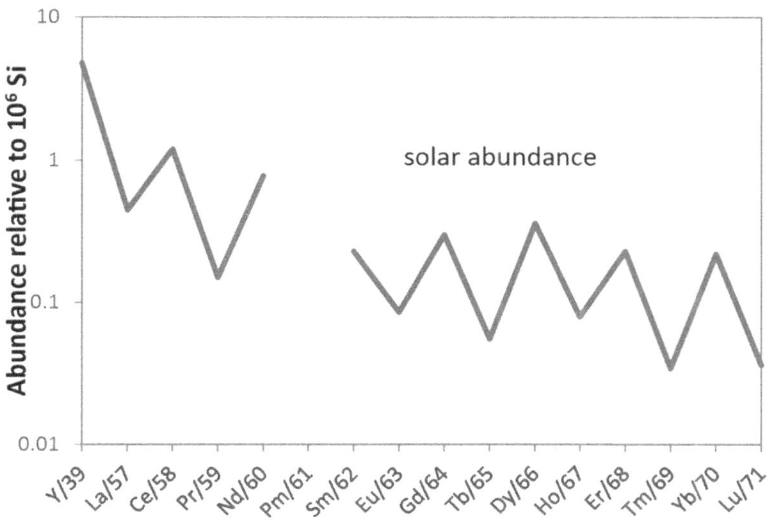

Figure 2.3 Sawtooth-like distribution of solar abundances of Y and REE (Camaron, 1973).

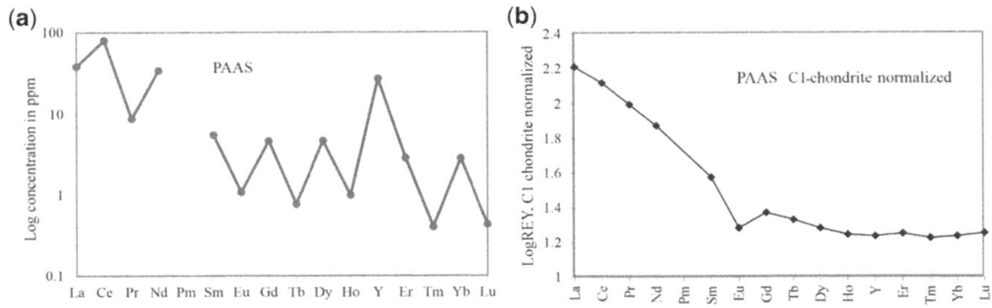

Figure 2.4 Effect of normalization of REY abundances. (a) logarithmic plot of abundances of REY in PAAS and (b) the same data of PAAS after normalization by C1-chondrite. Data are taken from Table 2.3.

(Anders & Grevesse, 1989), NASC a mixture of North American shale composite (Taylor & McLennan, 1995); PAAS post-Archean average shale (Gromet *et al.*, 1984) or UCC Upper Continental Crust (Rudnik & Gao, 2014) (Table 2.3).

Table 2.3 Composition of reference materials used in REY plots.

Compounds	C1 chondrite µg/g	PAAS µg/g	NASC µg/g	UCC µg/g
La	0.235	38.2	32	30
Ce	0.603	79.6	73	64
Pr	0.089	8.83	7.9	7.1
Nd	0.452	33.9	33	26
Pm				
Sm	0.147	5.55	5.7	4.5
Eu	0.056	1.08	1.24	0.88
Gd	0.197	4.66	5.2	3.8
Tb	0.036	0.774	0.85	0.64
Dy	0.243	4.68	5.8	3.5
Ho	0.056	0.991	1.04	0.8
Y	1.56	27		22
Er	0.159	2.85	3.4	2.3
Tm	0.024	0.405	0.5	0.33
Yb	0.163	2.82	3.1	2.2
Lu	0.024	0.433	0.48	0.32

C1 = C1 chondrite (Anders & Grevesse, 1989); PAAS = post-Archean Australian Shale (Taylor & McLennan, 1995); NASC = North American shale composite (Gromet *et al.*, 1984); UCC = upper crustal composition (Rudnick & Gao, 2014).

2.4 OCCURRENCE OF LANTHANOIDS AND Y

2.4.1 REY in rocks

In deep time, when the earth accreted, REE accumulated in the primary mantle of the earth. The fractionation of REY in the solid earth started by magmatic processes from solar composition and evolved to the composition of the deep mantle, depleted upper mantle, lower crust and upper crust (Figure 2.5). During this differentiation, REY were enriched in ascending alkaline melts due to their large ionic radii and high field strength, whereas the mantle was depleted preferably in LREY. Because LREE are more incompatible than HREE, the latter were enriched in rock-forming minerals from final melts such as garnets (Jébrak *et al.*, 2014). The consequence is that partial-melts show higher concentrations of LREE than HREE (Jébrak *et al.*, 2014). The least fractionated REE available materials on the earth are assumed to be C1-chondritic meteorites (Figure 2.5). Compared to HREE, LREE are enriched in the earth's crust because Y and HREE are preferably incorporated in minerals forming the mantle. LREE are preferably accumulated in the crust. The economic consequence is that large LREE ore bodies occur worldwide, whereas HREE ones are less common.

The compilation of analyses of 90 rocks suggested as reference materials in geoanalytical work (Dulski, 2001) yields a good overview of REE fractionation in rocks. Basalts show a wide spread of light and medium REE, whereas the heavy ones are rather similar in abundance (Figure 2.6a). The olivine rich dunite and peridotite show very low abundances and a minimum for the medium ones. The felsic rocks such as granites, granodiorites and rhyolites mostly show flat

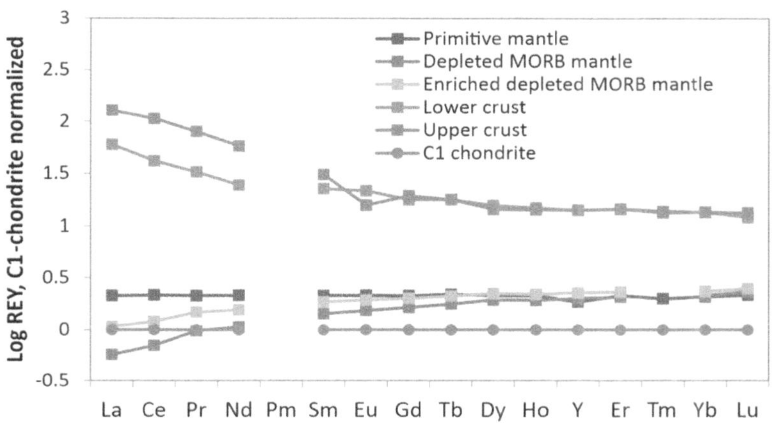

Figure 2.5 The fractionation of REY in solid earth. The primitive mantle (Palme *et al.* 2007) lost LREE (Workman & Hart, 2005), whereas REY in general but LREE in particular are enriched in the earth's crust (Taylor & McLennan, 1995). MORB = mid-ocean ridge basalt (Workman & Hart, 2005).

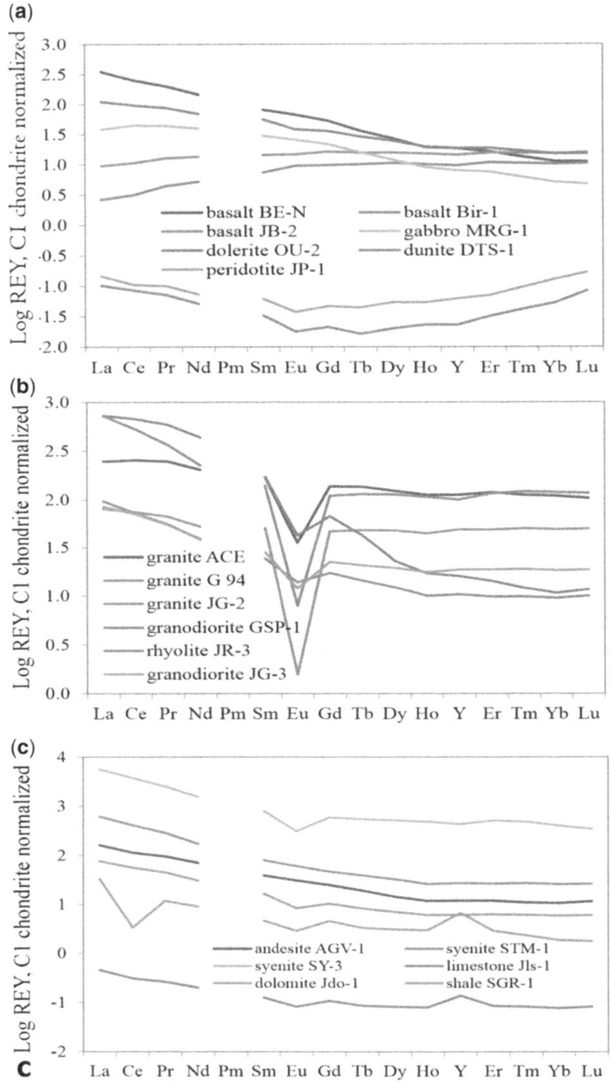

Figure 2.6 C1-chondrite –normalized REY abundances of diverse rocks. The analyses are taken from Dulski (2001). The abbreviations following every sample refer to the reference code of samples issued by international organizations such as USGS (United States Geological Survey): BIR-1, DTS-1, AGV-1, STM-1, SRG-1, GSP-1; GSJ (Geological Survey of Japan): JB-2, JP-1, JDo, JLs, JG-2, JR-3, JG-3; IAG (International Association of Geoanalysts): OU-2, G-94; GIT-IWG (Groupe Internationale de Travail – International Working Group): BE-N, ACE; CCRMP (Canadian Certified Reference Materials Project): MRG-1, SY-1.

patterns for the range of Gd to Lu, whereas the LREE decrease from La to Sm with strongly negative Eu anomalies (Figure 2.6b). Alkali feldspar rich plutons (syenite) and plagioclase rich plutons (andesite) show patterns with flat parts for the HREE but increasing ones from MREE to La (Figure 2.6c). This trend type is similar to sedimentites such as shale, limestones and dolostones. Limestones show negative Ce and positive Y anomalies.

2.4.2 Analyses of REY in abundant minerals

The lithophile REY form oxides, silicates and phosphates, but do not react with sulfur. Locally, abundant monazite, bastnaesite, parasite and xenotime formed and were accumulated during weathering in placer deposits (Figure 2.7). Due to isomorphous substitution, REY occur as traces in almost all minerals, particularly in Ca^{2+} bearing minerals such as calcite, fluorite and phosphates. During weathering of igneous and sedimentary rocks REY are partially released into the hydrosphere, where they are strongly adsorbed onto particulate matter or co-precipitated with calcite/aragonite (Möller & Siebert, 2016).

A selection of rare earth deposits either in production or under exploration is given in Table 2.4 indicating their worldwide distribution.

2.4.3 REY distribution in the hydrosphere

The REY distribution in seawater is very similar all over the world (Figure 2.8a). This is mainly caused by the movement of the seawater by the Great Ocean Conveyor Belt (Arsouce *et al.*, 2007). Groundwater composition is controlled by weathering of continental rocks and thereafter by the lithology of their aquifers.

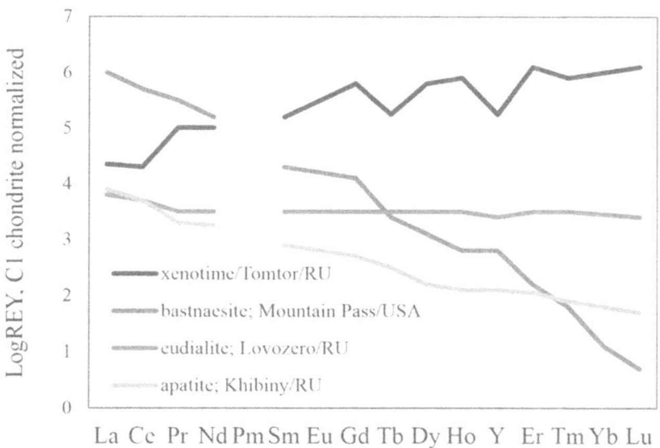

Figure 2.7 Examples of REY distribution patterns of REY ore minerals (after Chakhmouradian & Wall, 2012).

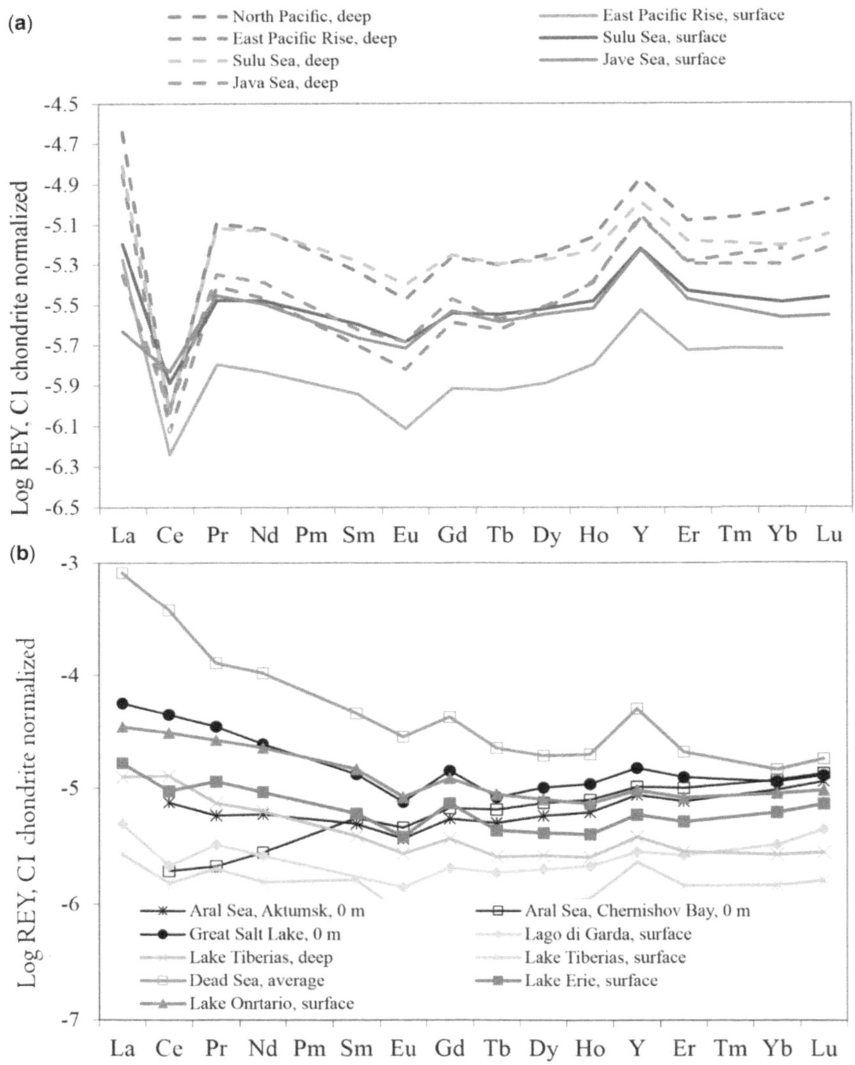

Figure 2.8 REY distribution patterns of seawater and continental lakes. Note the low spread in seawater (a) compared to the large spread of lake water (b). Data for water bodies are taken from: East Pacific Rise (Möller *et al.*, 1994); North Pacific (Nozaki *et al.*, 1999), Java Sea (unpubl.), Sulu Sea (Nozaki *et al.*, 1999), Gulf of Thailand (Nozaki *et al.*, 1997); Aral Sea (unpubl.), Great Salt Lake (unpubl.), Lago di Garda (unpubl.), Lake Tiberias (Siebert, 2006), Dead Sea (Möller *et al.*, 2009); Lake Erie and Ontario (Bau *et al.*, 2006).

Table 2.4 Selection of rare earth element deposits either in production or exploration.

Country/Locality	Type of deposit	Reference
Afghanistan: Khanneshin;	Carb, LREE	Tucker *et al.* (2012)
Argentinia: Rodeo des Los Molles;	Fenitization of monozogranite	Lira and Ripley (1992)
Australia: Mount Weld/W Australia;	Lateritic carb.; REY phosphates	Lottermoser (1990)
Olympic Dam/S Australia;	Fe-Cu-Au-U dep.; fluorobas, flor;	Schmandt *et al.* (2017)
Dubbo/New South Wales;	Altered carbonatite;	https://asm-au.com/projects/dubbo-project/
Yangibana/W Australia;	Mon in ironstone dykes/carb	https://www.mining-technology.com/projects/yangibana-rare-earth-project-gascoyne-region/
Nolans Bore/N Territories;	Vein dep.; fluoroapa, apa, mon	
Eneabba;	Coastal placer; mon, xen	
Chapel;	Coastal placer; mon	Huston *et al.* (2016)
Yoganup;	Coastal placer; Ti, Zr, LREE	
North Stadboke Island;	Coastal placer; mon	
Burundi: Gakara	Carbonatite	Ntiharirizwa *et al.* (2019)
Brazil: Salobo;	Cu-REE dep., allan., mon	Lindenmayer and Teixeira (1999)
Morro dos Seis Lagos;	Carb, laterite	Giovannini *et al.* (2017)
Camaratuba;	Placer deposit; mon	Morteani and Preinfalk (1996)
Araxa/Catalao;	Laterite on carb	
Canada: Thor Lake/NW;	IAC; HREE	Pedersen and LeCouteur (1991)
Hoidas Lake/Saskatchewan;	Veins; HREE, apa, all	Halpin (2010)
Strange Lake/Quebec-Labrador;	AIC; HREE;	Woolley (1987)
Elliot Lake/Ontario;	U, monazite in conglomerites	Karagiorgakis *et al.* (2018)
Oka;	Carb with REE enrichment	Chakhmouradian (1996)
China: Bayan Obo/Inner Mongolia;	Carb, Fe-REE-Nb; bas, mon	Yang and Le Bas (2004)
Maoniuping/Sichuan;	Barite vein type, carb-AIC,	Liu *et al.* (2018)
Miaoya/Hubei;	Syenite-carb; bast., mon, par	Wang *et al.* (2001, 2018)

(*Continued*)

Table 2.4 Selection of rare earth element deposits either in production or exploration (*Continued*).

Country/Locality	Type of deposit	Reference
Weishan/Shandong;	IAC; barite veins;	Shi (1980)
Xunwu/Jiangxi;	Laterite; ion adsorbtion on	Jia and Liu (2020)
Longnan/Jiangxi;	Weathered syenite	Wu and Bai (1996)
	Laterite; ion adsorption on granite;	Wu and Guo (1989)
India: Orissa;	Placer deposit; mon	Kumar (2011)
Chawara;	Placer deposit; mon	Kumar (2011)
Greenland: Kvanefjeld;	IAC; U, REE	Friis (2016)
Sarfartoq;	Carb; Nb, Ta, Zr, bas, mon	Hudson Resources Inc. (2020)
Kenya: Mrima Hill;	Carb., laterite; Nb REE	Coetzee and Edwards (1959)
Kyrgystan: Aktiuz, Kutessay;	Polymetallic dep; alkali granite;	Djenchuraeva *et al.* (2008)
Madagascar: Tantalus;	Laterite, HREE; skarn	Dolbear (2017)
Malawi: Chilwa;	Alkaline prov., carb, carbonatite	Broom-Fendley *et al.* (2017)
Kangankunde;		Otake *et al.* (2019)
Malaysia: Bukit Merah;	Xenotime placer	Sanusi *et al.* (2021)
Mongolia: Mushgai-Khudag;	Carb; magnetite/apa; REE veins	Kovalenko and Yarmolyuk (1995)
Luiigin Gol;	Carb; syn, bas	
Namibia: Lofdal;	HREE; xen	Wall *et al.* (2008)
Russia Chibiny;	Apatite body; carb, REE	Kovalenko and Yarmolyuk (1995)
Lovozero;	AIC; lop, eud, rin, ancyl	Kogarko *et al.* (1995)

Chadobetskoe/Siberia;	Laterite	Kogarko *et al.* (1995)
Tomtor/Jakutia;	HREE, Th deposit	Lapin *et al.* (2016)
South Africa: Palaborwa;	Carb, bas, par, syn	Giebel *et al.* (2017)
Steenskampskraal/W Cape Prov.;	REE-Th-Cu deposit	Andreoli *et al.* (1994)
Zandkopsdrift/N Cape Prov.;	Carb, mon	https://www.proactiveinvestors.com
Sri Lanka: Pulmoddai;	Alluvial placer; mon	Batapola *et al.* (2020)
Tanzania: Wigu Hill	Carb, REE in veins	Karpeta *et al.* (2012); Harmer and Nex (2016)
Turkey: Kizilkaören/Anatolia;	Bas, fluorite barite dep.	Morteani and Satir (1989)
Sweden: Ytterby;	Pegmatite	Sahlström *et al.* (2019)
Bastnäs;	Pegmatite	Sjöqvist *et al.* (2013)
Norra Kärr;	IAC; HREE, eud	
USA: Mountain Pass/Cal.;	Bastnaesite,	Castor (2008)
Bear Lodge/Wyoming;	Carb., veins; bur, bas	Moore *et al.* (2015)
Bokan Mountain/Alaska;	Peralkaline granite; HREE	Cafferty *et al.* (2014)
Green Cove Springs;	Placer deposit; mon	Jackson and Christiansen (1993)
Lemhi Pass/Idaho;	Th, mon veins	Staatz *et al.* (1972)
Elk Creek/Nebraska;	Carb., REE in beforsite unit	Drenth (2014)
Vietnam: Dong Pao/Lai Chau;	IAC; bas, par	Mariano (1997)

Abbreviation: all = allanite; anc = ancylike; apa = apatite; bas = bastnaesite; bra = brannerite; bur = burbankite; carb = carbonatite; dep = deposit; eud = eudialyte; flor = florencite; IAC = ion-adsorbed complex; lop = loparite; mon = monazite; par = parisite; rin = rinkite; syn = synchesite; xen = xenotime.

Brines are highly enriched in REY. The marine limestones are the most important sink of REY (Figure 2.6b). They are characterized by negative Ce and Eu as well as positive Gd and Y anomalies inherited from seawater (Figure 2.8a). The REY composition of primary carbonates is varied by diagenetic processes and thereby REY are enriched in limestones and marine precipitates such as phosphorites (Bau & Dulski, 1996) and Fe-Mn nodules (Ohta & Kawabe, 2001). By weathering of limestone small amounts of REY are recycled to oceans via rivers. This contribution is less than that released by weathering of limestone and igneous rocks because of scavenging of REY by Fe-Mn oxihydroxides and adsorption onto clay minerals.

2.5 REY DEPOSITS

2.5.1 Geopolitical sources of rare earth elements production

The top 10 countries for rare earth elements production are China, USA, Myanmar, Australia, India, Russia, Madagascar, Thailand, Brazil and Vietnam (Pistilli, 2021). In 2020, China alone produced 63% of the total, followed by USA, Myanmar and Australia with 12, 10 and 10%, respectively. The other countries produce about 1% each. The ion-adsorbed deposits (IADs) account for about 35% of China's total REE production and roughly 80% of the global HREE supplies (Yang *et al.*, 2013). Despite being low grade (0.05–0.2 wt-% total RE_2O_3, incl. Y_2O_3) and relatively low tonnage compared to hard-rock REE deposits associated with carbonatites and alkaline igneous complexes (Smith *et al.*, 2016), IADs are of high importance because of their HREE and Y contents. Exploration for this kind of deposit is actively taking place across the world. Currently, some potential deposits have been discovered in USA, Southeast Asia, Malawi, Brazil and Madagascar (Yan *et al.*, 2017).

A worldwide selection of REY deposits and occurrences are compiled in Table 2.4. Often, they are only by-products of other metals.

2.5.2 Endogenic enrichment of REY

REE deposits at the earth's surface are carbonatites and pegmatites related to either subduction or rifting. During rifting, small degrees of partial melting of garnet peridotites at several km depth produce alkaline magma enriched in incompatible elements such as REY. Along pre-existing fractures these magmas are emplaced in the crust or erupt at the surface. REY are enriched in A type granitoids denoting anorogenic and/or anhydrous granitoids occurring along rift zones and within stable continental blocks (Chappell & White, 1974; Loiselle & Wones, 1979; Whalen *et al.*, 1987) and M-type granitoids presumably deriving from the melting of subducted oceanic crust or the overlying mantle (White, 1979). Partial melting near subduction zones produces volatile rich magma within the asthenosphere with high concentrations of alkaline elements and REY (Jébrak *et al.*, 2014). Such melts ascent and either emplace above the subducting slab or erupt at the surface. REY occurrences also form from weathering S-type granitoids derived from metasedimentary sources (Chappell & White, 1974).

2.5.2.1 Carbonatites

Carbonatites are assumed to be derived from hydrous-carbonated lherzolite and originate from deep in the earth's mantle (Jébrak *et al.*, 2014). They are bound to extensional structures. These magmas develop in deep cratons such as the African and Canadian Shields. Large carbonatite REE deposits are Mount Weld (AUS), Thor Lake (Canada), Zandkopsdrift (South Africa) and Mountain Pass (USA).

2.5.2.2 Pegmatites

High abundances of REY are in pegmatites and their hydrothermal alteration zones such as in Bastnäs and Ytterby (Sweden) or Mountain Pass (USA). These resources owe their REY to hydrothermal mobilization of the latter (Sahlström *et al.*, 2019; Williams-Jones *et al.*, 2012). They are restricted to narrow zones of carbonatites and dolomitic marble, in which the carbonate is replaced by magnetite, tremolite, talc and REY minerals (Holtstam & Anderson, 2007). Granitic pegmatites are mostly emplaced along contact zones and axial faults in greenschist belts or higher-grade metamorphic terrains.

2.5.2.3 Mountain pass REY deposit

The Mountain Pass REY deposit is located in a block of Precambrian metamorphic rocks of garnet bearing gneisses, biotite granites and migmatites of granite, syenite and shonkinite bodies. These metamorphic complexes are subject to intrusion by a number of stocks and dykes of a K-feldspar rich igneous suite. They are non-foliated and therefore younger than the metamorphic complex itself. The most important intrusive carbonatite body is the Sulfide Queen carbonate body, which is related to the largest shonkinite-syenite stock. It contains breccias of all country rock types and shows discordant contacts and must be the last in the intrusion sequence. Other smaller carbonatite bodies cut older rocks discordantly. The Mountain Pass carbonatite differs from other carbonatites in many aspects: absence of feldspathoids in associated igneous rocks, related volcanic rocks, concentric ring structures and Ca-silicate minerals, and extreme concentrations of bastnaesite and barite, scarcity of Ti-Nb minerals, and relatively small amounts of magnetite. The mined ore contains 5–15% bastnaesite. Cross cutting dykes contain locally up to 30% bastnaesite (Robjohns, 1984).

Three types of rocks are distinguishable: brownish, iron bearing dolomite, grey- to rose-coloured calcite-barite and silicified carbonatite with up to 60% bastnaesite and parisite and small amounts of allanite, monazite, cerite and sahamalite. All rock types are unusually high in REY, Ba and Sr. Mountain Pass was once the most important source of light REE (La, Ce and Nd).

2.5.2.4 Bayan Obo REY deposit

The Bayan Obo giant REY deposit is related to primary mantle-derived carbonatite. Intensive magmatic differentiation seems to be the source of the enrichment of REY (Ling *et al.*, 2014; Yang *et al.*, 2019). Located in the Yin-schan Range, the deposit was formed in two steps: prior to the deposition of the Bayan-Obo Group the REY bearing pegmatites were emplaced in Precambrian rocks. In Late Paleozoic times, large masses of REY-bearing iron ore were emplaced in the carbonates of the Bayan-Obo Group and Precambrian schists, probably at high temperature. The associated magnetite deposit is rich in REY-bearing phosphates. The REY-rich iron ore is bound to Precambrian dolomites of the Bayan-Obo Group. The primary host minerals are magnetite, specularite and hematite. Common REY minerals are aeschynite, allanite, apatite, beiyenite (resembles bastnaesite), parisite, monazite, xenotime and others.

2.5.2.5 Peralkaline igneous deposits

The peralkaline igneous deposits associated with A-type granitoids (also called pegmatites because of their large crystal sizes) are emplaced in extensional zones, either as ring complexes, pipes, lenses, or massive bodies (Jébrak *et al.*, 2014). The subaluminous melts (Al_2O_3 is less than $Na_2O + K_2O$) are enriched in high field strength elements such as REY^{3+}, Nb^{5+}, Ti^{4+}, Zr^{4+} and Be^{2+} and form oxides and silicates. By interaction with carbonatites REY react with calcite forming Ca-phosphates and Ca-fluorocarbonates such as bastnaesite or yttrium members such as synchesite. These minerals are mostly formed by late hydrothermal or hypogenic replacement of minerals in alkaline syenitic rocks or in alkalic and subalkalic granitoids and their associated pegmatites. REE deposits spatially and genetically connected with nepheline syenite, alkali syenites, alkali granites and alkali ultrabasic complexes deserve special attention because of their concentration of both light and heavy REE. Although REE, Y and Ta (Nb) are enriched in complex granitic pegmatites and carbonatites, they do not always occur in the same pegmatite. REY mainly substitute mostly Ca^{2+} in Ca minerals.

M-type granitoids (nepheline syenite) consist of feldspars and feldspathoids, with up to 90% containing REY-bearing accessory minerals (Jébrak *et al.*, 2014). Examples include Illimaussaq-Kvanefeld (Greenland) and Lovozoro (Russia). Granitic pegmatites often contain gadolinite, fergusonite and xenotime, which are rich in HREE and Y.

2.5.2.6 Hydrothermal vein deposits

Hydrothermal vein deposits are often enriched in REE hosted by accessory minerals of magnetite or uranium deposits, or due to isomorphous replacement in the main mineral such as calcite, fluorite and apatite. Such REY rich vein deposits are often associated with carbonatites or alkali igneous intrusions.

2.5.3 Exogenic enrichment

Placer deposits are formed by the mechanical concentration of minerals from weathering debris. Heavy mineral accumulation occurs in weathering zones, on beaches, in rivers, streams, dunes and off-shore areas. This way, weathering-, ocean-, river- or wind actions result in the concentration of valuable heavy resistant minerals of economic importance such as monazite (Sengupta & van Gosen, 2016) or HREE, particularly in the placer deposits of South China.

2.5.3.1 *Regolith-hosted REY deposits*

The regolith-hosted rare earth element deposits occur along ridges in low-lying granitic hills in South China (Yan *et al.*, 2017). Weathering generates profiles consisting of soil on top followed by an accumulation of in situ weathered rocks particles, the less-weathered regolith, that is, saprolite, and beneath the bedrock. The orebody lies at lower layers of weathered soil (Chi & Jun, 2008; Wang *et al.*, 2018; Zhang, 1990). With respect to REY monazite is most common.

Regolith-hosted REY deposits are common in Southern China. These unconsolidated deposits consist of fragmented and decomposed rocks and include dust, soil, broken rock and other related materials. Two types of deposits exist: The 'light' rare earth element (LREE) (i.e., La, Ce, Pr, Nd) deposits dominate in the Heling and the Dingnan deposits in the Jiangxi Province (Simandl, 2014; Wu & Bai, 1996; Wu & Guo, 1989). The middle and heavy rare earth elements (i.e., Sm, Eu, Gd, Dy, Ho, Y, Er, Tm, Yb and Lu) are mined in the Zudong, Datian and Xiawentian deposits in the Jiangxi Province, South China. These two types of REY deposits are widespread in China's Jiangxi, Hunan, Guangdong and Fujian provinces.

2.5.3.2 *Ion-adsorbed deposits*

In ion-adsorbed deposits (IADs) REY are weakly adsorbed onto clay minerals (Borst *et al.*, 2020). These deposits are important resources of Gd-Lu and Y. This type of deposit occurs dominantly in southeast China, where they were formed by subtropical weathering of granitic rocks. Bauxite-laterite deposits are enriched in REY by secondary alteration by meteoric or hydrothermal fluids or erosion and transport of resistant REE minerals. Silica and other soluble elements are leached from primary minerals and clay minerals such as kaolinite and montmorillonite. This type of process produced bauxite and laterite deposits containing HREE such as those in Southern China which are the most important mines for HREE worldwide. Laterite also formed over the carbonatite at Mount Weld (Australia).

2.5.3.3 Ocean seabed mud

Undersea REE resources may be much more promising than on-land resources such as those mined in Southern China (Kato *et al.*, 2011; Powell, 2011). In seabed muds REE are accumulated over millions of years.

2.5.4 Anthropogenic REY enrichments

Anthropogenic deposits of REY are large stockpiles of mining wastes such as the red muds from aluminium production based on bauxite or the leftover materials of fertilizer production from phosphate rock, the so-called phosphogypsum. Their average REY contents vary between 300 and 500 ppm, and thousands of tons of REY metals could be produced from millions of tons of globally piled-up waste materials (Ritter, 2017). Another resource of REY may be coal ashes (Rozelle *et al.*, 2016).

2.6 SUMMARY

Historically, the most prominent source of discovered REY was the pegmatite in Ytterby, Sweden, which was the namesake for yttrium, terbium, erbium and ytterbium. Gadolinium, Ho, Tm and Sc were also extracted from *yttria*. All these elements were harboured in gadolinite. The elements La-Eu were isolated from *ceria* which was derived from different minerals such as cerite and samarskite. China is the most important REY producer followed by the USA, Myanmar and Australia.

The REY-mineral deposits may be grouped into:

- Endogenic enrichments of REE associated with alkali-ultrabasic complexes, carbonatites, alkaline pegmatites, metasomatites (albite and fenite), hydrothermal vein mineralizations, and contact metamorphic rocks, especially skarns.
- Exogenic accumulations of REE minerals in fluvial and coastal or alluvial sands and residual soils (laterites and bauxite) on top of carbonatites; sediments rich in organic material and REE. These rocks may become an interesting source in the future because of their availability in large quantities.
- Anthropogenic enrichments of REY exist in large stockpiles of mining wastes such as the red muds from aluminium production based on bauxite or the leftover materials of fertilizer production from phosphate rock, the so-called phosphogypsum. Another possible resource of REY may be coal ashes.

REFERENCES

Anders E. and Grevesse N. (1989). Abundances of elements: meteoric and solar. *Geochimica et Cosmochimica Acta*, **53**, 197–214.

Andreoli M. A. G., Smith C. B., Watkeys M., Moore L. D., Ashwal L. D. and Hart R. J. (1994). The geology of the steenkampskraal monazite deposit. South Africa; implications of REE-Th-Cu mineralization in charnockite-granulite terranes. *Economic Geology*, **89**(5), 994–1016.

Arsouce T., Dutay J.-C., Lacan F. and Jeandel C. (2007). Modeling the neodymium isotope composition with a global oceanic circulation model. *Chemical Geology*, **239**, 165–177.

Batapola N. M., Dushyantha N. P., Abeysinghe A. M. K. B., Rohitha L. P. S., Ratnayake N. P., Dissanayake D. M. D. O. K., Ilankoon I. M. S. K. and Dharmaratne P. G. R. (2020). A comparison of global rare earth element (REE) resources and their mineralogy with REE prospects in Sri Lanka. *Journal of Asian Earth Sciences*, **200**, 104475. doi: 10.1016/J.JSEAES.2020.104475

Bau M. and Dulski P. (1992). Y/Ho fraktionierung in geochemischen systemen: erste ergebnisse. *Beihefte, European Journal of Mineralogy*, **4**, 17.

Bau M. and Dulski P. (1996). Distribution of yttrium and rare-earth elements in the penge and kuruman iron-formation, transvaal supergroup, South Africa. *Precambrian Research*, **39**, 37–55.

Bau M., Knappe A. and Dulski P. (2006). Anthropogenic gadolinium as a micropullutant in river waters in pennsylvania and in lake erie, northwestern United States. *Chemie der Erde-Geochemistry*, **66**, 143–152.

Boisbaudran Lecoq de P. E. (1886). Le Yα de Marignac est définitevement nomme gadolinium. *Comptes Rendus*, **102**, 902.

Borst A. M., Smith M. P., Finch A. A., Estrade G., Villanova-de-Benavent C., Nason P., Marquis E., Horsburgh N. J., Goodenough K. M., Xu C., Kynicky J. and Graki K. (2020). Adsorption of rare earth element in regolith-hosted clay deposits. *Nature Communications*, **11**, 4386.

Brauner B. (1926). The new element of atomic number 61: illinium. *Nature*, **118**, 84–85.

Broom-Fendley S., Brady A. E., Horstwood M. S. A., Woodley A. R., Mtegha J., Wall F., Dawes W. and Gunn G. (2017). Geology, geochemistry and geochronology of the songwe hill carbonatite, Malawi. *Journal of African Earth Sciences*, **134**, 10–13.

Burt D. M. (1989). Compositional and phase relations among rare earth element minerals. In: Geochemistry and Mineralogy of Rare Earth Elements, B. R. Lipin and G. A. McKay (eds), Mineralogical Society of America; BookCrafter Inc, Chelsea, USA, pp. 259–307.

Bünzli J. -C. G. and McGill I. (2018). Rare earth elements. In: Wiley-VCH (ed.) Ullmann's Encyclopedia in Industrial Chemistry, Wiley. VCH, Weinheim. doi: 10.1002/14356007a.22_607.pub2.

Cafferty A. E., Stoeser D. B. and van Gosen B. S. (2015). Geophysical interpretation of U, Th, and rare earth element mineralization of the Bokan Mountain peralkaline granite complex, Prince of Wales Island, southeast Alaska. doi: 10.1190/INT-2014-0010.1

Camaron A. G. W. (1973). Abundances of element in the Solar System. *Space Science Reviews*, **15**, 121–146.

Castor S. B. (2008). The mountain pass rare-earth carbonatite and associated ultrapotassic rocks, California. *Canadian Mineralogist*, **46**(4), 779–806. doi: 10.3749/canmin.46.4.779

Chakhmouradian A. R. (1996). On the development of niobium and rare earth minerals in monticellite-calcite carbonatite of the Oka Complex, quebec. *Canadian Mineralogist*, **34**, 479–484.

Chakhmouradian A. R. and Wall F. (2012). Rare earth elements: minerals, mines and magnets (and more). *Elements*, **8**, 333–340.

Chappell B. W. and White A. J. R. (1974). Reprinted in – Chappell, B. W. and White, A. J. R. (2001). Two contrasting granite types: 25 years later. *Australian Journal of Earth Sciences*, **48**, 489–499.

Chi R. and Jun T. (2008). Weathered Crust Elution-Deposited Rare Earth Ores. Nova Science Publishers, New York, USA, ISBN 978-1-604-56387-0.

Chroustschoff K. D. (1989). New metals. *Chemical News*, **59**, 234.

Cleve P. T. (1879). Sur deux nouveaux éléments dans l'erbine. *Comptes Rendus*, **89**, 478–480.

Coetzee G. L. and Edwards C. B. (1959). The mrima hill carbonatite, coast province, Kenya. *South African Journal of Geology*, **62**(1), 373–397.

Crookes F. R. S. (1899). Photographic researches on phosphorescent Spectra: on victorium, a new element associated with yttrium. *Chemical News*, **80**, 49–51.

Demarçay E.-A. (1901). Sur un nouvel élément, europium. *Comptes Rendus*, **132**, 484–1486.

Djenchuraeva R. D., Borisov F. I., Pak N. T. and Malyukova N. N. (2008). Metallogeny and geodynamics of the Aktuz-Boordu mining district, Northern Tien Shan, Kyrgystan. doi: 10.1016/j.jseaes.2007.10.019

Drenth B. J. (2014). Geophysical expression of a buried niobium and rare earth element deposit: the Elk Creek carbonatite, Nebraska, USA. doi: 10.1190/INT-2014-0002.1

Dulski P. (2001). Reference materials for geochemical studies: new analytical data by ICP-MS and critical discussion of reference values. *Geostandards Newsletter*, **25**(1), 87–125.

Dolbear B. (2017). Tantalus Rare Earths Ionic Clay Project. Australian Pty Limited, North Sydney, New South Wales, 2060, Austarlia.

Emsley J. (2001). Natures Building Blocks. Oxford University Press.

Erämetsä O. (1965). Separation of Promethium from a natural lanthanide mixture. *Acta Polytechmica Scandinaviavica*, **37**, 21 pp.

Friis H. (2016). First occurrence of moskvinite-(Y) in the ilimaussaq alkaline complex, south Greenland-implications for rare-earth element mobility. *Mineralogical Magazine*, **80**(1), 31–41.

Giebel R. J., Gauert C. and Marks M. A. W. (2017). The multi-stage formation of REE minerals in the palabora carbonatite complex, South Africa. *American Mineralogist*, **102**(6), 1218–1233.

Giovannini A. L., Neto A. C. B., Porto V. P., Takehara L., Barbanson L. and Bastos P. H. S. (2017). Mineralogy and geochemistry of laterite from morro dos seis lagos Nb(Ti, REE) deposit (amazonas, *Brazil). Ore Geology Reviews*, **88**, 461–480.

Goldschmidt V. M. (1925). Geochemische Verteilungsgesetze der Elemente, Part V. Isomorphie und Polymorphie der Sesquioxyde. Die Lanthaniden-Kontraktion und ihre Konsequenzen, Oslo, Skrifter (Norske videnskaps-akademi. I-Mat.-naturv. Klasse) no. 7.

Gromet L. P., Haskin L. A., Korotev R. L. and Dymek R. F. (1984). The 'north American shale composite': its compilation, major and trace element characteristics. *Geochimica et Cosmochimica Acta*, **48**, 2469–2482.

Gschneidner K. A. (1966). Rare Earths. The Fraternal Fifteen. U.S. Atomic Energy Commission – Division of Technical Information, Oak Ridge, TN.

Gupta C. K. (1993). Extractive metallurgy of rare earths. *Journal of Alloys and Compounds*, **194**, 93–99.

Halpin K. M. (2010). The characteristics and origin of the Hoidas Lake REE deposit. MSc thesis, University Saskatchewan, Saskatoo, 84–118.

Harmer R. E. and Nex P. A. M. (2016). Rare earth deposits of Africa. *Episodes*, **39**(2), 381–406. doi: 10.18814/epiiugs/2016/v39i2/95784

Hofmann K. A. and Prandtl W. (1901). Ueber die zirkonerde im euxenit von brevig. *European Journal of Inorganic Chemistry*, **34**, 1064–1069.

Holtstam D. and Anderson U. (2007). The REE minerals of the bastnäs-type deposits, south central Sweden. *Canadian Mineralogist*, **45**, 1073–1114.

Hudson. (2020). Hudson Resources Inc. https://hudsonresourcesinc.com/ (last accessed 30 October 2021).

Huston D. L., Maas R., Cross A., Hussey K. J., Mernagh T. P., Fraser G. and Champion D. C. (2016). The noleans bore rare-earth-element-phosphorus-uranium mineral system: geology, origin and post depositional modifications. *Mineralium Deposita*, **51**, 797–822.

IUPAC (2005). Nomenclature of inorganic chemistry. In: N. G. Connelly, T. Damhus, R. M. Hartshorn and A. T. Hutton (eds) (pp. 51–52, 336), RSC Publishing, Biddles Ltd Norfolk.

Jackson W. D. and Christiansen G. (1993). International strategic minerals inventory report-Rare-earth oxides: U.S. Geological Survey Circ. 930-N.

James C. (1911). Thulium I. *Journal of the American Chemical Society*, **33**(8), 1332–1344. doi: 10.1021/ja02221a007

Jébrak M., Marcoux E., Laithier M. and Skipwith P. (2014). Geology of Mineral Resources, 2nd edn. St. John's, NL. *Geological Association of Canada*, p. 662. ISBN 9781897095737.

Jia Y. and Liu Y. (2020). REE Enrichment during magmatic–hydrothermal processes in carbonatite- related REE deposits: a case study of the weishan REE deposit. *China Minerals*, **10**(1), 25. doi: 10.3390/min10010025

Karagiorgakis A. L., Schindler M. and Spiers G. A. (2018). Retention of rare earth elements in authigenic phases following biochemical dissolution of ore from elliot lake, Ontario. *Hydrometallurgy*, **177**, 9–20.

Karpeta W. P., Harwood R., Bucibo S., pKulwa A. and Musengi S. (2012). The Geology of Wigu Hill Rare Earth Carbonatite Complex. Internal report: Montero Mining, 42 p.

Kato Y., Fujinaga K., Nakamura K., Takaya Y., Kitamura K., Ohta J., Toda R., Nakashima T. and Iwamori H. (2011). Deep-sea mud in the Pacific Ocean as a potential resource for rare-earth elements. *Nature Geocience*, **4**, 535–539.

Kogarko L. N., Kononova V. A., Orlova M. P. and Wooley A. R. (1995). Alkaline Rocks and Carbonatites of the World. Part 2. Former USSR. Chapman & Hall, London.

Koppenol W. H. (2002). Naming of new elements (IUPAC recommendations 2002). *Pure and Applied Chemistry*, **74**, 787–791. doi: 10.1351/pac200274050787

Kovalenko V. I. and Yarmolyuk V. V. (1995). *Endogenous Rare Metal ore Formations and Rare Metal Metallogeny of Mongolia: Economic Geology*, **90**(3), 520–529.

Kumar A. (2011). A Review of Seabed and Placer Mining Deposits in India (Doctoral Dissertation). National Institute of Technology Rourkela, Odisha, India.

Lapin A. V., Tolstov A. V. and Kulikova I. M. (2016). Distribution of REE, Y, Sc, and Th in the unique complex rare-metal ores of the tomtor deposit. *Geochemistry International*, **54**, 1061–1078. doi: 10.1134/S0016702916120065

Lindenmayer Z. G. and Teixeira J. B. G. (1999). Ore genesis at the Salobo copper deposit, Serra dos Carajas. https://www.researchgate.net/publication/291904547 (last accessed 30 October 2021).

Linnemann E. (1886). Austrium, ein neues metallisches element. *Monatshefte für Chemie*, **7**, 121–123.

Ling M.-X., Zhang H., Li H., Liu Y.-L., Liu J., Li L.-Q., Li C.-Y., Yang X.-Y. and Sun W. (2014). The permian-triassic granitoids in bayan Obo, north China craton: a geochemical and geochronological study. *Lithos*, **190–191**, 430–439.

Lira R. and Ripley E. M. (1992). Hydrothermal alteration and REE-Th mineralization at the rodeo de los molles deposit, Las chacras batholith, central argentinia. *Contributions to Mineralogy and Petrology*, **110**, 370–386.

Liu R. and Wang R. C. (2016). Nano-sized rare earth minerals from granite-related weathering-type REE deposits in southern jiangxi. *Acta Petrologica et Mineralogica*, **35**, 617–626.

Liu S., Fan H. R., Yang K. F., Hu F. F., Rusk B., Liu X., Li X. C., Yang Z. F., Wang Q. W. and Wang K. Y. (2018). Fenitization in the giant bayan Obo REE-Nb-Fe deposit: implication for REE mineralization. *Ore Geology Reviews*, **94**, 290–309. doi: 10.1016/j.oregeorev.2018.02.006

Loiselle M. C. and Wones D. R. (1979). Characteristics and origin of anorogenic granites. *Geological Society of America*, Abstracts with Programs **11**(7), 468.

Lottermoser B. (1990). Rare earth mineralization within Mt. Weld carbonatite laterite, western Australia. *Lithos*, **24**, 151–167.

Mariano A. N. (1997). Mineralogical and textural analysis of bastnaesite ore from Dong Pao, northern Vietnam: confidential report to AMR Technologies Inc.

Marinsky J.-A. and Glendenin L. E. (1948). Promethium, the new name for element 61. *Nature*, **162**, 175.

Marinsky J.-A., Glendenin L. E. and Coryell C. D. (1947). The chemical identification of radioisotopes of neodymium and of element 61. *Journal of the American Chemical Society*, **69**(11), 2781–2785. doi: 10.1021/ja01203a059

Marshal J. L. and Marshal V. R. (2016). Rediscovery of the elements: the rare earths-the confusing years. *The Hexagon Spring*, **2016**, 72–77.

Meier H., Zimmerhackl E., Albrecht W., Bösche D., Hecker W., Menge P., Unger E. and Zeitler G. (1970). *Zeitschrift für Naturforschung*, **25a**, 1945–1953.

Moeller T. (1973). The lanthanides. In: Comprehensive Inorganic Chemistry, J. C. Bailer, H. J. Emeleus, R. Nyholm and A. F. Trotman-Dickenson (eds), Pergamon Press, New York, Vol. **3**, pp. 1–101.

Möller P. and Siebert C. (2016). Cycling of calcite and hydrous metal oxides and chemical changes of mayor element and REE chemistry in monomictic hardwater lake: impact on sedimentation. *Chemie der Erde – Geochemistry*, **76**, 133–148.

Möller P., Dulski P. and Bau M. (1994). Rare-earth element adsorption in a seawater profile above the east pacific rise. *Chemie der Erde – Geochemistry*, **54**, 129–149.

Möller P., Rosenthal E., Dulski P. and Geyer S. (2009). Characterization of recharge areas by rare earth elements and stable isotopes. In: The Water of the Jordan Valley, H. Hoetzl, P. Möller and E. Rosenthal (eds), Springer, Berlin Heidelberg, pp. 13–147.

Moore M., Chakhmouradian A. R., Mariano A. N. and Sidhu R. (2015). Evolution of rare earth mineraliztion in bear lodge carbonatite, wyoming: mineralogical and isotopic evidence. *Ore Geology Reviews*, **64**, 499–521.

Morteani G. and Preinfalk C. (1996). REE Distribution and REE carriers in laterites formed on the alkaline complexes of Araxa and Catalao (Brazil). In: Rare Earth Minerals-Chemistry, Origin and Ore Deposits, A. P. Lones, F. Wall and C. T. Williams (eds), Chapman & Hall, London, pp. 227–255. ISBN 0412610302.

Morteani G. and Satir M. (1989). The bastnaesite-fluorite-barite deposit of the kizilcaören district, eskisehir, Turkey. In: Lanthanides, Tantalum and Niobium, P. Möller, P. Černý and F. Saupé (eds), Special Publication No. 7 of the Society for Geology Applied to Mineral Deposits. Springer, Berlin, Heidelberg, Vol. **7**, pp. 189–194, doi: 10.1007/978-3-642-87262-4_7

Mosander C. G. (1843). On the new metals lanthanum and didymium, which are associated with cerium; and on erbium and terbium, new metals associated with yttria. *Philosophical Magazine Series* **3**, 23, 241–254.

Norton T. and Hillebrand W. F. (1875). Ueber metallisches Ce, La und didym. *Poggendorfs Annalen*, **156**, 466–476.

Nozaki Y., Zhang J. and Amakawa H. (1997). The fractionation between Y and Ho in the marine environment. *Earth and Planetary Science Letters*, **148**, 329–340.

Nozaki Y., Alibo D.-S., Amakawa H. M., Gamo T. and Hasumoto H. (1999). Dissolved rare earth elements and hydrography in the sulu Sea, *Geochimica et Cosmochimica Acta*, **63**, 2171–2181.

Ntiharirizwa S., Boulvais P., Poujol M., Branquet Y., Midende G., Morelli C. and Ntungwanayo J. (2019). The Gakara rare earth deposit, Burundi: geology, mineralogy, U-Th-Pb dating. 15th Biennial Meeting of the Society for Geology Applied to Mineral Deposits at Glasgow, Scotland. https://www.researchgate.net/publication/335382249 (last accessed 30 October 2021).

Ohta A. and Kawabe I. (2001). REE(III) adsorption onto Mn dioxide (δ-MnO_2) and Fe oxyhydroxide: Ce(III) oxidation by δ-MnO_2. *Geochimica et Cosmochimica Acta*, **65**, 695–703.

Otake F. T., Ito O. Y., Yokoyama T. D. and Chikanda S. T. (2019). Magmatic-hydrothermal processes associated with rare earth element enrichment in the kangankunde carbonatite complex, *Malawi. Minerals*, **9**(7), 442. doi: 10.3390/min9070442

Palme H., O'Neill H. S. C., Heinrich D. H. and Karl K. T. (2007). Cosmochemical estimates of mantle composition. In: Treatise on Geochemistry, K. K. Turekian and H. D. Holland (eds), Pergamon, Oxford, Vol. **2**, pp. 1–38.

Pedersen J. C. and Lecouteur P. C. (1991). The thor lake beryllium-rare metal deposit, northwest territories. In: Mineral Deposits of the Slave Province, Northwest Territories (Field Trip 13). Open-file Report-Geological Survey of Canada, Calgary, pp. 128–136.

Pistilli M. (2021). 10 Top Countries for rare earth metal production. https://investingnews.com/daily/resource-investing/critical-metals-investing/rare-earth-investing/rare-earth-producing-countries/ (last accessed 30 October 2021).

Powell D. (2011). Rare earth elements plentiful in ocean sediments. Science News, July 3, https://www.sciencenews.org/article/rare-earth-elements-plentiful-ocean-sediments (last accessed 30 October 2021).

Ritter S. (2017). A whole world for rare earths. *Chemical and Engineering News*, **95**(34), 30–34.

Robjohns N. (1984). Rare earths. *Mining Annual Review*, **1984**, 91–92.

Rollinson H. R. (1993). Using Geochemical Data: Evaluation, Presentation, Interpretation. Longman Scientific & Technical, Harlow, Essex, England. ISBN 9780582067011.

Rozelle P. L., Khadilkar A. B., Pulati N., Soundarrajan N., Klima M. S., Mosser M. M., Miller C. E. and Pisupati S. V. (2016). A study on removal of rare earth elements from U. *S. Coal Byproducts by ion Exchange. Metallurgical and Materials Transaction E*, **3**, 6–17. doi: 10.1007/s40553-015-0064-7

Rudnick R. L. and Gao S. (2014). Composition of the continental crust. In: Treatise of Geochemistry, 2nd edn, K. K. Turekian and H. D. Holland (eds), Pergamon, Oxford, Vol. **3**, pp. 1–64. doi: 10.1016/B0-08-043751-6/03016-4

Sahlström F., Jonsson E., Högdahl K., Troll V. R., Harris C., Jolis E. M. and Weis F. (2019). Interaction between high-temperature magmatic fluids and limestone explains 'bastnäs-type' REE deposit in central Sweden. *Scientific Reports*, **9**, 15203. doi: 10.1038/s41598-019-49321-8

Sanusi M. S. M., Ramli A. T., Hashim S. and Lee M. H. (2021). Radiological hazard associated with amang processing industry in peninsular Malaysia and its environmental impacts. *Ecotoxicology and Environmental Safety*, **208**, 111727.

Schmandt D. S., Cook N. J., Ciobanu C. L., Ehrig K., Wade B. P., Gilbert S., Vadim S. and Kamenetsky V. S. (2017). Rare earth element fluorocarbonate minerals from the olympic Dam Cu-U-Au-Ag deposit, south Australia. *Minerals*, **7**(10), 202. doi: 10.3390/min7100202

Sengupta D. and Van Gosen B. S. (2016). Placer-type rare earth element deposits. *Reviews in Economic Geology*, **18**, 81–100.

Siebert C. (2006). Saisonale chemische Variationen des See Genezareth, seiner Zuflüsse und deren Ursachen. PhD thesis, Free University Berlin, ISSN 1860-0387.

Shannon R. D. (1976). Revised effective ionic radii and systematic studies of interatomic distances in halides and chalcogenides. *Acta Crystallographica Section A*, **32**, 751–767. doi: 10.1107/S0567739476001551

Shi L. (1980). Geochemical features and petrogenesis of miayoa carbonatites, *Hupeh. Geochimica*, **9**(4), 345–355.

Simandl G. J. (2014). Geology and market-dependent significance of rare earth element resources. *Mineralium Deposita*, **49**(8), 889–904. doi: 10.1007/s00126-014-0546-z

Sjöqvist A. S. L., Cornell D. H., Andersen T., Erambert M., Ek M. and Leijd M. (2013). Three compositional varieties of rare earth element ore: eudialyte-group minerals from norra kärr alkaline complex, southern Sweden. *Minerals*, **3**(1), 94–120. doi: 10.3390/min3010094

Smith M. P., Moore K., Kavecsanszki D., Finch A. A. and Wall F. (2016). From mantle to critical zone: a review of large and giant-sized deposits of rare earth elements. *Geoscience Frontiers*, **7**(3), 315–334.

Spedding F. H., Voigt A. F., Gladrow E. M., Sleight N. R., Powell J. E., Wright J. M., Butler T. A. and Figard P. (1947). The separation of rare earths by ion exchange. II. Neodymium and praseodymium. *Journal of the American Chemical Society*, **69**(11), 2786–2792.

Spencer J. F. (1919). The Metals of the Rare Earths. Longmans, Green, and Co., New York, p. 135.

Staatz M. H., Shaw V. E. and Wahlberg J. S. (1972). Occurrence and distribution of rare earths in the lemhi pass thorium veins, idaho and Montana. *Economic Geology*, **67**, 72–82.

Taylor S. R. and McLennan S. M. (1995). The geochemical evolution of the continental crust. *Reviews of Geophysics*, **33**, 241–265. doi: 10.1029/95RG00262

Thomsen T. (1811). XLVII. Experiments on allanite, a new mineral from Greenland. *Philosophical Magazine*, **37**, 278–288.

Tucker R., Belkin H. E., Schulz K. J. and Peters S. G. (2012). A light rare-earth element (LREE) resource in the khanneshin carbonatite complex of southern Afghanistan. *Economic Geology*, **107**, 197–208.

Urbain M. G. (1908). Un nouvel élément, le lutécium, résultant du dédoublement de l'ytterbium de Marignac. *Comptes Rendus*, **145**, 759–762.

Urbain M. G. (1910). Lutetium und neoytterbium oder cassiopeium und aldebaranium. *Monatshefte für Chemie*, **31**, I–VI. doi: 10.1007/BF01530262

Wall F., Niku-Paavola V. N., Storey C., Müller A. and Jeffries T. (2008). Xenotime-(Y) from carbonatite dykes at lofdal, Namibia: unusual low LREE: HREE ratio in carbonatite, and first dating of xenotime overgrowths in zircon. *Canadian Mineralogist*, **46**, 861–877.

Wang D.-H., Yang J., Yan S., Xu J., Chen Y., Pu G. and Luo Y. (2001). A special orogenic-type rare earth element deposit in maoniuping, sichuan, China: geology and geochemistry. *Resource Geology*, **51**(3), 177–188.

Wang D.-H., Zhao Z., Yu Y., Dai J.-J., Deng M.-C., Zhao T. and Liu L.-J. (2018). Exploration and research progress on ion-adsorption type REE deposit in south China. *China Geology*, **1**(3), 415–424. doi: 10.31035/cg2018022

Weeks M. E. (1932). The discovery of the elements. XVI. The rare earth elements. *Journal of Chemical Education*, **9**(10), 1751–1773. doi: 10.1021/ed009p1751

Weigel F. (1963). Darstellung von metallischem promethium. *Angewandte Chemie*, **75**, 451. doi: 10.1002/ange.19630751009

Welsbach C. A. V. (1908). Die zerlegung des ytterbiums in seine elemente. *Monatshefte für Chemie*, **29**, 181–225. doi: 10.1007/BF01558944

Whalen J. B., Currie K. L. and Chappell B. W. (1987). A-type granites: geochemical characteristics, discrimination and petrogenesis. *Contributions to Mineralogy and Petrology*, **95**, 407–419.

White A. J. R. (1979). Sources of granite magmas. *Geological Society of America, Abstracts with Programs* **11**(7), 539.

Williams-Jones A. E., Migdisov A. A. and Samson I. M. (2012). Hydrothermal mobilisation of the rare earth elements-A tale of 'ceria' and 'yttria'. *Elements*, **2012**(8), 355–360.

Woolley A. R. (1987). Alkaline Rocks and Carbonatites of the World; Part 1, North and South America. British Museum of Natural History, London, UK. https://mrdata.usgs.gov/ree/show-ree.php?rec_id=250 (last accessed 30 October 2021).

Workman R. K. and Hart S. R. (2005). Major and trace element composition of the depleted mantle. *Earth and Planetary Science Letters*, **231**, 53–72.

Wu C. H. D. and Guo Z. (1989). Geochemistry of REE in the weathering crust of granites in longnan, jiangxi. *Acta Geologica Sinica*, **63**(4), 349–362.

Wu C. Z. and Bai G. (1996). Rare earth deposits in China. In: Rare Earth Minerals, Origin, and Ore Deposits, A. P. Jones, F. Fall and C. T. Williams (eds), Mineralogical Society Series 7. Chapman and Hall, London, pp. 281–310.

Yan H. M., Zhao W. W. and Zhou M.-F. (2017). Nature of parent rocks, mineralization styles and ore genesis of regolith-hosted REE deposits in south China: an integrated genetic model. *Journal of Asian Earth Sciences*, **148**, 65–95. doi: 10.1016/j.jseaes.2017.08.004. ISSN 1367-9120.

Yang X.-M. and Bas M. J. (2004). Chemical compositions of carbonate minerals from bayan Obo, inner Mongolia, China: implications for petrogenesis. *Lithos*, **72**(1–2), 97–116. doi: 10.1016/j.lithos.2003.09.002

Yang X. J., Lin A., Li X-L. and Wu Y. (2013). China's ion-adsorption rare earth resources, mining consequences and preservation. *Environmental Development*, **8**(1), 131–136.

Yang K., Fan H., Piajno F. and Li X. (2019). The bayan Obo (China) giant REE accumulation conundrum elucidated by intense magmatic differentiation of carbonatite. *Geology*, **47** (12), 11981202.

Zhang Z. (1990). A study on weathering crust ion adsorption type REE deposits, proterozoic Fe–Cu metallogeny and supercontinental cycles of the southwestern China. *Contribution to Geology and Mineral Resources Research*, **5**, 57–71 (in Chinese with English abstract).

Chapter 3

Occurrence and detection of the rare earth elements

Simon M. Jowitt

3.1 INTRODUCTION

The rare earth elements (REE) play a vital role in modern society and have a variety of uses in modern technology, low- and zero-CO_2 energy and transport, and in the defense sector (e.g. Jowitt *et al.*, 2018a; Table 3.1). However, although knowledge of the role of these elements in human society has increased rapidly in the last two decades their supply remains problematic for a number of reasons. This has resulted in their designation as a group of critical elements in the vast majority of criticality studies and reports (e.g. Hayes & McCullough, 2018; Jowitt *et al.*, 2018b). The REE are also not all that rare, having average crustal abundances that are higher than a range of other commonly used and economically important elements such as Cu, Au and Pt (Table 3.1). This chapter aims to provide a basic overview of the REE, their mineralogy, the mineral deposits that commonly host these critical elements, and the implications for future supply of these elements.

The REE consist of the 15 lanthanide elements as outlined in Table 3.1 and Figure 3.1 (IUPAC, 2005). Each of these elements have distinctive characteristics and usages that reflect their chemical characteristics, with the REE typically divided by electron shell configuration into the light REE (LREE; La to Gd) and the heavy REE (HREE; Tb to Lu; Table 3.1 and Figure 3.1). However, the global mining industry often uses a slightly differing classification (as also shown in Table 3.1 and Figure 3.1 for clarity), where the HREE includes Pm, Sm, Eu and

doi: 10.2166/9781789062236_0045

Table 3.1 Outline of the rare earth elements, their chemical compositions, average crustal abundances, and common uses.

Element	Atomic Number	Element Name	Classification[1]		Average Crustal Abundance (ppm)[2]	Average 2020 Price (US$/kg oxide)[3]	Typical Uses
			IUPAC	Minerals Industry			
La	57	Lanthanum	Light	Light	31	2	Optics, batteries, catalysis
Ce	58	Cerium	Light	Light	63	2	Chemical applications, coloring, catalysis
Pr	59	Praseodymium	Light	Light	7.1	—	Magnets, lighting, optics
Nd	60	Neodymium	Light	Light	27	47	Magnets, lighting, lasers, optics
Pm	61	Promethium	Light	Heavy	N/A[4]	—	Limited use due to radioactivity, used in paint and atomic batteries; very rare in nature
Sm	62	Samarium	Light	Heavy	4.7	—	Magnets, lasers, masers
Eu	63	Europium	Light	Heavy	1	31	Lasers, lighting, medical applications
Gd	64	Gadolinium	Light	Heavy	4	—	Magnets, glassware, lasers, X-ray generation, computer applications, medical applications

Tb	65	Terbium	Heavy	Heavy	0.7	628	Lasers, lighting
Dy	66	Dysprosium	Heavy	Heavy	3.9	258	Magnets, lasers
Ho	67	Holmium	Heavy	Heavy	0.83	—	Lasers
Er	68	Erbium	Heavy	Heavy	2.3	—	Lasers, steelmaking
Tm	69	Thulium	Heavy	Heavy	0.3	—	X-ray generation
Yb	70	Ytterbium	Heavy	Heavy	2	—	Lasers, chemical industry applications
Lu	71	Lutetium	Heavy	Heavy	0.31	—	Medical applications, chemical industry applications
Sc	21	Scandium	N/A	N/A	14	—	Alloys in aerospace engineering, lighting
Y	39	Yttrium	Heavy	Heavy	21	—	Lasers, superconductors, microwave filters, lighting

Adapted from Weng *et al.* (2014, 2015) and Jowitt *et al.* (2013, 2018a, 2018b).

[1]The chemical classification of the REE uses the configuration of electrons in the outer shell of the element where the LREE have an increasing number of unpaired electrons in their 4f shells, starting at lanthanum, which has zero unpaired electrons, through to gadolinium, which has seven unpaired electrons. In comparison, the HREE have both clockwise and counterclockwise spinning electrons. Sc and Y are chemically similar to these elements and are also included, with Y classified as an HREE. However, the properties of Sc are not similar enough to either LREE or HREE to allow further chemical classification.

[2]From Rudnick and Gao (2003).

[3]Prices are for REE oxides and are from United States Geological Survey (USGS), 2021.

[4]Concentration too low to assess, reflecting the short radioactive half-life of this element.

'–' = not available. IUPAC = International Union of Pure and Applied Chemistry.

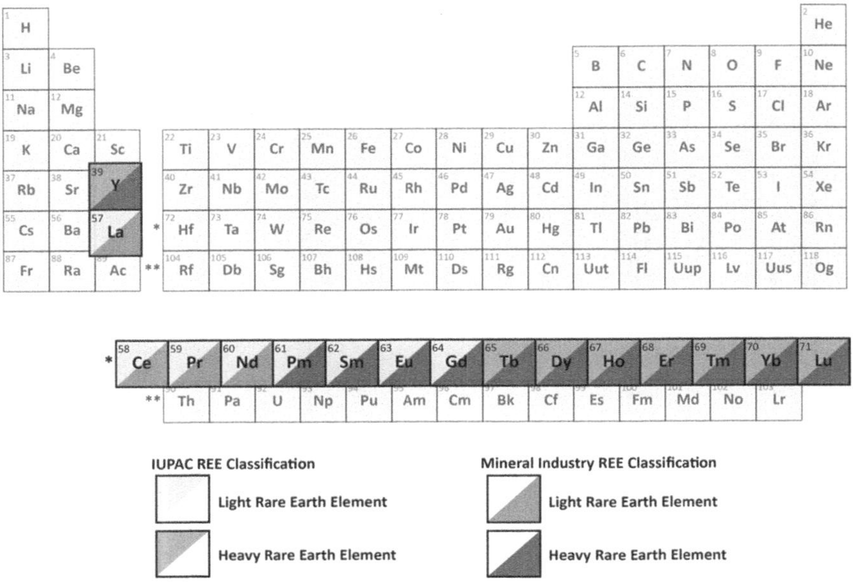

Figure 3.1 Periodic table of elements highlighting the rare earth elements and the two different classification schemes as discussed in the text.

Gd. This chapter uses the IUPAC classification for consistency with the majority of the scientific literature focused on the REE. The chemically similar elements Sc and Y are also frequently grouped with the REE, although the geochemical behavior of Sc differs from the rest of the REE and Y. This means that although the REE and Y commonly occur together in mineral deposits, Sc is not enriched by the same processes and hence is not commonly extracted from REE $\pm$ Y mineralization. Both Sc and Y are also incorporated into the global mining industry grouping, with Y classified as an HREE despite having a relatively low molecular weight, and Sc not formally classified as either an LREE or an HREE but often considered together with the REE as a whole.

The REE have a wide variety of uses in modern society and are frequently used in compounds or alloys where they impart specific characteristics to end-products, such as enhanced magnetism (Table 3.1). This means that the REE are inherently unsuitable for substitution, meaning that it is not possible to substitute another element for one of the REE in an end-product without some loss in usability or performance (e.g. Graedel *et al.*, 2015). Their increased use in modern society is demonstrated by the rapid increase in REE production in the last 70 years (Figure 3.2). This diagram clearly shows the different episodes of REE production, from early very minor production, predominantly from monazite extracted from heavy mineral sands (HMS), to the era of carbonatite-dominated

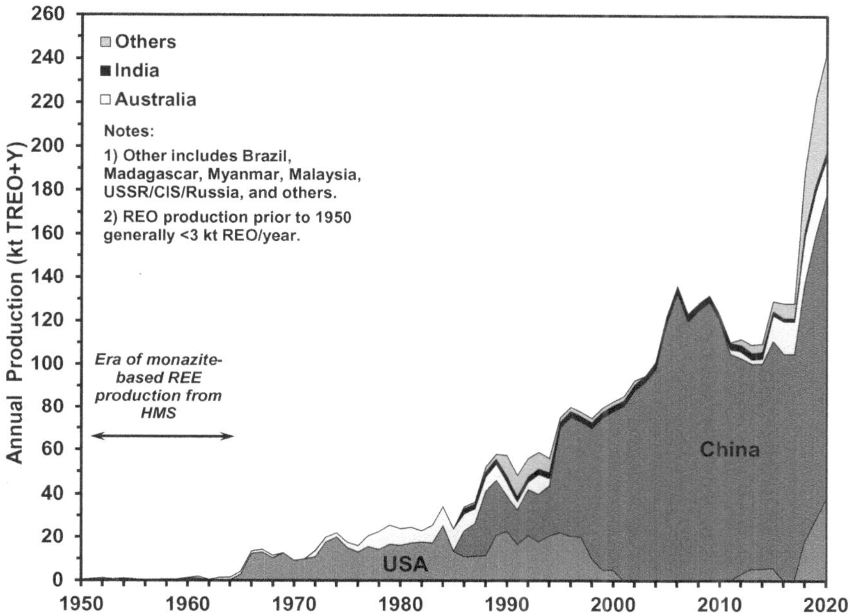

Figure 3.2 Global REE production from 1950 to 2020; adapted from Weng *et al.* (2015) with additional data from USGS (2021) and Jowitt *et al.* (2020). REO: rare earth oxides; TREO: total rare earth oxides; REE: rare earth elements; HMS: heavy mineral sand.

REE extraction, first from Mountain Pass in California and then from Bayan Obo in China. The latest phase in REE production started around 2010, with (i) increased production from a number of countries outside of the USA such as Australia, (ii) more transparent reporting of production from countries such as Myanmar where production was formerly attributed to China, and (iii) the eventual reopening of the mine at Mountain Pass.

This increased production also coincided with an increased amount of mineral exploration and research into the REE, both in terms of the processes associated with REE mineral deposit formation as well as improvements in approaches to extract the REE from the minerals that host these elements and the further splitting of the resulting REE concentrates produced during mining into individual REE phases (Table 3.2). The increase in demand evident from the increased production shown in Figure 3.2 also demonstrates the reliance that the REE supply chain has on primary production: very little REE production is derived from secondary sources such as mining waste or tailings and REE recycling rates are very low (typically $<1\%$; Jowitt *et al.*, 2018b) for a number of reasons. These include the complexity of end-uses of the REE, the inherent

Table 3.2 Outline of major REE-bearing mineral phases, their REO contents, their discovery history, and their potential for REE extraction.

Mineral	Composition	REO Content (wt.%)	First Discovered	Mineral Processing Routes Established?	Use for REE Production?
Aeschynite	$(Ce,Ca,Fe,Th)(Ti,Nb)_2(O,OH)_6$	36	Norway; Ural Mountains, Russia		
Allanite (orthite)	$(Ce,Ca,Y)_2(Al,Fe)_3(SiO_4)_3(OH)$	3 to 51	1810: Aluk Island, Greenland		
Anatase	$(Ti,REE)O_2$	3	1801: St Christophe-en-Oisans, France		
Ancylite–(Ce)	$SrCe(CO_3)_2(OH) \cdot H_2O$	46 to 53	1899: Narssârssuk, Narsaq, Greenland	Yes	
Apatite	$Ca_5(PO_4,CO_3)_3(F,Cl,OH)$	19	Bronze Age (3300–1200 BC); widespread		
Bastnäsite– (Ce)	$(Ce,La)(CO_3)F$	70 to 74	1838: Bastnäs mine, Västmanland, Sweden	Yes	Yes
Brannerite	$(U,Ca,Y,Ce)(Ti,Fe)_2O_6$	6	1920: Kelly Gulch, Stanley, Idaho, USA		
Britholite–(Ce)	$(Ce,Ca)_5(SiO_4,PO_4)_3(OH,F)$	56	1901: Naujakasik, Narsaq, Greenland		
Brockite	$(Ca,Th,Ce)(PO_4) \cdot H_2O$		1962: Bassick mine, Colorado, USA		
Calcio–ancylite (Ce)	$(Ca,Sr)Ce_3(CO_3)_4(OH)_3 \cdot H2O$	60	Kola Peninsula, Russia		

Cerianite–(Ce)	$(Ce_4+,Th)O_2$	81	1955: Firetown, Sudbury, Ontario, Canada	
Cerite–(Ce)	$Ce_9^{3+}Fe^{3+}(SiO_4)_6[SiO_3(OH)](OH)_3$	60	1751: Bastnäs mine, Västmanland, Sweden	
Cheralite–(Ce)	$(Ce,Ca,Th)(P,Si)O_4$	5	Chera, Travancore, India	
Chevkinite	$(Ca,Ce,Th)_4(Fe^{2+},Mg)_2(Ti,Fe^{3+})_3Si_4O_{22}$		1839: Ural Mountains, Russia	
Churchite–(Y)	$YPO_4 \bullet 2H_2O$	44	1923: Maffei mine, Bavaria, Germany	
Crandallite	$CaAl_3(PO_4)_2(OH)_5 \bullet H_2O$		1917: Brooklyn mine, Silver City, Utah, USA	
Doverite (synchysite–(Y)	$YCaF(CO_3)_2$		1951: Scrub Oak mine, New Jersey, USA	
Eudialyte	$Na_4(Ca,Ce)_2(Fe^{2+},Mn^{2+},Y);ZrSi_8O_{22}(OH,Cl)_2$	1 to 10	1819: Kangerdluarssuq Firth, Narsaq, Greenland	Yes
Euxenite–(Y)	$(Y,Ca,Ce,U,Th)(Nb,Ta,Ti)_2O_6$	<40	1840: Jølster, Sogn og Fjordane, Norway	
Fergusonite–(Ce)	$(Ce,La,Y)NbO_4$	47	1806: Kikertaursak, Greenland; Ukraine	Yes
Fergusonite–(Y)	$YNbO_4$		1826: Kikertaursak, Greenland	Yes
Florencite–(Ce)	$CeAl_3(PO_4)_2(OH)_6$	32	pre–1951: Mata dos Criolos, Minas Gerais, Brazil	
Fluocerite–(Ce)	$(Ce,La)F_3$		1845: Broddbo and Finnbo, Dalarna, Sweden	

(Continued)

Table 3.2 Outline of major REE-bearing mineral phases, their REO contents, their discovery history, and their potential for REE extraction (*Continued*).

Mineral	Composition	REO Content (wt.%)	First Discovered	Mineral Processing Routes Established?	Use for REE Production?
Fluocerite–(La)	$(La,Ce)F_3$		1969: Zhanuzak, Kazakhstan		
Fluorapatite–(Ce)	$(Ca,Ce)_5(PO_4)_3F$	0 to 21	1860: Greifenstein Rocks, Saxony, Germany		
Fluorite	(Ca,REE)F		1530: England; Czech Republic; Germany		
Gadolinite	$(Ce,La,Nd,Y)_2Fe^{2+}Be_2Si_2O_{10}$	40	1788: Ytterby mine, Resarö, Sweden	Yes	
Gagarinite–(Y)	$NaCaY(F,Cl)_6$		pre–1961: Akzhaylyautas Mountains,Kazakhstan		
Gerenite–(Y)	$(Ca,Na)_2(Y,REE)_3Si_6O_{18} \cdot 2H_2O$		1998: Strange Lake, Quebec–Labrador, Canada		
Gorceixite	$(Ba,REE)Al_3(PO_4)(PO_3OH)(OH)_6$		1906: Ouro Preto, Minas Gerais, Brazil		
Goyazite	$SrAl_3(PO_4)_2(OH)_5 \cdot H_2O$		1884: Diamantina, Minas Gerais, Brazil		
Hingganite–(Y)	$(Y,Yb,Er)_2Be_2Si_2O_8(OH)_2$		1984: Greater Hinggan Mountains, China		
Huanghoite–(Ce)	$BaCe(CO_3)_2F$	38	1960s: Bayan Obo deposit, Inner Mongolia	Yes	

Hydroxylbastnäsite–(Ce)	$(Ce,La)(CO_3)(OH,F)$	75	Kola Peninsula and Ural Mountains, Russia		
Iimoriite–(Y)	$Y_2(SiO_4)(CO_3)$		pre–1970: Honshu Island, Japan		
Kainosite–(Y)	$Ca_2(Y,Ce)_2Si_4O_{12}(CO_3) \bullet H_2O$	38	1885: Hidra, Vest-Agder, Norway		
Loparite–(Ce)	$(Ce,Na,Ca)(Ti,Nb)O_3$	32 to 34	1925: Maly Mannepakhk, Kola Peninsula, Russia	Yes	Yes
Monazite–(Ce)	$(Ce,La,Nd,Th)PO_4$	35 to 71	1823: Ilmen Mountains, Ural Mountains, Russia	Yes	Yes
Mosandrite	$(Na,Ca,Ce)_3Ti(SiO_4)_2F$	<65	1841: Låven Island, Larvik, Norway		
Parisite–(Ce)	$Ca(Ce,La)_2(CO_3)_3F_2$	59	Muzo mine, Colombia	Yes	
Perovskite	$(Ca,REE)TiO_3$	≤ 37	1839: Achmatovsk mine, Ural Mountains, Russia		
Pyrochlore	$(Ca,Na,REE)_2Nb_2O_6(OH,F)$		1826: Stavern, Larvik, Vestfold, Norway		
Rhabdophane–(Ce)	$(Ce,La)PO_4 \bullet H_2O$		pre–1992: Fowey Consols, Cornwall, England		
Rhabdophane–(La)	$(La,Ce)PO_4 \bullet H_2O$		1883: Salisbury Iron mines, Connecticut, USA		
Rinkite	$(Ca,Ce)_4Na(Na,Ca)_2Ti(Si_2O_7)_2F_2(O,F)_2$		1884: Kangerdluarssuk, Narsaq, Greenland		
Samarskite–(Y)	$(Y,Ce,U,Fe^{3+})_3(Nb,Ta,Ti)_5O_{16}$	12	1839: Blyumovskaya pit, Ural Mountains, Russia		

(Continued)

Table 3.2 Outline of major REE-bearing mineral phases, their REO contents, their discovery history, and their potential for REE extraction (*Continued*).

Mineral	Composition	REO Content (wt.%)	First Discovered	Mineral Processing Routes Established?	Use for REE Production?
Steenstrupine–(Ce)	$Na_{14}Ce_6Mn^{2+}Mn^{3+}Fe_2^{2+}(Zr,Th)(Si_6O_{18})_2(PO_4)_7 \cdot 3H_2O$		1853–54: Kangerdluarssuk, Narsaq, Greenland	Yes	
Synchysite–(Ce)	$Ca(Ce,La)(CO_3)_2F$	49 to 52	1953: Narsarsuk, Greenland	Yes	
Thalénite–(Y)	$Y_3Si_3O_{10}(OH)$	63	Österby, Dalama, Sweden		
Titanite (sphene)	$(Ca,REE)TiSiO_5$	≤ 3	1795: Hauzenberg, Bavaria,Germany		
Uraninite	$(U,Th,Ce)O_2$		1772: Jáchymov, Bohemia, Czech Republic		
Vitusite–(Ce)	$Na3(Ce,La,Nd)(PO_4)_2$		1980: Kola Peninsula, Russia; Narsaq, Greenland		
Xenotime–(Y)	YPO_4	52 to 67	1824: Ytterby mine, Sweden; Hidra, Norway	Yes	Yes
Yttrofluorite	$(Ca,Y)F_2$		1911: Hundholmen, Tysfjord, Norway		
Yttrotantalite–(Y)	$(Y,U,Fe^{2+})(Ta,Nb)O_4$	<24	Ytterby mine, Resarö, Sweden		
Zircon	$(Zr,REE)SiO_4$	<5	1783: Sri Lanka; 1789: Germany	Yes	

Adapted from Weng *et al.*, 2013 and Wall, 2014.

difficulty involved in separating the individual REE from each other to yield pure single elements, and the long lifetimes involved in certain end-uses of the REE (e.g. permanent magnets; Jowitt *et al.*, 2020). All of this means that improved knowledge of the processes involved in REE deposit formation and the approaches used to detect and discover these resources is key.

3.2 MINERALOGY OF THE REE

The REE are generally never present as native metals within the geological environment but instead typically combine with other elements to form a variety of carbonate, phosphate and silicate minerals (Table 3.2). However, REE-focused beneficiation and processing approaches have only been developed for a few of these minerals, with monazite, bastnäsite and loparite being the only commercially important sources of the REE (Table 3.2). It is also important to note that REE processing approaches are often specific to individual mineral deposit types or even ore deposits, contrasting with other sectors of the mining industry where mineral processing approaches can be widely applied to a number of different types of deposit (e.g. Au heap leaching or Cu solvent extraction and pyrometallurgy). This is one of the key issues surrounding expansion of the REE mining sector and makes the development of new mines and projects typically more complex than is the case in other parts of the minerals industry.

The often expensive processing facilities needed for REE extraction also increase capital costs for new projects, again slowing down or delaying the development of new REE deposits. This is compounded by the fact that, unlike porphyry Cu (Sillitoe, 2010), orogenic Au (Bierlein *et al.*, 2006) or volcanogenic massive sulfide Cu-Pb-Zn systems (Mortensen *et al.*, 2015) which often produce multiple, spatiotemporally related deposits that define individual mineral provinces, REE deposits are often solitary and may be located distal from other similar REE deposits. This makes it unlikely that a single processing plant could use REE concentrates derived from several different mines although REE concentrates from the mines at Mountain Pass in the USA and Mount Weld in Australia are exported for refining to China and Malaysia, respectively. This contrasts with the base metal industry, where for example refineries and smelters often process concentrates derived from several different mines (e.g. Molymet in Chile; Naumov, 2007). Detailed analysis of the processing of different types of REE mineralization is beyond the scope of this chapter, but the reader is referred to useful reviews in Wall (2014) for more details of this.

3.3 PRIMARY SOURCES OF THE RARE EARTH ELEMENTS

Global REE production remains dominated by the exploitation of carbonatite or weathered carbonatite mineralization along with relatively minor volumes (but of major importance in terms of heavier REE production) of production from ionic clay and other sources. Carbonatite REE producers include the major Bayan Obo mine in China, Mountain Pass in the USA, Mount Weld in Australia, and Araxa in Brazil. Other producers include the hydrothermal Browns Range deposit in Australia, ionic clay deposits in southern China and poorly quantified production of the REE from probably carbonatitic sources in Myanmar (Figure 3.2). Each of these deposits contains different concentrations and abundances of the REE as well as different proportions of the LREE and HREE (even between carbonatites, which are typically LREE enriched), complicating issues around mineral processing and the production of refined REE products. Increasing demand for the REE has also accelerated the exploration for REE deposits of all types, including potential production as a by-product of other minerals and metals (e.g. Mudd & Jowitt, 2016).

Despite increased interest and investment in REE exploration, global REE production remains dominated by Chinese operations. This is clearly evident in Figure 3.2, where Chinese production has dominated global REE production statistics since the early 1990s, if not before. The relatively recent identification of production from Myanmar that formerly flowed through China (and hence was attributed to Chinese mineral deposits) is also evident in Figure 3.2 from around 2010 onwards, as shown by the increased proportion of REE production from other countries in this diagram. The relatively recent reopening of the Mountain Pass mine in California (USA) is also evident in Figure 3.2, although it should be noted that although this mine at the time of writing does produce REE concentrates, this material is exported to China for further processing rather than being processed on site (as was the case during the majority of the Molycorp era of production from Mountain Pass). This means that although the USA was self-reliant in terms of REE supply and demand during the majority of the 20th century, the situation that developed by the late 1990s, where USA REE demand was 100% reliant on imports (Humphries, 2012), continues to the present day. Similar situations exist within countries that either formerly or currently produce the REE, with production from Mount Weld in Australia being shipped to Malaysia for processing rather than being processed within Australia as mentioned above.

The increase in demand for the REE that has been met by the increased primary production of these elements (Figure 3.2) reflects the growing use of these elements in a variety of areas, including electronics, oil refineries, low- and zero-CO_2 energy generation and storage and medical and defense technologies, including fluid cracking catalysts, REE alloys in batteries, specialist magnets for wind turbines or hybrid and full electric cars and personal electronic devices and phosphors in liquid crystal displays (LCDs; e.g. Jowitt *et al.*, 2018a). This pattern of increased demand is likely to increase further given the need for REE in the global transition to low- and zero-CO_2 energy generation and storage (among other uses; e.g. Lee *et al.*, 2020), meaning in turn that further REE mines are likely to be needed to ensure supply meets demand for these critical elements.

As outlined in Table 3.2, the REE within REE mineral deposits are typically hosted by a limited range of minerals such as bastnäsite ($(Ce,La,Y)CO_3F$), monazite ($(La–Gd,Th)PO_4$) and xenotime (YPO_4), with even fewer of these minerals having identified processing routes that enable the extraction of the REE. This is not to say that REE minerals are uncommon or are not diverse: more than 200 minerals are known to contain the REE in significant amounts (Table 3.2; e.g. Christie *et al.*, 1998; Hoatson *et al.*, 2011). These minerals are formed as a result of a diverse range of geological processes and hence are found in a wide range of igneous, sedimentary and metamorphic rocks, although typically not at concentrations sufficient to warrant economic evaluation. Equally importantly, the REE can be adsorbed onto other minerals, as is the case for ionic clay-type REE deposits. Here, the REE are bound onto rather than incorporated into the associated minerals, meaning they are typically present at low concentrations but can be liberated relatively easily by leaching, a common approach to REE extraction from ionic clay deposits (e.g. Sanematsu & Watanabe, 2016).

Although a number of different schemes can be used to classify REE deposits (e.g. Chakhmouradian & Wall, 2012; Kanazawa & Kamitani, 2006; Long *et al.*, 2010; Mariano & Mariano Jr, 2012; Walters *et al.*, 2010), this chapter uses the approach developed by Weng *et al.* (2013, 2015) to metallogenically subdivide REE deposits based on the geological processes that form or concentrate the REE minerals within these systems. As with any classification, this is naturally a simplification of the complexity present within the geological systems that concentrate the REE although this approach is more comprehensive than some of the other classifications outlined above. The approach used divides REE deposits into three main families of processes, namely igneous, hydrothermal and sedimentary, before subdividing these families further into groups and then individual deposit types (Table 3.3). However, as is the case with all mineral deposit classification schemes, the classification of REE deposits using this approach is reliant on available data, typically obtained from scientific publications or publicly available reporting from mining and exploration companies.

In general terms, primary REE mineral deposits can also be split into two main groups. These are main sources such as carbonatites that have formed the basis of the global REE industry for the past few decades (e.g. Mountain Pass and Bayan Obo), and potential sources that represent opportunities to expand the REE mining sector (e.g. Round Top) but as yet have not produced significant amounts of REE (sometimes as a result of hurdles relating to the often difficult processing of these deposits). This section outlines the key processes involved in the formation of REE deposits before providing more specific details on important groups of REE deposits as previously developed by Weng *et al.* (2013, 2015).

Table 3.3 Process-based classification of major REE deposits (adapted from Weng *et al.*, 2013) and selected key examples of individual REE deposit types.

Family	Mineral Deposit Type		Key Examples
Igneous	Silica Undersaturated	Carbonatite	Bayan Obo, China; Araxá, Brazil; Karonge, Burundi; Mountain Pass, USA; Nolans Bore, Australia; Steenkampskraal, South Africa
		Alkaline Complexes and Alkaline Pegmatites	Khibina and Lovozero, Russia; Norra Kärr, Sweden; Bokan, USA; Thor Lake, Canada; Kipawa Lake, Canada; Kola Peninsula, Russia
	Silica Saturated to Oversaturated	Felsic Volcanic/Rhyolite	Round Top, USA; Foxtrot, Canada
		Granites and Granitic Pegmatites	Khibina Massif, Russia; Motzfeldt, Greenland; Ytterby, Sweden
Hydrothermal	Iron Oxide Copper Gold (IOCG)		Olympic Dam, Australia; Milo, Australia
	Skarn	Granite-related	Mary Kathleen, Australia
		Carbonatite-related	Saima, China
	Unconformity-related		Browns Range, Australia
Secondary/ Sedimentary	Heavy Mineral Sands		WIM150, Australia; Monazite stockpile in India
	Laterite/Soil/Clay		Tantalus, Madagascar
	Tailings		Steenkampskraal, South Africa; Port Pirie, Australia; Mary Kathleen, Australia
	Shale-hosted		Buckton, Canada;
	Alluvial/Placer		Charley Creek, Australia; India; Sri Lanka; Florida, USA

3.4 PROCESSES INVOLVED IN THE FORMATION OF REE DEPOSITS

There are a variety of different geological processes that can generate REE enrichments up to economic levels and/or generate specific mineralogical assemblages that are well-suited for the extraction of the REE. The formation of individual REE deposits may involve one or more of these types of processes (e.g. igneous then hydrothermal, or igneous then sedimentary). This is especially true for the economically important carbonatite family of REE deposits, which almost always involve some type of late hydrothermal activity in and around carbonatites as a result of ubiquitous sub-solidus alteration (e.g. Anenburg *et al.*, 2020). As such, these descriptions should be considered a general introduction to the geological processes involved in REE deposit formation.

3.4.1 Igneous processes involved in REE deposit formation

The majority of the REE behave incompatibly within most igneous systems (with a few key exceptions). This means that they are concentrated in melts that are generated by low degree partial melting of crustal or mantle material, especially metasomatically enriched regions of the mantle that contain elevated concentrations of the REE. This also includes the generation of unusual but important melts such as carbonatites that are ubiquitously enriched in the REE, especially the LREE. Igneous processes after melting, such as differentiation and fractionation, can also significantly increase the REE concentration of the remaining melt within the system (e.g. Chakhmouradian & Zaitsev, 2012). This holds true as long as the REE continue to behave incompatibly; this may change after prolonged fractionation, a process that can lead to some of the LREE becoming compatible in very evolved felsic igneous systems (e.g. Cameron & Cameron, 1986; Jowitt *et al.*, 2017; Miller & Mittlefehldt, 1982).

The same applies to carbonatite-related REE deposits where REE mineralization is often associated with the most evolved parts of carbonatite systems. This may also relate to the behavior of fluids within magmatic systems, where the later stages of evolution of a magma can involve fluid exsolution, a process invoked in the formation or enhancement of REE mineralization within carbonatite and rhyolite-associated REE deposits (e.g. Jowitt *et al.*, 2017). All of these processes operating either in isolation or in combination can generate REE-enriched magmas and eventually REE-enriched igneous rocks. This indicates that the majority of igneous-type REE deposits are related to either rocks formed from magmas generated by very low degree partial melting or by extreme fractionation, along with other associated processes such as undercooling and zone refining (e.g. London, 2018), both of which are thought to be key steps in the formation of REE-bearing pegmatites.

The evolved nature of the magmas and rocks resulting from these processes, such as carbonatites or alkaline igneous complexes, also form ideal sources for REE-enriched hydrothermal fluids. As mentioned above, this has generated several REE deposits that contain both primary igneous and hydrothermal REE mineralization (e.g. Bayan Obo), suggesting that combinations of geological processes are vital to the formation of economic REE mineral deposits.

3.4.2 Hydrothermal processes involved in REE deposit formation

Although the REE are thought to be generally immobile during the majority of hydrothermal activity (e.g. Anenburg & Mavrogenes, 2018), this certainly does not mean that hydrothermal processes are not important in REE deposit formation. Instead, the mobilization, deposition and concentration of the REE during hydrothermal activity are typically thought to require somewhat unusual styles of fluid compared to the vast majority of other hydrothermal mineral deposits. Hydrothermal REE systems can be split into two end-members, albeit with some overlap between these two groups. These are high temperature magmatic-hydrothermal systems and systems involving F- and Cl-bearing hydrothermal fluids, with the latter being somewhat similar to some hydrothermal fluids involved in the formation of non-REE hydrothermal mineral deposits (e.g. Spandler *et al.*, 2020).

In general terms, higher temperature hydrothermal fluids have higher REE solubilities than lower temperature fluids, as evidenced by experimental work by Williams-Jones *et al.* (2012) who determined that Nd solubilities in moderate salinity brines (10 wt.% NaCl) increased from ~1 ppm to ~200 ppm over a temperature change from 300°C to 400°C. Importantly, the presence of ligands such as F, Cl and Li can also increase the solubility of the REE as well as elements such as U that are also associated with the REE (e.g. McGloin *et al.*, 2016; McPhie *et al.*, 2011; Williams-Jones *et al.*, 2012; Xing *et al.*, 2019). McPhie *et al.* (2011) and Xing *et al.* (2019) also suggest that F-bearing hydrothermal fluids can break down the often refractory minerals that host the REE. This dissolution increases the concentrations of the REE in the associated hydrothermal fluids, enabling their transport and eventual precipitation within REE-enriched mineral deposits (although this process may not always generate economic REE mineralization). This or similar models have been suggested for Olympic Dam (McPhie *et al.*, 2012) and other REE-enriched IOCG deposits of the Gawler Craton (Skirrow *et al.*, 2007) and elsewhere (e.g. Ernst & Jowitt, 2013). This model is also analogous to the model of McGloin *et al.* (2016) for uranium mineralization, which also involves F-rich fluids.

The F-rich fluids invoked in the formation of these mineral deposits are thought to be derived from either evolved igneous or carbonatitic magmas or rocks. The former is exemplified by the fact that the REE mineralization associated with carbonatites is also often accompanied by hydrothermal high-F REE minerals in surrounding skarns or alteration haloes (e.g. Anenburg & Mavrogenes, 2018). In addition, acidic hydrothermal fluids exsolved from granites can interact with the high-F igneous rocks that form during the same silicic large igneous province events that form the granitic source of the fluids (e.g. Ernst & Jowitt, 2013; Pankhurst *et al.*, 2011). This interaction can remove the F within fluorite (CaF_2) in the solid granites, generating the high-F hydrothermal fluids mentioned above that could act as transportation agents for the REE within hydrothermal mineralizing systems (e.g. Ernst & Jowitt, 2013; McPhie *et al.*, 2012). However, the fact that Cl may be a more important ligand in these fluids (e.g. Williams-Jones *et al.*, 2012) suggests that the presence of F may not be the only factor in the transportation of the REE in hydrothermal fluids. As mentioned above, this is especially true at high temperatures where REE-F complexes decrease in stability at temperatures $> \sim350°C$ (Williams-Jones *et al.*, 2012), although F complexes may effectively transport the REE at temperatures below this. This tentatively suggests that lower temperature hydrothermal REE deposits may be associated with F-bearing fluids, whereas higher temperature hydrothermal REE deposits are more likely to be associated with Cl-bearing fluids. Both of these fluids most likely deposit their REE during interaction with cooler and pH-neutralizing rocks or fluids during fluid-rock or fluid-fluid interaction (Williams-Jones *et al.*, 2012), although variations in solubilities and complexing behavior means that the deposition of the REE may not always be contemporaneous or even spatially associated with the deposition of other metals, such as Cu, Au or U.

3.4.3 Sedimentary, secondary and placer processes

The majority of economically important REE minerals (e.g. Table 3.2) are refractory, meaning they are resistant to alteration and weathering. This, coupled with the fact that these minerals are often denser than the majority of silicate minerals, means that transportation by water (or more rarely as a result of aeolian processes) after their erosion from their original hosting rock can lead to the development of sedimentary, secondary or placer REE enrichments and deposits. This means that REE minerals are generally concentrated within sedimentary (e.g. shale-hosted deposits such as at Buckton (Canada); although these may also be associated with hydrothermal REE mineralization), secondary (e.g. laterite and ionic clay deposits) and placer (e.g. heavy mineral sands; Mudd & Jowitt, 2016) type REE deposits, where the latter two deposit types host REE mineralization that is often associated with other dense, refractory minerals (e.g. zircon and ilmenite; Mudd & Jowitt, 2016). The preservation of eroded and transported deposits within the geological record as paleoplacer deposits means that the REE within these systems could be further upgraded by post-depositional hydrothermal or metamorphic activity. Anthropogenic processes that concentrate the REE can also lead to the formation of secondary REE mineralization, as exemplified by the REE-enriched tailings generated as a result of mining of the primary Mary Kathleen skarn near Mount Isa (Queensland, Australia; e.g. Weng *et al.*, 2013).

The next section provides an overview of the major deposit types as summarized in Table 3.3. Describing the processes involved in the formation of currently minor and unexploited REE deposit types such as skarn (Mary Kathleen, Australia), shale (Talvivaara, Finland; Buckton, Canada) and quartz-pebble conglomerate U deposits (Eco-Ridge, Canada) is beyond the remit of this chapter. The reader is referred to descriptions of these deposits in Weng *et al.* (2013, 2015) and references therein given that the future importance of these deposits is unknown.

3.5 RARE EARTH ELEMENT MINERAL DEPOSIT TYPES

3.5.1 Carbonatites

The majority of carbonatites are enriched in REE when compared to crustal abundances. However, carbonatites with economic enrichments of any metals or commodities are comparatively rare, with carbonatites containing economic concentrations of the REE in minerals amenable for processing forming a subset of economic carbonatites. This reflects the fact that carbonatite systems contain economic or anomalously high concentrations of a wide variety of elements, including the REE, P, Nb, U, Th, Cu, Fe, Ti, Ba, F, Zr and other incompatible elements (Modreski *et al.*, 1995). This means that carbonatites are not just sources of the REE (as is the case at Bayan Obo, Mountain Pass and Mount Weld e.g.) but of other elements, as exemplified by mineralization within the Palabora/Phalaborwa Cu-Fe-P and Catalao Nb mines, both of which formed as a result of carbonatite magmatism (e.g. Groves *et al.*, 2010). Economic REE enrichments are typically associated with the latest and most highly evolved sections of intrusive carbonatite systems (Mariano & Mariano Jr., 2012), and (as mentioned above) the majority of the world's REE production to date has been from carbonatites.

Carbonatites are classified as igneous rocks with high modal abundances of carbonates ($>50\%$) that are typically enriched in Sr, Ba, P and the LREE (Jones *et al.*, 2013; Nelson *et al.*, 1988). They can be intrusive or extrusive and can be associated with metasomatic and hydrothermal alteration (sometimes termed antiskarn; Anenburg & Mavrogenes, 2018; Jones *et al.*, 2013). The rocks formed by carbonatite magmatism are texturally variable as well as potentially having a wide range of grainsizes. The processes that form carbonatite magmas remain controversial but the main models can be summarized (e.g. Jones *et al.*, 2013 and references therein) as genesis from (a) residual melts associated with carbonated nephelinite or melilitite magmas, (b) immiscible carbonatite melts formed by the saturation of a silicate magma with CO_2 and (c) the partial melting of CO_2-bearing peridotite. Whatever the processes involved, the carbonatites formed from these magmas are found either as individual intrusions or more commonly with rocks formed from other SiO_2-undersaturated magmas (e.g. syenites, nepheline syenites and nephelinites) in alkaline–carbonatite complexes that are suggestive of a common origin for the magmas that formed these rocks (e.g. Christie *et al.*, 1998). The most common (and important) REE-bearing minerals in carbonatites include bastnäsite and monazite (Table 3.2), with xenotime generally rare, barring examples such as the xenotime-bearing carbonatite dykes within the Lofdal REE deposit in Namibia (Siegfried & Hall, 2012).

3.5.2 Alkaline rocks

The SiO_2-undersaturated alkaline felspathoid (e.g. nepheline)-bearing igneous rocks that are often (but not always) associated with carbonatites can also host potentially economic concentrations of the REE. These rocks form from alkali-rich, silica-unsaturated magmas that contain high concentrations of K and Na. This magmatism typically occurs in rift-type tectonic settings and involves generally deep and always low degree partial melting of mantle material that has often undergone previous metasomatism. This metasomatism sometimes increases the concentration of the REE within the source mantle material associated with these magmas. This, combined with the low degree partial melting involved in their petrogenesis, means that alkaline magmas are generally enriched in the incompatible REE (e.g. Chakhmouradian & Zaitsev, 2012).

However, the REE mineralization within alkaline igneous complexes is not just a function of the petrogenesis of these magmas but also their magmatic evolution prior to their emplacement. The REE enrichments in these complexes reflect this evolution in that they are associated with highly fractionated magmas, where the REE are essentially doubly concentrated as a result of the magma petrogenetic processes and the incompatibility of the REE (and other elements such as Zr, Hf, Nb, Ta, Be) during fractionation and differentiation (e.g. Chakhmouradian & Zaitsev, 2012). This means that the REE enrichments and mineralization present in alkaline igneous systems are predominantly associated with highly evolved sections or sequences of alkaline lavas, tuffs, mafic volcanics and subvolcanic intrusives (especially the upper sections of layered intrusions). This mineralization may also be associated with magmas that are volatile-enriched, as any late-stage magmato-hydrothermal F- and Cl-bearing fluid activity associated with these magmas may also increase the concentrations of the REE (Hoatson *et al.*, 2011). Typical deposits of this type include the Khibini and Lovozero complexes in Russia, the Norra Kärr complex in Sweden and Thor Lake in Canada.

3.5.3 Granites and rhyolites

Intrusive (e.g. granites) and extrusive (e.g. rhyolites) SiO_2-enriched felsic magmas can also be associated with REE enrichments up to potentially economic levels. The magmas that form these felsic intrusions and volcanic units can be generated by the melting of igneous or sedimentary rocks during metamorphism, by the anhydrous melting of the lower crust (e.g. during mantle plume-related underplating) or by the extreme fractionation of mafic magmas (e.g. Jowitt *et al.*, 2018a, 2008b). Granites and rhyolites can be compositionally identical and are discussed together as the processes that control the REE abundances within these systems are similar. However, the REE deposits associated with these systems are split into two separate categories in the classification used in this chapter (Table 3.3), reflecting the fact that the REE mineralization associated with these different types of felsic igneous rock have very different forms.

The most important form of granite-related igneous REE mineralization is pegmatites, very coarse-grained igneous intrusions that form from volatile-rich magmas derived and/or exsolved from highly fractionated felsic magmas that undergo processes such as zone-refining that enrich elements such as the REE into concentrations that allow the form of distinct REE mineral phases. In general, pegmatitic magmas contain elevated concentrations of elements (e.g. the REE, Be, Li), fluxes (e.g. B, F and P) and volatiles (e.g. H_2O and Cl), all of which are incompatible during the fractionation of quartz and feldspar, the two minerals that typically dominate the eventual pegmatite assemblage formed by these systems (e.g. London & Kontak, 2012). The concentrations of the incompatible elements in these systems increase as a result of fractional crystallization, eventually generating highly REE-enriched pegmatitic magmas that then undergo processes such as zone-refining, concentrating elements such as the REE even further. This process produces pegmatites that are much smaller in volume than the parental pluton (or source region if generated by anatexis) but are often highly enriched in elements such as the REE, meaning pegmatite REE deposits can contain significant tonnages of the REE (and other associated elements). Examples of this include the Khibina Massif in Russia, the Motzfeldt deposit in Greenland and the Ytterby REE deposit in Sweden.

Rhyolite-related REE deposits are generally rarer than pegmatite-hosted REE deposits, with only one single REE-dominated resource delineated to date (the Round Top REE rhyolite deposit in Texas, USA; Jowitt *et al.* 2017). However, other REE-enriched rhyolites are known and there is significant potential for these systems to represent bulk-tonnage but low grade sources of the REE and other critical metals (e.g. Jowitt *et al.*, 2017). The REE enrichments in rhyolites are thought to at least partly form as a result of extensive fractional crystallization, although these systems may also involve magmato-hydrothermal activity and hydrothermal fluids, a process that appears to be an important step in generating leachable REE mineral assemblages (such as those at Round Top) that allow the REE to be extracted. The large volumes of rhyolites compared to pegmatites means that they typically contain lower concentrations of the REE but are far larger in size, representing voluminous but low grade REE mineralization that is exemplified by Round Top but is also present in highly evolved rhyolites elsewhere (e.g. Christiansen *et al.*, 1983, 2007; Jowitt *et al.*, 2017).

3.5.4 Iron oxide-copper-gold (IOCG)

Iron oxide-copper-gold (IOCG) deposits are frequently enriched in the REE among a number of other elements. They are exemplified by the giant Olympic Dam Cu–U–Au–Ag deposit, which contains significant amounts of the LREE (e.g. Schmandt *et al.*, 2017) but also includes a broad range of somewhat loosely grouped mineral deposit types (e.g. Groves *et al.*, 2010; Mudd *et al.*, 2013). Strictly speaking, IOCG deposits are structurally controlled magmato-hydrothermal mineral deposits that are often thought to form as a result of the mixing of basinal brines and fluids derived from magmas or igneous rocks. They are often LREE-enriched with elevated concentrations of La, Ce and Nd and are associated with significant amounts of brecciation and Na-Ca alteration. These systems also typically contain high modal abundances of low-Ti Fe oxides as well as low-S sulfides that host Cu and Au mineralization within these systems.

The REE enrichments within IOCG deposits most likely reflect the involvement of high-F hydrothermal fluids derived from magmas or igneous rocks that can effectively mobilize significant concentrations of these elements as mentioned above (e.g. McPhie *et al.*, 2011; Xing *et al.*, 2019). These fluids can also mobilize and transport elements such as Fe, Cu, Au and U before the REE is precipitated within these systems as minerals such as bastnäsite, apatite and allanite, potentially as a result of fluid mixing. However, the REE enrichments within IOCG systems may not be spatially related to the economically important Cu, Au and U mineralization often within these systems, reflecting the different solubilities of these elements. In addition, this separation of the REE from Cu, Au and U within these systems has hampered the extraction of the REE from IOCG systems to date. A number of REE-enriched IOCG deposits are known, including Olympic Dam and the Mount Cobalt deposit of the Eastern Fold Belt of the Mount Isa Inlier, both of which contain Fe, Cu, Au and U in addition to being enriched in the LREE (predominantly La, Ce and Nd; Weng *et al.*, 2013).

3.5.5 Unconformity-related

The relatively recent opening of the Browns Range HREE-enriched mineral deposit in northwestern Australia has highlighted the potential importance of unconformity-related REE mineralization, especially as this style of REE deposit is enriched in Dy and Tb relative to other REE deposits, such as carbonatites (e.g. Nazari-Dehkordi *et al.*, 2018, 2020; Spandler *et al.*, 2020). The REE mineralization present within unconformity-related systems is dominated by xenotime with minor florencite, is structurally controlled and is sediment-hosted, with some similarities to the important unconformity-related U deposits of Canada. In addition to the Browns Range deposit and other mineralization in NW Australia, similar types of mineralization have been identified within the Athabasca Basin of Canada (which also hosts major unconformity-related U systems; Rabiei *et al.* 2017) and in the Mount Isa area of Australia (Jaireth *et al.*, 2014).

The Browns Range deposit is the best example of this REE deposit type and contains several HREE-enriched ore bodies associated with a regional unconformity between Archean metasedimentary and Proterozoic sedimentary units (Spandler *et al.*, 2020). This type of mineralization is thought to have formed as a result of large-scale regional fluid flow driven by distal orogenic (mountain building) events, with the latter also generating faulting and shearing that allowed the movement of fluids and the eventual precipitation of REE minerals. These systems source the REE from the underlying metamorphosed sedimentary units as a result of interaction with saline brines, with REE mineral precipitation as a result of the mixing of these brines with P-bearing acidic fluids from the sandstones overlying the unconformity (e.g. Spandler *et al.*, 2020). This suggests that the targeting of regional unconformities and associated fault zones that may allow the mixing of REE- and P-bearing fluids may be an effective approach to exploring for these systems.

3.5.6 Placer and heavy mineral sands

As briefly described above, placer deposits form as a result of the mechanical concentration of minerals (typically by fluids) by sedimentary processes, resulting in the downstream transport, concentration and deposition of denser minerals, with less dense minerals carried further downstream. The majority of placer or heavy mineral sands deposits are located along modern or ancient marine coastlines that contain sands typically derived from heavy mineral (e.g. zircon, rutile, ilmenite, xenotime and magnetite)-bearing granites. These heavy mineral sands deposits represent the world's most important source of Ti along with by-product zircon, monazite, xenotime and other minerals (e.g. Perks & Mudd, 2021). These deposits also contain significant amounts of REE, as demonstrated by Orris and Grauch (2002), who provide details on more than 360 REE-bearing placer deposits, the REE potential of which is discussed in detail by Mudd and Jowitt (2016).

The preservation of these deposits within the geological record can also form paleoplacer REE mineralization that can subsequently be upgraded by other processes. This is similar to the processes associated with the generation of paleoplacer-hosted Au and U mineralization and is exemplified by the Eco Ridge REE deposit in Ontario (Canada). This deposit consists of two separate paleoplacer, conglomerate units that have been upgraded by interaction with hydrothermal fluids, depositing secondary U-minerals and generating higher grades of the REE (Weng *et al.*, 2013). The upgrading of one of these units, the basal conglomerate bed (BCB), generated higher grade U mineralization than within the overlying main conglomerate bed (MCB), whereas the reverse is true for the REE as the MCB contains higher concentrations of the REE than the BCB (Pele Mountain Resources Inc [PMR], 2012).

3.5.7 Laterite and ionic clay deposits

Lateritic REE deposits are formed in the same way as other lateritic deposits (e.g. Freyssinet *et al.*, 2005) by deep surficial weathering of a protolith typically under tropical conditions, although laterites can be found in higher latitudes where there are preserved records of lateritic weathering during earlier, warmer and more humid climates. Laterite REE systems involve the weathering of REE-enriched rocks such as carbonatites that are naturally susceptible to these processes as a result of their carbonate-rich mineralogy. This, combined with deep weathering and associated karst development means that significant thicknesses (>200 m in some cases; Lottermoser, 1990) of REE-enriched laterite horizons can develop above carbonatite rocks in tropical regions.

The refractory nature of many of the REE-minerals (such as apatite; Lottermoser, 1990) can lead to the generation of enrichments of these minerals (and an increase in REE concentrations) during weathering. This weathering can also vary the relative abundances of the REE within the carbonatite in question. One key example of this is Mount Weld in Australia, where weathering has modified the concentrations and mineralogy of the REE within the originally LREE-dominated carbonatite protolith. This increased the concentrations of the HREE within the laterite-altered carbonatite relative to the protolith (and to other carbonatites globally), generated a monazite, churchite, plumbogummite-group and rhabdophane REE mineral assemblage, and lowered the concentrations of the LREE in the resulting laterite developed from the underlying carbonatite (e.g. Spandler *et al.*, 2020). Sourcing of the REE from this type of deposit could well provide a possible solution to the balance problem of oversupply of the LREE and undersupply of the HREE and other key REE such as Nd and Dy (e.g. Binnemans *et al.*, 2013; Elshkaki & Graedel, 2014).

The second class of lateritic deposits that host important amounts of the REE are ionic clay deposits, as exemplified by the Chinese Long Nan and Yian Xi HREE laterite deposits (e.g. Sanematsu & Watanabe, 2016). These deposits form as a result of the mobilization of the REE by weathering and the breakdown of REE-bearing minerals within a protolith. If the right weathering conditions are present the REE can be generally immobile during this process, meaning they can be concentrated within distinct horizons in the laterite profile as a result of their adsorption onto particular clay minerals. This can cause lateral variations in REE grades, with saprolite grades often higher than ferruginous laterite horizons (e.g. Lottermoser, 1990).

3.6 EXPLORATION FOR REE DEPOSITS

It is difficult to identify a generic approach to the discovery of REE deposits given the variety of different mineralizing systems that can concentrate these elements as outlined above. Equally importantly, there is nothing specific to the REE mineral systems outlined above that means that they are more detectable than (for example) carbonatites that are enriched in Nb (e.g. Araxa in Brazil) or Cu and other elements (e.g. Palabora in South Africa) beyond their above average concentration of the REE. As such, it is outside of the scope of this review to focus on the exploration approaches suitable for all of the deposit systems outlined above. However, it is certainly important to understand the environments these different deposits form in, and the processes that generate these systems as outlined above. There are also two other key (and linked) factors that need to be considered for effective exploration for REE mineral deposits. The first of these is the development of REE enrichments to levels considered economic for a given mineral deposit type and 'basket' of the REE, reflecting the price variations for the individual REE shown in Table 3.1. The other crucial factor is the extractability and processability of the REE in given mineral deposits. For example, the REE in some rhyolites may be hosted by phases that are not amenable to REE extraction, whereas other rhyolites may contain REE-bearing minerals that are eminently suitable for the production of the REE (e.g. Jowitt *et al.*, 2017; Table 3.2). This is compounded by the difficulty in quantifying the concentrations of individual REE within minerals as a result of analytical costs and challenges, thus potentially leading to issues associated with REE deportment and recovery. The controls on these factors as well as the factors that generate the different REE deposits outlined above as well as concentrating the REE within these mineral systems are the focus of a significant amount of recent research (e. g. Anenburg & Mavrogenes, 2018; Anenburg *et al.*, 2020; Dostal, 2016; Jowitt *et al.*, 2017; Sanematsu & Watanabe, 2016; Verplanck *et al.*, 2016).

3.7 CONCLUSIONS

This chapter provides an overview of the geological processes that concentrate the REE, leading to the development of a variety of different REE deposits as a result of igneous, hydrothermal and sedimentary processes. Global REE resources continue to be dominated by carbonatite and carbonatite-associated deposits such as Bayan Obo in China, Mountain Pass in the USA and Mount Weld in Australia, all of which contain REE resources predominantly formed by igneous processes. Sedimentary and surficial processes are responsible for the formation of important ionic clay REE deposits in China as well as the important development of laterites above the primary Mount Weld carbonatite, changing the mineralogy of this deposit and increasing the abundance of the HREE within this system. Only limited development of hydrothermal REE resources has occurred to date, although these deposits represent potentially important future sources of the REE. All of this means that it is unclear whether carbonatites will continue to dominate the long term supply of the REE or whether other deposit types of classes will come to dominate (or at least rival) carbonatite-related REE supply. Adding to this complexity is the difficulty in identifying prospective REE deposits that not only contain elevated concentrations of the REE but also contain the REE in a processable form, the main barrier to increasing REE production. The latter is especially problematic given the energy and material intensity of REE extraction and processing, which is also often associated with the generation of U- and Th-bearing mine waste formed during REE refining and extraction. All of this means that securing future supplies of the REE requires not only the identification and delineation of new REE resources, but also the economical and sustainable extraction and processing of these resources while limiting significant environmental and social impacts.

ACKNOWLEDGEMENTS

Brian McNulty is thanked for comments and assistance with the writing of this chapter.

REFERENCES

Anenburg M. and Mavrogenes J. A. (2018). Carbonatitic versus hydrothermal origin for fluorapatite REE-Th deposits: experimental study of REE transport and crustal "antiskarn" metasomatism. *American Journal of Science*, **318**(3), 335–366.

Anenburg M., Mavrogenes J. A., Frigo C. and Wall F. (2020). Rare earth element mobility in and around carbonatites controlled by sodium, potassium, and silica. *Science Advances*, **6**(41). doi: 10.1126/sciadv.abb6570

Bierlein F. P., Groves D. I., Goldfarb R. J. and Dubé B. (2006). Lithospheric controls on the formation of provinces hosting giant orogenic gold deposits. *Mineralium Deposita*, **40**(8), 874–886.

Binnemans K., Jones P. T., Blanpain B., Van Gerven T., Yang Y., Walton A. and Buchert M. (2013). Recycling of rare earths: A critical review. *Journal of Cleaner Production*, **51**, 1–22.

Cameron K. L. and Cameron M. (1986). Whole-rock/groundmass differentiation trends of rare earth elements in high-silica rhyolites. *Geochimica et Cosmochimica Acta*, **50**(5), 759–769.

Chakhmouradian A. R. and Wall F. (2012). Rare earth elements: minerals, mines, magnets (and more). *Elements*, **8**(5), 333–340.

Chakhmouradian A. R. and Zaitsev A. N. (2012). Rare earth mineralization in igneous rocks: sources and processes. *Elements*, **8**(5), 347–353.

Christiansen E. H., Burt D. M., Sheridan M. F. and Wilson R. T. (1983). The petrogenesis of topaz rhyolites from the western United States. *Contributions to Mineralogy and Petrology*, **83**(1–2), 16–30.

Christiansen E. H., Haapala I. and Hart G. L. (2007). Are Cenozoic topaz rhyolites the erupted equivalents of Proterozoic rapakivi granites? Examples from the western United States and Finland. *Lithos*, **97**(1–2), 219–246.

Christie T., Brathwaite B. and Tullock A. (1998). Mineral commodities report 17: rare earth and related elements. *New Zealand Mining*, **24**(7), 1–13.

Dostal J. (2016). Rare metal deposits associated with alkaline/peralkaline igneous rocks. *Reviews in Economic Geology*, **18**, 33–54.

Elshkaki A. and Graedel T. E. (2014). Dysprosium, the balance problem, and wind power technology. *Applied Energy*, **136**, 548–559.

Ernst R. E. and Jowitt S. M. (2013). Large igneous provinces (LIPs) and metallogeny. In: Tectonics, Terranes, Metallogeny and Discovery in the Northern Circum-Pacific Region, T. Bissig, M. Colpron, B. Rusk and J. Thompson (eds), Society of Economic Geologists, Littleton, CO, USA, pp. 17–51.

Freyssinet P., Butt C. R. M., Morris R. C. and Piantone P. (2005). Ore-forming processes related to lateritic weathering. *Economic Geology*, 100th Anniversary Volume, 681–722.

Graedel T. E., Harper E. M., Nassar N. T., Nuss P. and Reck B. K. (2015). Criticality of metals and metalloids. *Proceedings of the National Academy of Sciences of the United States of America*, **112**(14), 4257–4262.

Groves D. I., Bierlein F. P., Meinert L. D. and Hitzman M. W. (2010). Iron oxide copper-gold (IOCG) deposits through earth histoiy: implications for origin, lithospheric setting, and distinction from other epigenetic iron oxide deposits. *Economic Geology*, **105**(3), 641–654.

Hayes S. M. and McCullough E. A. (2018). Critical minerals: A review of elemental trends in comprehensive criticality studies. *Resources Policy*, **59**, 192–199.

Hoatson D. M., Jaireth S. and Miezitis Y. (2011). The major rare-earth-element deposits of Australia: geological setting, exploration, and resources. Geoscience Australia, Canberra, Australia. http://www.ga.gov.au/webtemp/image_cache/GA19659.pdf (last accessed 1 November 2021)

Humphries M. (2012). The Global Supply Chain. CRS Report for Congress R41347.

IUPAC (2005). Nomenclaurte of inorganic chemistry IUPAC recommendations 2005. In: International Union of Pure and Applied Chemistry, N. G. Connelly, T. Damhusm, R. M. Hartshorn and A. T. Hutton (eds), RSC Publishing, London, UK, 377 p.

Jaireth S., Hoatson D. M. and Miezitis Y. (2014). Geological setting and resources of the major rare-earth-element deposits in Australia. *Ore Geology Reviews*, **62**, 72–128.

Jones A. P., Genge M. and Carmody L. (2013). Carbonate melts and carbonatites. *Reviews in Mineralogy and Geochemistry*, **75**, 289–322.

Jowitt S. M., Weng Z. and Mudd G. (2013). Rare earth elements: deposits, uncertainties and wasted opportunities. *Materials World*, **21**, 22–24.

Jowitt S. M., Medlin C. C. and Cas R. A. (2017). The rare earth element (REE) mineralisation potential of highly fractionated rhyolites: A potential low-grade, bulk tonnage source of critical metals. *Ore Geology Reviews*, **86**, 548–562.

Jowitt S. M., Mudd G. M., Werner T. T., Weng Z., Barkoff D. W. and McCaffrey D. (2018a). The critical metals: An overview and opportunities and concerns for the future. *Metals. Minerals, and Society: Society of Economic Geologists Special Publication*, **21**, 25–38.

Jowitt S. M., Werner T. T., Weng Z. and Mudd G. M. (2018b). Recycling of the rare earth elements. *Current Opinion in Green and Sustainable Chemistry*, **13**, 1–7.

Jowitt S. M., Mudd G. M. and Thompson J. F. H. (2020). Future availability of non-renewable metal resources and the influence of environmental, social, and governance conflicts on metal production. *Communications Earth & Environment*, **1**, 1–8. doi: 10.1038/s43247-020-0011-0

Kanazawa Y. and Kamitani M. (2006). Rare earth minerals and resources in the world. *Journal of Alloys and Compounds*, **408**, 1339–1343.

Lee J., Bazilian M., Sovacool B., Hund K., Jowitt S. M., Nguyen T. P., Månberger A., Kah M., Greene S., Galeazzi C., Awuah-Offei K., Moats M., Tilton J. and Kukoda S. (2020). Reviewing the material and metal security of low-carbon energy transitions. *Renewable and Sustainable Energy Reviews*, **124**, 109789.

London D. (2018). Ore-forming processes within granitic pegmatites. *Ore Geology Reviews*, **101**, 349–383.

London D. and Kontak D. J. (2012). Granitic pegmatites: scientific wonders and economic bonanzas. *Elements*, **8**(4), 257–261.

Long K. R., Van Gosen B. S., Foley N. K. and Cordier D. (2010). The principal rare earth elements deposits of the United States: A summary of domestic deposits and a global perspective. In: Non-renewable Resource Issues, Springer, Dordrecht, pp. 131–155. doi: 10.1007/978-90-481-8679-2_7

Lottermoser B. G. (1990). Rare-earth element mineralisation within the Mt. Weld carbonatite laterite, western Australia. *Lithos*, **24**(2), 151–167.

Mariano A. N. and Mariano A., Jr. (2012). Rare earth mining and exploration in North America. *Elements*, **8**(5), 369–376.

McGloin M. V., Tomkins A. G., Webb G. P., Spiers K., MacRae C. M., Paterson D. and Ryan C. G. (2016). Release of uranium from highly radiogenic zircon through metamictization: The source of orogenic uranium ores. *Geology*, **44**(1), 15–18.

McPhie J., Kamenetsky V., Allen S., Ehrig K., Agangi A. and Bath A. (2011). The fluorine link between a supergiant ore deposit and a silicic large igneous province. *Geology*, **39**(11), 1003–1006.

McPhie J., Kamenetsky V., Allen S., Ehrig K., Agangi A. and Bath A. (2012). The fluorine link between a supergiant ore deposit and a silicic large igneous province. *Geology*, **40**(8), 3543931.

Miller C. F. and Mittlefehldt D. W. (1982). Depletion of light rare-earth elements in felsic magmas. *Geology*, **10**(3), 129–133.

Modreski P. J., Armbrustmacher T. J. and Hoover D. B. (1995). Carbonatite deposits. In: E. A. du Bray (ed.), Preliminary Compilation of Descriptive Geoenvironmental Mineral Deposits, Geological Survey and Department of the Interior, Reston, VA, pp. 47–49.

Mortensen J. K., Gemmell J. B., McNeill A. W. and Friedman R. M. (2015). High-Precision U–Pb zircon chronostratigraphy of the mount read volcanic belt in western tasmania, Australia: implications for VHMS deposit formation. *Economic Geology*, **110**, 445–468.

Mudd G. M. and Jowitt S. M. (2016). Rare earth elements from heavy mineral sands: assessing the potential of a forgotten resource. *Applied Earth Science*, **125**(3), 107–113.

Mudd G. M., Weng Z., Jowitt S. M., Turnbull I. D. and Graedel T. E. (2013). Quantifying the recoverable resources of by-product metals: The case of cobalt. *Ore Geology Reviews*, **55C**, 87–98.

Naumov A. V. (2007). Rhythms of rhenium. *Russian Journal of Non-Ferrous Metals*, **48**(6), 418–423.

Nazari-Dehkordi T., Spandler C., Oliver N. H. S. and Wilson R. (2018). Unconformity-related rare earth element deposits: A regional-scale hydrothermal mineralization type of Northern Australia. *Economic Geology*, **113**(6), 1297–1305.

Nazari-Dehkordi T., Spandler C., Oliver N. H. S. and Wilson R. (2020). Age, geological setting, and paragenesis of heavy rare earth element mineralization of the tanami region, western Australia. *Mineralium Deposita*, **55**(1), 107–130.

Nelson D. R., Chivas A. R., Chappell B. W. and McCulloch M. T. (1988). Geochemical and isotopic systematics in carbonatites and implications for the evolution of ocean-island sources. *Geochimica et Cosmochimica Acta*, **52**(1), 1–17.

Orris G. J. and Grauch R. I. (2002). Rare earth element mines, deposits and occurrences (Vol. 2, No. 189). US Department of the Interior, US Geological Survey.

Pankhurst M. J., Schaefer B. F. and Betts P. G. (2011). Geodynamics of rapid voluminous felsic magmatism through time. *Lithos*, **123**(1–4), 92–101.

Perks C. and Mudd G. (2021). Soft rocks, hard rocks: the world's resources and reserves of Ti and Zr and associated critical minerals. *International Geology Review*, 1–22, doi: 10.1080/00206814.2021.1904294

PMR (2012). Technical Report on the Eco Ridge Mine Project, Elliot Lake, Ontario, Canada, NI43–101 Report. Pele Mountain Resources Inc (PMR), Canada.

Rabiei M., Chi G., Normand C., Davis W. J., Fayek M. and Blamey N. J. F. (2017). Hydrothermal rare earth element (xenotime) mineralization at Maw Zone, Athabasca Basin, Canada, and its relationship to unconformity-related uranium deposits. *Economic Geology*, **112**(6), 1483–1507.

Rudnick R. L. and Gao S. (2003). Composition of the continental crust. In H. D. Holland and K. Turekian, (ed.), Treatise on Geochemistry, **3**, 1–64, https://sec.report/otc/financial-report/60060/Technical-Report-on-the-Eco-Ridge-Mine-Rare-Earths-and-Uranium-Project-Elliot-Lake-Ontario-Canada.pdf (last accessed 1 November 2021).

Sanematsu K. and Watanabe Y. (2016). Characteristics and genesis of ion adsorption-type rare earth element deposits. *Reviews in Economic Geology*, **18**, 55–79.

Schmandt D. S., Cook N. J., Ciobanu C. L., Ehrig K., Wade B. P., Gilbert S. and Kamenetsky V. S. (2017). Rare earth element fluorocarbonate minerals from the olympic dam Cu-U-Au-Ag deposit, South Australia. *Minerals*, **7**(10), 202. doi: 10.3390/min7100202

Siegfried P. and Hall M. (2012). NI-43-101 Technical Report and Mineral Resource Estimate for Area 4 of the Lofdal Rare Earth Element (REE) Project, *Khorixas District*, *Republic of Namibia. The MSA group*.

Sillitoe R. H. (2010). Porphyry copper systems. *Economic Geology*, **105**, 3–41.

Skirrow R. G., Bastrakov E. N., Barovich K., Fraser G. L., Creaser R. A., Fanning C. M., Raymond O. L. and Davidson G. J. (2007). Timing of iron oxide Cu-Au-(U) hydrothermal activity and Nd isotope constraints on metal sources in the Gawler craton, South Australia. *Economic Geology*, **102**(8), 1441–1470.

Spandler C., Slezak P. and Nazari-Dehkordi T. (2020). Tectonic significance of Australian rare earth element deposits. *Earth-Science Reviews*, **207**, 103219.

USGS (2021). Mineral Commodity Summaries., U.S. Geological Survey (USGS), Reston, VA; https://pubs.usgs.gov/periodicals/mcs2021/mcs2021.pdf (last accessed 1 November 2021).

Verplanck P. L., Mariano A. N. and Mariano A., Jr. (2016). Rare earth element ore geology of carbonatites, In: Rare Earth and Critical Elements in ore Deposits, Society of Economic Geologists, Inc., Littleton, CO, pp. 5–32.

Wall F. (2014). Rare earth elements. In: Critical Metals Handbook, G. Gunn (ed.), John Wiley & Sons, Ltd, Chichester, UK, pp. 312–339.

Walters A., Lusty P., Chetwyn C. and Hill A. (2010) for British Geological Survey. Rare Earth Elements. Keyworth, Nottingham. See http://nora.nerc.ac.uk/id/eprint/12583/1/Rare_Earth_Elements_profile.pdf (last accessed 1st November 2021).

Weng Z. H., Jowitt S. M., Mudd G. M. and Haque N. (2013). Assessing rare earth element mineral deposit types and links to environmental impacts. *Transactions of the Institutions of Mining and Metallurgy, Section B: Applied Earth Science*, **122**(2), 83–96.

Weng Z., Jowitt S. M., Mudd G. M. and Haque N. (2015). A detailed assessment of global rare earth element resources: opportunities and challenges. *Economic Geology*, **110** (8), 1925–1952.

Williams-Jones A. E., Migdisov A. A. and Samson I. M. (2012). Hydrothermal mobilisation of the rare earth elements-a tale of "ceria" and "yttria". *Elements*, **8**(5), 355–360.

Xing Y., Etschmann B., Liu W., Mei Y., Shvarov Y., Testemale D., Tomkins A. and Brugger J. (2019). The role of fluorine in hydrothermal mobilization and transportation of Fe, U and REE and the formation of IOCG deposits. *Chemical Geology*, **504**, 158–176.

Chapter 4

Sources and applications of rare earth elements

V. Balaram

4.1 INTRODUCTION

The International Union of Pure and Applied Chemistry (IUPAC) has defined the rare earth elements (REE) [REE: lanthanum (La), cerium (Ce), praseodymium (Pr), neodymium (Nd), promethium (Pm), samarium (Sm), europium (Eu), gadolinium (Gd), terbium (Tb), dysprosium (Dy), holmium (Ho), erbium (Er), thulium (Tm), ytterbium (Yb), lutetium (Lu), yttrium (Y) and scandium (Sc)] as a group of 17 chemically similar metallic elements that comprise the 15 lanthanide elements (La-Lu), Y and Sc. Due to the similar geochemical behavior of Y and Sc to the lanthanides in most environments in the earth's crust, they are very often grouped with the lanthanides and referred to as REE (Taylor & McLennan, 1985). The lower atomic weight elements La to Sm, with atomic numbers 57 to 62, are referred to as the light REE (LREE); while Eu to Lu, with atomic numbers 63 to 71, are known as the heavy REE (HREE) (Walters *et al.*, 2010). Despite their low atomic weights, Y and Sc are included with the HREE subgroup because of their co-occurrence, ionic radius and closer behavioral properties to the HREE than to the LREE (Table 4.1).

Another related consequence is the so-called *'lanthanide contraction'* in which the ionic radius progressively decreases from La^{3+} (1.06 Å) to Lu^{3+} (0.85 Å) with the increase of atomic number. In lanthanide atoms, the configuration of the valence electrons of the outermost shell is the same for all the species while the

doi: 10.2166/9781789062236_0075

Table 4.1 Average abundance (in μg/g) of REE in the earth's crust in comparison with chondritic abundance.

Element	Symbol	Average Crustal Abundance (Lide, 1997)	Chondritic Abundance (Pourmand *et al.*, 2012)
LREE – Light REE			
Lanthanum	La	39	0.2469
Cerium	Ce	66.5	0.6321
	Pr	9.2	0.0959
Praseodymium			
Neodymium	Nd	41.5	0.4854
Promethium	Pm	—	—
Samarium	Sm	7.05	0.1556
Europium	Eu	2	0.0599
Gadolinium	Gd	6.2	0.2093
HREE – Heavy REE			
Terbium	Tb	1.2	0.0378
Dysprosium	Dy	5.2	0.2577
Holmium	Ho	1.3	0.0554
Erbium	Er	3.5	0.1667
Thulium	Tm	0.52	0.0261
Ytterbium	Yb	3.2	0.1694
Lutetium	Lu	0.8	0.0256
Yttrium	Y	33	1.395
Scandium#	Sc	22	5.493
Total (μg/g)		242.17	9.5118

#Scandium's properties are different enough from those of the other REE that most of the scientific and general literature excludes it and focuses on the lanthanides and yttrium.

4f orbitals are progressively filled with increasing atomic numbers. It occurs when electrons in the 'f' orbital do not screen the other electrons from the positive pull towards the nucleus. Screening of the 4f orbitals leads to the extremely similar physical and chemical properties of the elements.

REE are not radioactive except promethium, which doesn't occur in any significant abundance in the earth's crust as it has the most stable isotope with a half-life of only 17.7 years (Castor & Hendrik, 2006). Promethium is formed in nature as a result of fission or decay of heavier elements, with only trace amounts found in naturally occurring ores. That is why it is extremely rare, with only less than 500 grams thought to be on the planet at any given time. However, five of the REE (excluding promethium) also each contain a proportion of a radioactive isotope: ^{138}La, ^{144}Nd, ^{147}Sm, ^{152}Gd and ^{176}Lu (Henderson *et al.*, 2011).

4.1.1 Occurrence in different geological systems, mineralogy and demand

All REE occur in nature but not in the pure metal form as they are found in the form of minerals, such as phosphates, silicates, carbonates, oxides and halides (Jordens *et al.*, 2013; Van Gosen *et al.*, 2017). These metals occur together in various geological deposits and are found worldwide in more than 250 minerals, and are diverse and often complex in composition. Out of >250 known REE-bearing minerals, only three are economically viable and exploited on a commercial scale: bastnäsite, monazite and xenotime (Jordens *et al.*, 2013). Bastnäsite is likely the primary valuable mineral for REE in the world. These minerals are generally beneficiated using gravity, magnetic, electrostatic and flotation separation techniques.

In fact, REE are currently governing our lifestyle as they are used in the manufacturing of all types of personal gadgets such as mobile phones, laptops, television sets and smartwatches. REE and the alloys that contain them are used in a variety of high-tech applications, such as the national defense and satellite systems, radar systems, wind turbines, electric vehicles, rechargeable batteries, laser crystals and robots. The 2010 'Rare Earth' crisis of the world, following China's monopoly with over 80% share and export restrictions in the REE market, led to an exploration boom for REE all over the world, including India (Krishnamurthy, 2020).

4.1.2 Behavior of REE in different geological systems

Geologically, the REE are not especially rare in terms of average crustal abundance, but the concentrated deposits of REE are limited in number (Chakhmouradian & Wall, 2012). In the earth's crust, the estimated average concentration of the REE is ~250 μg/g (Table 4.1) which exceeds that of many other familiar metals that are mined. Cerium is the most abundant of the REE. For example, Ce being the 25th most abundant element at 67 μg/g average abundance is more abundant than copper (55 μg/g). They are in fact quite common just like most other metals like copper or lead and are found on every continent and even on the ocean floor. But REE are never found in very high concentrations and are usually found mixed together with other elements or with radioactive elements, such as uranium and thorium. The unique properties of the REE have led them to solve numerous geological research problems. The distribution of REE in a series of igneous rocks, for example, can indicate details of their origin and history, as well as their relationship to one another. The same techniques can be applied to cosmic materials. It has been shown that some meteorites (achondrites) were formed by processes very similar to those by which many igneous rocks on the earth are formed (fractional crystallization).

The existence of multiple valencies of some REE and their different chemical behavior, can also help to reveal some geological processes. Most of the lanthanides exhibit a trivalent state (i.e. Ln^{3+}, where Ln is the generic symbol for lanthanides) but cerium can also be quadrivalent (Ce^{4+}) and europium is often divalent (Eu^{2+}) and can behave differently in different geological systems (Balaram, 1996). The anomalous behavior of cerium and europium, together with the coherent behavior of the rest of the series, is used extensively as a tool to study geochemical processes within the earth's mantle and crust, as well as in natural waters. These alternative redox states can generate concentration anomalies for these two elements because REE are fractionated during mineralogical and petrological processes. For example, under reducing conditions, such as those found within the mantle or the lower crust, europium may exist in the divalent state (Eu^{2+}). This results in an increase in ionic radius of about 17% (from 1.066 to 1.25 Å) making it essentially identical to Sr^{2+} (1.26 Å). The consequence of this is that Eu substitutes freely in place of Sr in feldspars, notably Ca-plagioclase, resulting in a strong positive Eu-anomaly which leads to distinctive geochemical behavior compared to other REE. An instance of this is on the moon – strong but opposite europium anomalies in the rocks of the lunar highlands and the mare basalts have provided useful information on their origin (Brophy & Basu, 1990).

Processes like partial melting, fractional crystallization, assimilation and mixing of magmas leave distinct signatures on REE fractionation as far as igneous rocks are concerned (Henderson, 1983). The REE abundances in sedimentary, metamorphic rocks and ore deposits provide clues to the sources and the processes related to their occurrence. Let us consider an interesting example from the marine environment. Under oxidizing conditions, Ce^{3+} may be oxidized to Ce^{4+}, leading to a decrease in the ionic radius of about 15% (from 1.1143 to 0.97 Å). The only place on Earth where this reaction can occur on a large scale is the marine environment (Balaram *et al.*, 2012). For example, seawater is considered to be a major contributor of REE in manganese nodules. Redox conditions can lead to fractionation of Ce relative to other REE, resulting in significant variations in REE distributions. The abundance of Ce is high relative to that of other REE in most manganese nodules. The preferential incorporation of Ce in manganese nodules is generally considered to be due to the oxidation of Ce^{3+} to Ce^{4+} under oxidizing conditions, with a consequent reduction in its ionic radius. This results in the precipitation of cerium hydroxide and phosphates in manganese nodules. This also explains the dramatic depletion of Ce in seawater compared to other lanthanides (Balaram *et al.*, 2015). This property has been used, for example, to estimate paleo redox conditions in ancient oceans. Hence, for geochemical studies related to all kinds of rock formations, the REE data is vital in providing information on the geochemical processes within the mantle and crust.

The REE radioactive isotopes have long half-lives. This makes some of them (with their daughter products) very suitable for geochronological studies, especially ^{138}La-^{138}Ce, ^{147}Sm-^{143}Nd and ^{176}Lu-^{176}Hf (Schnabel *et al.*, 2017). They are used extensively in the dating of rocks, particularly ancient ones and high-grade metamorphic rocks. They are also used in investigating problems such as ocean mixing, the mixing of magmas, or contamination of the earth's mantle by rocks from the crust. Understanding geological concepts like this and further developments in geochemical exploration techniques such as soil surveys, bedrock surveys, stream sediment surveys, biogeochemistry, hydro-geochemistry and soil gas studies, in conjunction with the traditional geological and geophysical methods have allowed their exploration activity to expand in previously underexplored terrains (Balaram, 2021).

4.2 BRIEF HISTORY OF REE

The term rare earth was coined when an unusual black rock was unearthed by a miner in Ytterby, Sweden, in 1787 (Gschneidner & Cappellen, 1987). In 1794 the chemist Johan Gadolin named this previously unknown 'earth' yttria, after the town where it was discovered. Over time, the mines around Ytterby extracted rocks that yielded four elements named after the town (Y, Yb, Tb and Eb). During the 19th century, among the many minerals and elements studied, Swedish chemist Jöns Jacob Berzelius is credited with discovering cerium. Berzelius isolated and named cerium in 1803. Another Swedish chemist, Carl Gustaf Mosander, discovered the element lanthanum in 1839 while studying a compound of cerium. He pursued his investigations of the REE and in 1843 reported discovery of the elements erbium, terbium and didymium. In 1885, the Austrian chemist Baron Carl Auer von Welsbach found that didymium was in reality a mixture of two elements: neodymium and praseodymium (Welsbach & Auer, 1885).

The invention of the spectroscope by Gustav Kirchhoff and Robert Bunsen in 1859 and the development of spectral analysis as a technique for identifying elements by examining emission spectral lines, along with the development of the periodic table by Mendeleev and Meyer in 1869, provided valuable tools for the study of rare earths. Carl Auer von Welsbach was a student of Robert Bunsen, the inventor of the Bunsen burner at the University of Heidelberg (Germany) in 1880. Welsbach began to work with the REE, identified and named Nd and Pr, and became the first person to develop a commercial use for the REE. Welsbach found a way to alloy these REE wastes with iron, creating a 'flintstone' that sparked when struck, which he named ferrocerium. This material was widely used in cigarette lighters, as well as ignition devices in automobiles. These were some of the earliest industrial applications of REE.

The last naturally occurring REE, lutetium was independently discovered in 1907 by Carl Auer von Welsbach, Charles James and Georges Urbain. However, research into the chemistry of these elements was difficult mainly because of their chemical similarity and no one knew how many true REE existed (Emsley, 2003). Fortunately, in 1913–14 the research of the Danish physicist Niels Bohr and English physicist Henry Gwyn Jeffreys Moseley resolved this situation.

Bohr's theory of the hydrogen atom enabled theoreticians to show that only 15 lanthanides exist. Moseley's experimental studies verified the existence of 14 of these elements and showed that the 15th lanthanide must be element 61 (promethium) and lie between neodymium and samarium. In the old days, the ores to supply these REE came largely from Brazil, India and North Carolina (USA) thus creating the first international trade in REE.

4.3 TYPES OF REE DEPOSITS

REE ore deposits are relatively abundant in the earth's crust and found all over the world, but unlike most other metals, they are rarely concentrated into mineable ore deposits. Minable deposits are hard to locate with concentrations high enough for economical extraction. The major deposits are in China, the US, Australia and Russia, while other viable deposits are found in Canada, India, South Africa and Southeast Asia. In fact, by the early 21st century, China had become the world's largest producer of REE. Figure 4.1 depicts the REE reserves worldwide and the estimated statistical data for their production in 2020. It can be seen that China dominates in both reserves and production (United States Geological Survey [USGS], 2021). Australia, Brazil, India, Kazakhstan, Malaysia, Russia, South Africa and the United States also extract and refine significant quantities of these materials. Table 4.2 presents the REE resources and their global distribution. Singh (2020) gave a detailed description of the REE resources from the Indian context. Although the number of identified deposits in the world is close to a thousand, there are only a handful of actual operating mines. Prominent currently operating mines are Bayan Obo in the Inner Mongolia region of China, Mountain Pass in the US, the recently opened Mount Weld in Australia (Haque *et al.*, 2014) along with a few more mines in other countries.

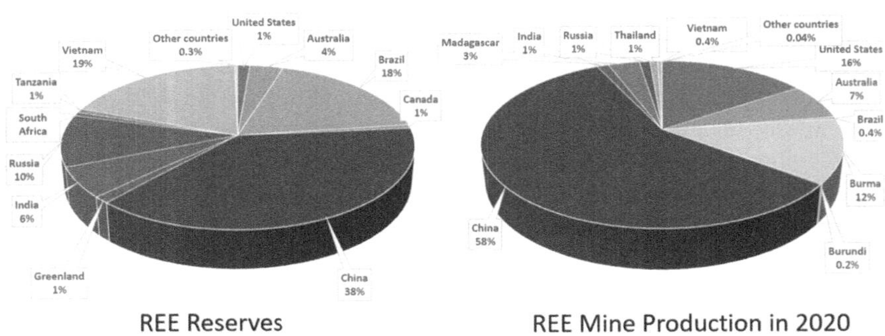

Figure 4.1 REE reserves worldwide (120 million tons); and estimated statistical data for their production in 2020 (240,000 tons). It can be seen that China dominates in both reserves and production (*Source*: USGS, 2021).

Table 4.2 REE resources and their global distribution (*Sources*: Bhushan & Kumar, 2013; Voncken, 2016; Dushyantha *et al.*, 2020).

Deposit	Location	Type	Main REE	REE Mineral(s)
Bayan Obo	China	Carbonatite/ hydrothermal	La, Ce, Nd	Bastnäsite, parisite, monazite
Mountain Pass	USA	Carbonatite	LREE	Bastnäsite
Mount Weld	Australia	Laterite/carbonatite	LREE	Apatite, monazite, synchysite, churchite
Illimaussaq	Denmark	Peralkaline igneous	La, Ce, Nd, HREE	Eudialyte, steenstrupine
Pilanesberg	South Africa	Peralkaline igneous	Ce, La	Eudialyte
Steenkampskraal	South Africa	Vein	La, Ce, Nd	Monazite, apatite
Hoidas Lake	Canada	Vein	La, Ce, Pr, Nd	Apatite, allanite
Thor Lake	Canada	Alkaline igneous	La, Ce, Pr, Nd, HREE	Bastnäsite
Strange Lake and Misery Lake	Canada	Alkaline igneous/hydrothermal	La, Ce, Nd, HREE	Gadolinite, bastnäsite
Nolans Bore	Australia	Vein	La, Ce, Nd	Apatite, allanite
Norra Karr	Sweden	Peralkaline igneous	La, Ce, Nd, HREE	Eudialyte
Khibina and Lovenzero	Russia	Peralkaline igneous	LREE + Y, minor HREE	Eudialyte, apatite
Nkwombwa Hill	Zambia	Carbonatite	LREE	Monazite, bastnäsite
Kagankunde	Malawi	Carbonatite	LREE	Monazite-Ce, bastnäsite-Ce
Tundulu	Malawi	Carbonatite	LREE	Synchesite, parisite, bastnäsite
Songwe	Malawi	Carbonatite	LREE, Nd	Synchesite, apatite
Chinese ion-adsorption deposits	China	Soils	La, Nd, HREE	Clay minerals

(*Continued*)

Table 4.2 REE resources and their global distribution (*Sources*: Bhushan & Kumar, 2013; Voncken, 2016; Dushyantha *et al.*, 2020) (*Continued*).

Deposit	Location	Type	Main REE	REE Mineral(s)
Maoniuping	China	Carbonatite	LREE	Bastnäsite
Dong Pao	Vietnam	Carbonatite	LREE	Bastnäsite, parisite
Kamthai, Rajasthan	India	Carbonatite	LREE	Bastnäsite-La, bastnäsite-Ce, synchysite, carbocernaite, cerianite, ancylite, parisite

The global distribution of REE deposits is shown in Figure 4.2. REE deposits are associated with igneous, sedimentary and metamorphic rocks in a wide range of geological environments. The concentration and redistribution are influenced by rock-forming and hydrothermal processes, including enrichment in magmatic or hydrothermal fluids, separation into mineral phases and precipitation and subsequent redistribution and concentration through weathering and other surface processes (Jaireth *et al.*, 2014). As REE deposits appear in a wide variety of geological environments, it's not very easy to classify them into different categories. However, using the evolutionary framework, Walters *et al.* (2010)

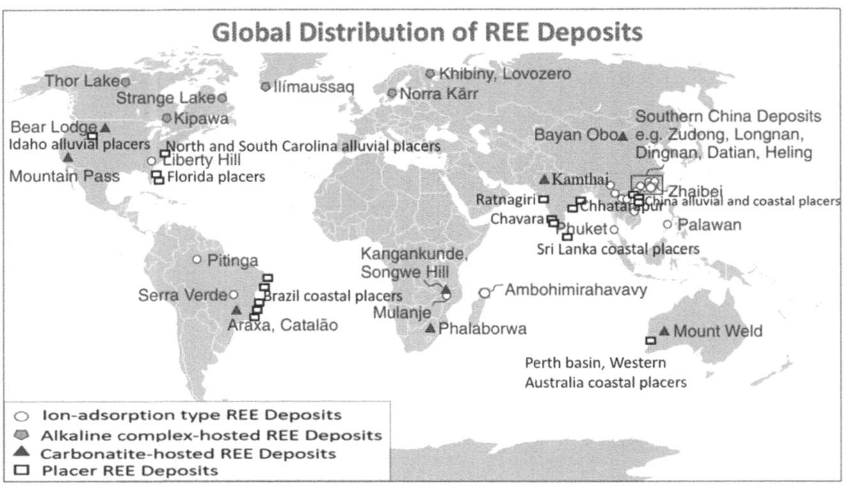

Figure 4.2 Global distribution of REE deposits (adopted from Sengupta and Gosen, 2016 and Borst *et al.*, 2020).

broadly divided the REE deposits into: (i) primary deposits associated with igneous and hydrothermal processes and (ii) secondary deposits concentrated by sedimentary processes and weathering.

Primary REE deposits are the formations associated with metamorphic rocks, pegmatite, carbonatite and hydrothermal veins, enriched by primary processes. In contrast, secondary deposits, namely, placers, laterites, bauxite and ion-adsorption clays are formed by weathering and erosion of the primary sources (Balaram, 2019). Within these two groups, REE resources can be further subdivided depending on their genetic association, mineralogy and form of occurrence.

These classifications are helpful in prospecting new deposits and determining the feasibility of mining, processing and refining them. Jaireth *et al.* (2014) provided more comprehensive information on the classification of REE deposits. Sowerbutts (2017) classified the economically important REE deposits into four categories. However, in view of the significant quantities of these elements in coal and marine sediment, Balaram (2019) divided the REE deposits into the following five categories: (i) alkaline igneous rocks: pegmatites and carbonatites, (ii) residual deposits, (iii) heavy mineral placers, (iv) REE in coal and (v) REE in ocean-bottom sediments.

Based on the information currently available in the literature, REE resources can be divided into the following six categories (Figure 4.3): (i) primary REE deposits

Figure 4.3 Illustrative view of the different types of REE resources (*Source*: author's design).

including carbonatites, alkaline igneous rocks, pegmatites, iron oxide copper-gold deposits, vein and skarn deposits, (ii) secondary deposits including residual deposits laterites, ion adsorption clays, (iii) offshore sources like phosphorites, ferromanganese crusts, manganese nodules and marine mud, (iv) REE resources in coal, coal ash, red mud and waste rocks from old and closed mines, (v) extraterrestrial REE resources and (vi) REE resources e-waste. Figure 4.3 provides an illustrative view of these different types of REE resources currently available.

4.3.1 Primary REE deposits

Primary REE deposits include carbonatites, alkaline igneous rocks, pegmatites, iron oxide copper-gold deposits, vein and skarn deposits. Many of the world's significant REE deposits occur in carbonatites as REE have a strong genetic association with alkaline magmatism (Figure 4.4). Alkaline igneous rocks are formed by the partial melting of deep mantle rocks, which subsequently rise and cool within the earth's crust. Typically, an alkaline magma is enriched in not just REE, but also zirconium (Zr), niobium (Nb), strontium (Sr), barium (Ba) and lithium (Li). As the magma ascends, it changes chemically in response to a complex interaction of factors including temperature, pressure and the chemistry of the surrounding rocks. This complex interaction results in the formation of a great variety of REE deposits. The hosts are differentiated rocks ranging from nepheline syenites and trachytes to peralkaline granites. The carbonatites and alkaline igneous rocks are characteristically found in the interiors of tectonic plates, that is away from the active plate margins where volcanic activity is at its greatest (Mehmood, 2018).

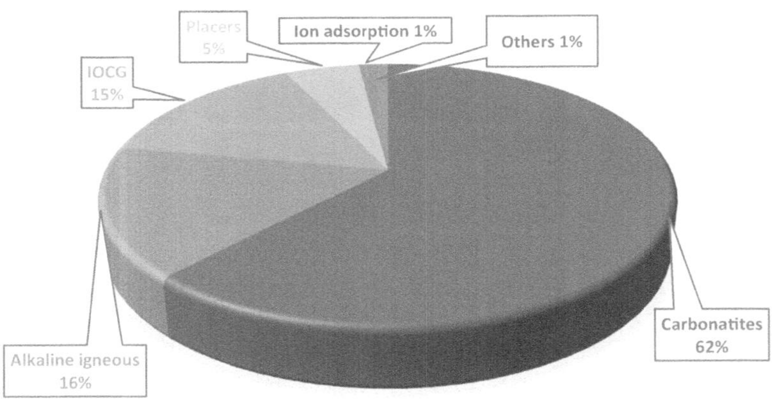

Figure 4.4 Distribution of global REE by principal deposit type. It can be seen that the majority of the world's significant REE deposits occur in carbonatites (Zhou *et al.*, 2017).

In addition, among several important earth's processes such as hazardous volcanic eruptions, the growth of the lithosphere, build-up of mountains, earthquakes, modulating earth's climate by releasing mantle-derived gases and rifting of continents, volcano-tectonic systems are also responsible for providing natural resources of REE minerals.

The carbonatites are mostly confined to continental areas, while the alkaline rocks also occur over much of the world's oceanic intraplate areas on volcanic islands. Although more than two-thirds of carbonatites that have been dated are Phanerozoic in age (less than 500 million years, e.g., Bayan Obo REE deposit corresponds to 442 ± 42 Ma), they are also overwhelmingly concentrated in areas comprising older Precambrian rocks (Balashov *et al.*, 1964; Hu *et al.*, 2009). These are igneous rocks comprising more than 50% carbonate minerals, principally calcite, but in some cases magnesium-bearing carbonates (dolomite and magnesite) or iron-bearing carbonates (siderite), that crystallized from a high-temperature liquid from deep in the earth. The REE mineralization is also found in layered alkaline complexes, granitic stocks and late-stages dikes as well as rarely trachytic volcanic and volcaniclastic deposits (Dostal, 2017). Though bastnäsite, apatite, monazite, allanite, parisite and synchysite are the most common REE-bearing minerals found in carbonatites, bastnäsite is considered as the primary REE mineral in carbonatite deposits. LREE are relatively more abundant in carbonatites than HREE. The Bayan Obo REE deposit in Inner Mongolia (China) is the largest REE deposit and is a high-grade, igneous related carbonatite deposit that sources 80% of the world's LREE requirements (Verplanck *et al.*, 2014). Other prominent examples of REE deposits associated with alkaline igneous rocks include Mountain Pass (California, USA), Saima (China) and Ytterby (Sweden). In a recent study, Bhushan and Kumar (2013) discovered carbonatite plugs, sills and dykes hosting REE deposits at Kamthai (Barmer district, Rajasthan, India) with potential REE resources of 4.91 million tons at estimated depths of up to 84 m. In a subsequent study of these carbonatites as well as the host ijolite plug of Kamthai, Bhushan (2015) attributed the genesis of this deposit to mantle 'hot spot' activity contemporary with the Deccan volcanism.

The Amba Dongar carbonatite complex, the first carbonatite discovery in India in 1963 with its rich fluorite deposit (over 10 million tons of resources), and a number of other complexes in India have been explored for their REE resources since the 1970s (Krishnamurthy, 2019 and references therein). The Amba Dongar complex has been explored for fluorite since 1969 by opencast mining. The complex itself and also the overburden and the mine-tailing dumps have long been recognized as a potential REE resource by the Atomic Minerals Directorate (AMD) for Exploration and Research, Hyderabad (India). Seeking non-placer REE sources, in the carbonatites, AMD's evaluation drilling at Amba Dongar has estimated there are substantial reserves in this complex. A total of 346–462 tons of rare-earth element oxide (REO) reserve has been established (Sinha, 2020).

The iron-oxide copper-gold (IOCG) type deposit contains large amounts of REE and uranium. Examples include the Olympic Dam deposit in South Australia and Eagle Mountain deposit in California, US (Long *et al.*, 2012). Many other deposits of this type have been identified around the world. IOCG may contain REE enrichments and it has the potential to produce REE as a by-product, but recovering REE from this deposit is very difficult and expensive.

There are also some REE mineral concentrations of economic interest in granite pegmatites and some hydrothermal (hot water) vein systems. Pegmatites are course-grained felsic igneous intrusive rocks that are rich in incompatible elements including REE, most commonly enriched in apatite $Ca_5(PO_4)_3$. Granitic pegmatites, coarse-grained igneous rocks with an overall granitic composition, are of varied origins and can show a similarly varied range of REE enrichments (Ercit, 2005; London, 2018). In addition to REE, deposits of zirconium and niobium are also found in pegmatitic zones of syenites. Some skarn deposits contain economic concentrations of REE and uranium (Lentz, 1991). Uneconomic quartz-phosphate hydrothermal vein occurrences improve our understanding of large REE ore deposits, as well as the behavior of REE-phosphate phases in hydrothermal processes.

4.3.2 Secondary REE deposits

These deposits include residual deposits such as laterites, ion adsorption clays and placer deposits. Residual deposits are formed from deep weathering of igneous rocks, pegmatites, iron-oxide copper-gold deposits. In tropical environments, rocks are deeply weathered to form a unique soil profile consisting of laterite, an iron- and aluminum-rich soil, which may be many tens of meters thick. When REE deposits undergo such weathering, they may be further enriched in REE up to concentrations of economic interest. Laterite deposits result from the enrichment of REE by *in situ* chemical alteration. Examples include REE-laterite in south China resulting from weathering of tin granite. Nugraheni *et al.* (2020) describe the mechanism of enrichment of REE concentrations in granitoids and bauxite laterite in the Bukit Tinggi (Pahang State, Malaysia) as well as Bangka Island (Indonesia). The weathering process, including higher mechanical weathering, contributes to ionic bonds destruction of the host minerals leading to a higher concentration of REE-bearing minerals. Intense weathering of carbonatite and peralkaline intrusive rocks may also form concentrated residual deposits of REE minerals. Laterite REE deposits are formed by weathering of carbonatites in tropical or forested warm environments (Richardson & Birkett, 1996). Substantial deposits of laterite REE are found in Brazil and southern Africa. North East India experiences climatic conditions favorable for the development of thick lateritic profiles (Singh *et al.*, 2014).

4.3.2.1 Ion-adsorption deposits

This particular type of REE deposits, the ion-adsorption type, is formed by subtropical weathering of igneous rocks and by leaching of REE from seemingly common igneous rocks such as REE-rich granites and fixing the elements into clay minerals like kaolinite, halloysite and illite in soil (Borst *et al.*, 2020; Wu *et al.*, 1996). REE-bearing ion-adsorption clay deposits form in tropical regions with moderate to high rainfall through the following general processes: (i) the REE are leached by groundwater from granites (the bedrock), (ii) thick zones of laterite soils develop above the granites; this intensely weathered zone contains an abundance of clays and (iii) the mobilized REE become weakly fixed (ion adsorption) onto the clays (kaolinite and halloysite) in the soils.

Ion-adsorption deposits are mainly located in southern China, although they are relatively low-grade, generally with only 0.05% to 0.5% REE, easily mined and contain a high content of the more valuable HREE. These ion-adsorption-type REE deposits dominantly occur in the tropics and subtropics (Figure 4.2) and currently provide most of the world's heavy REE supply. These deposits are mostly known to be in southern China and Kazakhstan. In deposits of this type in southern China, REE concentrations range from approximately 300 μg/g to 2000 μg/g and appear to vary with the parent rocks (Bao & Zhao, 2008). These modest REE concentrations are economic for the southern China deposits because (a) the REE can be easily extracted from the clays with weak acids, (b) the deposits are often enriched in the high-value HREE, (c) the area has low labor costs and (d) there has been a localized lack of environmental protection (Liang *et al.*, 2014). Temga *et al.* (2021) have reported high REE-enrichment in soils controlled by ion-adsorption clays, carbonate and Fe-Mn oxides in tropical Cameroonian soils. The soils developed on alkaline igneous rocks usually show great or high REE contents, ranging from 678.73 to 1148.02 μg/g, whereas on limestones the REE contents vary from 152 to 1927 μg/g. On the other hand, the soils from igneous rocks show lower REE contents (24 to 312 μg/g).

4.3.2.2 Heavy mineral placer deposits

Potentially useful concentrations of REE-bearing minerals are also found in placer deposits. The weathering of all types of rocks yields sediments that are deposited in a wide variety of environments, such as streams and rivers, shorelines, alluvial fans and deltas. Depending on the source of the erosion products, certain REE-bearing minerals, such as monazite and xenotime, can be concentrated along with other heavy minerals such as ilmenite, rutile, magnetite and zircon. The source does not need to be an alkaline igneous rock or a related REE deposit. Most placer accumulations with significant amounts of REE minerals are tertiary or quaternary deposits derived from source areas that include granitic rocks or high-grade metamorphic rocks. Many common igneous, metamorphic and even older sedimentary rocks contain enough monazite to produce a monazite-bearing placer. As a result, monazite is almost always found in any placer deposit (Long *et al.*, 2012). Ancient and modern types of sedimentary placer deposits formed in both alluvial and coastal environments have been significant sources of REE (Ghosal *et al.*, 2020; Sengupta & Gosen, 2016). Depending on source rock (provenance) characteristics, beach placers can be particularly enriched not only in REE, but also in other critical metals (e.g. Co, Nb, Ta).

These REE placer deposits are widely distributed in the world on all continents (Figure 4.2). Examples of placer deposits of REE include: tin-rich river placers in Malaysia; paleo-placers in Witwatersrand (South Africa) and Elliot Lake (Ontario, Canada); and beach sand deposits of Kerala, Andhra Pradesh and Orissa in India. There are also plenty of placer deposits available in other countries such as Australia, Madagascar, Brazil, Thailand, China, New Zealand, Sri Lanka, Indonesia, Zaire, Korea, Greece and the US (Batapola *et al.*, 2020; Dushyantha *et al.*, 2020). India has nearly 35% of the world's beach and sand mineral deposits, which are significant sources of REE. Placer deposits are usually associated with a high content of radioactive elements (e.g. Th and U). India is the most important thorium producer, through the exploitation of the monazite-rich beach placers of Orissa in India. With ~6000 km of coastal length, India has the world's richest shoreline heavy mineral placer deposits (dominated by ilmenite, rutile, sillimanite and zircon) which also include REE-bearing minerals such as monazite, xenotime and garnet (Singh, 2020). Palaparthi *et al.* (2017) have carried out a comprehensive study to determine the radioelement and REE concentrations in beach placer deposits at selected locations along the eastern coast of Andhra Pradesh in India to understand their economic potential. Singh *et al.* (2020) made a detailed exploration study revealing new placer deposits with radioactive and non-radioactive mineral assemblages formed from diverse sources in northeastern Odisha, India.

4.4 ALTERNATE SOURCES FOR REE

As the gap between their global demand and supply increases, the search for alternative resources of REE becomes more and more important, especially for the countries which depend highly on their import. In recent years sources like marine sediments, coal and fly ash, waste rocks in closed mines, mine drainage, river sediments, and industrial wastes like red mud, have been found to contain significant amounts of REE which can be utilized. These aspects will be discussed in detail in the following section, in addition to extraterrestrial and e-waste REE resources.

4.4.1 REE in coal and coal fly ash

In recent years, coal deposits have attracted much attention as promising alternative raw materials for REE because the REE concentrations in many coals or coal ashes are equal to or higher by many times than those found in conventional REE ores. Coal-related materials, including coal refuse and coal mine drainage have been identified as a potentially promising resource (Zhang *et al.*, 2020). Besides coal, fly ash which until now was considered as a waste material has also been studied for its potential for REE extraction.

The ΣREE was found to be 159.9 μg/g in some Polish fly ash samples, though the economically justified level of the REE content in power plant wastes is above 1000 μg/g (Baron, 2020). The average REE contents in lignite and bituminous coals, as well as lignite and bituminous coal ashes worldwide, have been estimated to be around 69, 72, 378 and 469 μg/g, respectively (Ketris & Yudovich, 2009). On a worldwide basis, coal and coal ash contain approximately 68 and 404 μg/g of REE, respectively. In general, the world's coal fly ash has 445 μg/g REE and has the potential to be a source of REE when the amount is comparable to conventional ore (Franus *et al.*, 2015). The REE content of coal ash is close to the REE content of the ion-adsorbed clays that are currently mined in China. For example, in coal samples from the Mazino Coal Mine (Tabas Coalfield, Iran), ΣREE range from 16.4 to 184 μg/g with an average of 88.9 μg/g (Pazand, 2015). On the other hand, the ΣREE value is ~ 404 μg/g in the world coal ash (Dai *et al.*, 2016). In fact, even much higher values up to 1358 μg/g of REE (including Y) have been reported in the literature (Hower *et al.*, 2016). This would offer the potential to reduce our dependence on other countries for these critical materials and also create new industries in regions where coal has played an important economic role.

Zou *et al.* (2020) carried out a detailed study to understand the controlling geological factors for the enrichment of not only REE but also another demanding element, Li, in coal from the Donggou Mine (southeastern Chongqing Coalfield, China). The possible recovery of REE from abundant coal and coal ash is an exciting new research area, representing a dramatic paradigm shift for coal, innovations to REE manufacturing and bypass mining. Some industries that rely on REE are looking for ways to bypass mining entirely by extracting REE from other materials. For example, countries like the US could someday obtain these elements as by-products from power plant coal ash, coal mining waste and mine drainage. Acid mine drainage from coal mines often contains considerable concentrations of REE as these elements are released in low pH conditions (Binnemans *et al.*, 2013a). Costis *et al.* (2021) very recently indicated that REE can be economically recovered from the residues coming from the exploitation of uranium, mine tailings and mine drainage, which contain significant quantities of REE.

4.4.2 REE in ocean-bottom sediments

Recent studies (Balaram *et al.*, 2012; Francesca *et al.*, 2020; Hein *et al.*, 2010; Kato *et al.*, 2011) indicated that in addition to land resources, significant REE resources

Table 4.3 Overview of ΣREE range (µg/g) in marine sediments from different oceans (modified after Balaram, 2019).

Ocean	ΣREE/ΣREY Range (µg/g)	Matrix	Reference
Afanasy Niktin Seamount (ANS) in the Eastern Equatorial Indian Ocean.	1727–2511	Cobalt crust	Balaram *et al.* (2012)
Mid-Pacific seamount	2085	Cobalt-rich crust	Cui *et al.* (2009)
Indian Ocean	928–1570	Ferromanganese crust	Nath *et al.* (1992)
Scotia Sea	3400	Ferromanganese crust	Gonzalez *et al.* (2010)
Eastern South Pacific	1000–2230	Deep sea mud	Kato *et al.* (2011)
North Pacific (east & west of Hawaiian Islands)	400–1000	Deep sea mud	Kato *et al.* (2011)
Mid Pacific Ocean	1178–1434	Fe-Mn nodules	Bu *et al.* (2003)
Pacific Ocean	1326	Deep nodules	Piper (1974)
Pacific Ocean	1398	Shallow nodules	Piper (1974)
Pacific Ocean	22 000	Ocean-floor sediments	Milinovic *et al.* (2021)

are available in ocean-bottom sediments (marine mud, ferromanganese crusts and polymetallic nodules). Table 4.3 presents comparative data of ΣREE from different marine sediments (cobalt crust) of world oceans, including that of the Indian Ocean. Figure 4.5 depicts the occurrence of manganese nodules, cobalt crusts, marine mud and phosphorites at the ocean basins, seamounts and continental margins in the oceans (Balaram, 2019). Deep-sea sediments with high REE and yttrium ranging from 1727 to 2511 µg/g are mainly concentrated around seamounts (e.g. Afanasy Niktin Seamount, Indian Ocean). Ocean-floor sediments containing ΣREE concentrations as high as 22,000 µg/g are reported from the Pacific Ocean (Milinovic *et al.*, 2021).

Though several types of sea-floor sediments harbor high concentrations of these elements, cobalt crusts (more precisely referred to as 'cobalt-rich ferromanganese crusts') and manganese nodules are considered to be the most important deposits of REE including other metals in the sea today. The composition of these ocean-bottom sediments reflects the proportions of influxes of genetically different materials: hydrogenic, biogenic, hydrothermal, volcanogenic or lithogenic (Banakar & Borole, 1991; Nath *et al.*, 1992). Out of these two types, cobalt crusts assume a lot of significance from the REE point of view. Cobalt crusts are found throughout the ocean basins on seamounts, ridges and plateaus on hard-rock substrates. They form by precipitation from cold ambient bottom waters at depths typically of about 2000–5000 m. The crusts grow at extremely slow rates, of about 1–7 mm/my and contain high concentrations of strategically and economically important metals (e.g. Co, Ti, Ce, Zr, Ni, Pt, Mo, Te, Cu and W) besides containing very high amounts of REE. There are essentially two types of ferromanganese crusts: hydrogenous (normally continuous) or hydrothermal (mostly episodic) in origin. REE patterns are useful in distinguishing these two types of deposits. These ocean-bottom sediments are found to contain surprising amounts of REE (ΣREE <2511 μg/g). Hot plumes from hydrothermal vents brought these elements out of the seawater and deposited them on ocean floors, bit by bit, over tens of millions of years (Sander & Koschinsky, 2011). At places, the Pacific mud deposit contains 100–1000 times more REE than the world's presently known land reserves of 120 million tons of REE.

According to some studies, REE resources undersea are much more promising than on-land resources. One square patch of metal-rich mud 2.3 km wide might contain enough REE to meet most of the global demand for a year. Kato *et al.* (2011) showed that REE and yttrium are readily recovered from the mud by simple acid leaching and suggested that deep-sea mud constitutes a highly promising resource for these elements. In an attempt to prove that the Indian Ocean also bestowed huge resources of these valuable metals in its bottom sediments, Balaram *et al.* (2012) explored the seamount crust at the Afanasy Niktin Seamount (ANS) in the Indian Ocean. Subsequently, the study was extended to detailed multi-beam swath bathymetric and geochemical investigations of seamount cobalt crust.

4.4.3 Phosphorite deposits

Considering other ocean sediments, phosphorite deposits have a higher potential as a resource for REE than most conventional REE deposits. Phosphorite deposits are formed in the continental margins at relatively shallower depths in oceans (Mazumdar *et al.*, 1999). Phosphorite deposits form as chemical precipitates on continental shelves (Figure 4.5). Upwelling of cold phosphate-rich waters causes warming and a decreasing solubility leading to their precipitation. It has been recognized that REE are enriched in the francolite mineral phase of phosphorites where REE substitute for Ca in the francolite lattice (Jarvis *et al.*, 1994). Some of the REE deposits contain ΣREE up to 2000 μg/g (Emsbo *et al.*, 2015), although the composition of these rocks mostly depends on their type and origin. The ΣREE of Northern African (Tunisian and Algerian) phosphorites exhibit a median of 314.5 μg/g (maximum value 1018.34 μg/g) and 289.21 μg/g (maximum value 1759.41 μg/g), respectively, and fall in the second highest range after the US

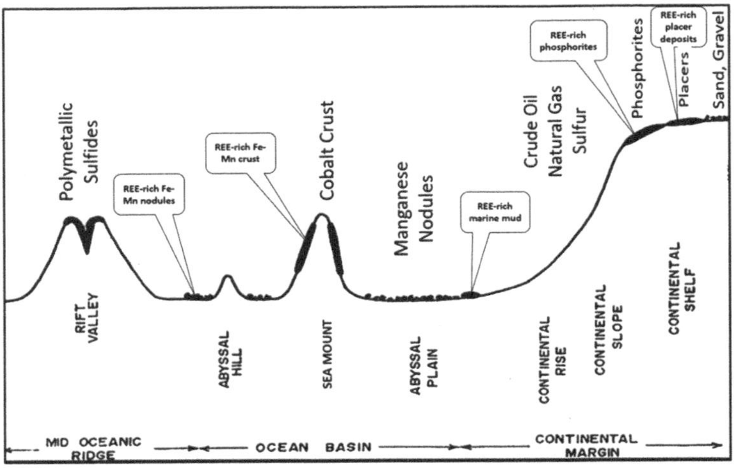

Figure 4.5 Occurrence of manganese nodules, cobalt crusts, marine mud and phosphorites at the ocean basins, seamounts and continental margins in the oceans (Balaram, 2019).

deposits that are the most REE-enriched (average = 4394 μg/g) (Buccione *et al.*, 2021).

However, with the current state of ocean mining technology, the economic viability of deep-sea mining is questionable. If the environmental and financial factors were cleared, then deep-sea mining would definitely be a feasible option for the long term.

4.4.4 REE in river sediments

According to the study of Ramesh *et al.* (2000), some of the Himalayan river sediments are showing ΣREE (excluding Y & Sc) as high as 402 μg/g with the highest average concentrations of 152 μg/g, and the Amazon reservoir reporting 217 μg/g. The physical weathering process seems to be a major controlling factor for the distribution of REE and trace metals in the sediments of the Himalayan rivers. According to Zhu *et al.* (1997), ΣREE in Chinese river sediments ranged between 44.5 and 315.8 μg/g. Recovery as well as extraction of REE from river sediments must be easier and less expensive than from ocean sediments. A comparative study of the REE contents of Chinese and Korean river sediments indicated that the Korean river sediments are enriched in REE contents with an average ΣREE concentration of 244 μg/g, against a concentration of 189 μg/g in Chinese river sediments (Xu *et al.*, 2009).

4.4.5 Waste rock sources from old and closed mines

Because of their essential role in our modern life, the demand for REE is continuously growing, pressuring the mining industry to fulfill these needs. The possibility of accessing REE from mine waste and mill tailings is attractive partly because the minerals have already been excavated from the ground which would make it easier to access the minerals and hence reduce mining costs. For example, REE are mostly contained in millimeter-sized crystals known as fluorapatite that are found in deposits of iron ore. When the iron was mined, the fluorapatite was usually left behind as waste because it was considered an unwanted impurity. As the need for certain mineral resources has changed with time, these waste products are now of interest. The fluorapatite has elevated levels of heavy REE such as Gd. It is estimated that acid mine drainage sludge from Pennsylvania and West Virginia in the United States alone could provide 45,000 tons of REE per year (Ganguli & Cook, 2018). The possibility of efficiently extracting REE from acid mine drainage has also caught people's attention. The mine drainage and the acidic wastewaters of abandoned REE mines can be potential sources for REE provided the technology of extracting the REE is economical.

4.4.6 Red mud

Red mud is a waste product from the Bayer process of producing alumina (aluminum oxide) from bauxite and is generated at a rate of up to 175.5 million tons per year. The global stockpile of red mud is near 4 billion (10^9) tons. REE have been found in significant quantities in red mud. REE concentrations in red mud samples can reach up to 500–1700 μg/g (Akcil *et al.*, 2018). Table 4.4 presents individual REE concentrations (μg/g) of red mud from different countries in comparison with the earth's average crust composition (Archambo & Kawatra, 2020). During a time when there is an intensive search for finding alternative sources for these vital elements, red mud may become a promising feedstock for their secondary production.

4.4.7 Extraterrestrial REE resources

Accessible terrestrial supplies of REE are becoming limited as a result of their high demand in advanced technology applications, which provides a motivation to investigate possible sources in extraterrestrial environments. According to some recent studies (McLeod & Krekeler, 2017), the REE reserves currently available across the world will be sufficient for our day-to-day needs for <2500 years. Based on current demand, mining operations and technologies, with an expected increase in the population and technology needs, exploration of potential extraterrestrial REE resources is inevitable, with the earth's moon being a logical first target. While REE abundances of trace phases in lunar rocks are high, their abundances are low compared to terrestrial ores which means that there is, to

Table 4.4 REE concentrations (μg/g) of red mud from different countries in comparison with the earth's average crust composition (Archambo & Kawatra, 2020).

Location (Reference)	REE (μg/g)												
	La	Ce	Pr	Nd	Sm	Gd	Tb	Dy	Ho	Er	Yb	Y	Sc
Average in Earth's crust (Balaram, 2019)	39	66	9	41	7	6	1	5	1	3	3	33	22
Chinalco, China (Ujaczki *et al.*, 2018)	416	842	95	341	64	56	184	48	25	28	28	266	158
Australia (Wang *et al.*, 2013)	—	—	—	—	—	—	—	—	—	—	—	68	54
Brazil (Botelho *et al.*, 2019)	—	—	—	—	—	—	—	—	—	—	—	24	43
India (Abhilash *et al.*, 2014)	110	70	0.5	—	—	—	—	—	—	—	—	1	5
India (Singh *et al.*, 2019)	58	98	—	—	—	—	—	—	—	—	—	—	48
India (Ujaczki *et al.*, 2018	112	191	18	48	9	7	—	4	—	1	2	13	58
Jamaica (Narayanan *et al.*, 2019)	287	366	74	69	0	37	0	37	5	21	16	373	55
Greece (Borra *et al.*, 2015)	114	386	28	98	21	22	—	16	4	13	14	75	121
Alumine de Greece, Greece (Vind *et al.*, 2018)	130	480	29	107	19	22	3	20	4	13	13	108	—
Greece (Deady *et al.*, 2016)	127	409	28	103	20	18	2	19	3	11	13	98	—
Ajka, Hungary (Ujaczki *et al.*, 2015)	114	368	—	—	—	—	—	—	—	—	—	68	—
Turkey (Deady *et al.*, 2016)	169	480	47	161	32	4	26	23	4	13	14	113	—
Russian Federation (Martoyan *et al.*, 2016)	—	—	—	—	—	—	—	—	—	—	—	53	25
Russian Federation (Ujaczki *et al.*, 2018)	—	—	—	—	—	—	—	—	—	—	—	—	90
Iran (Martoyan *et al.*, 2016)	—	—	—	—	—	—	—	—	—	—	—	1	19

date, no geological evidence to support mining the Moon for REE. While the relative concentrations and abundances of the REE in different mineral phases of the moon, Mars and chondritic meteorites that have been characterized to date provide crucial insights into the differentiation history of planetary objects, current and future missions to the moon, Mars and other nearby objects in our solar system may yet reveal one or more extraterrestrial REE resources that one day will be utilized (Haque *et al.*, 2014; Heggy *et al.*, 2020). Already, mining experiments on the International Space Station (ISS) have been conducted to test hypotheses on the bioleaching of REE from basaltic rock in microgravity and simulated Mars and Earth gravities, using three microorganisms and a purposely designed biomining reactor (Cockell *et al.*, 2020).

4.4.8 REE from electronic and industrial waste

Overwhelming challenges in conventional REE explorations and mining are currently indicating that secondary REE resources, such as electrical and electronic waste (e-waste) and mine tailings are promising resources. The world is moving towards a cleaner and greener future and it is becoming extremely difficult to meet the growing demand for the REE as most of the production is located in only a few countries such as China, the US, Australia and India, although lots of research is going on to search for replacements for REE in critical technologies such as super magnets. Currently, mountains of e-waste, rich in REE and other valuable metals are indeed growing across the globe. When this waste is turned into a valuable resource, this will protect the planet's increasingly strained REE resources.

Recycling of electronic and industrial waste such as printed circuit boards, scrap automobile magnets and hard drives and the extraction of REE from this scrap have become attractive. However, recycling activity is a threat to both the health of the individuals working on the recovery of the elements and to the environment. In addition, REE are often used in small quantities, making their recycling challenging. Therefore, it is estimated that less than 20% of the e-waste is recycled and documented, while the rest (80%) is deposited in landfills without further treatment (Szucs *et al.*, 2021). The recovered REE represents a very small proportion (estimated to be less than 1%) of the 20% recycled e-waste (Binnemans *et al.*, 2013b). Each year, the electronics industry generates up to 50 million tons of e-waste, but as the number of consumers rises and the lifespan of devices shrinks in response to demand for updated equipment, this figure could exceed 50 million tons in the coming years. There is also another argument that the amount of electronic waste generated is shrinking due to the phasing out of bulky products, such as large cathode-ray tube televisions and computer monitors (Althaf *et al.*, 2021). In addition to a decline in total e-waste mass, the sheer number of electronic devices entering the waste stream is also leveling off or shrinking.

Despite these arguments, the amount of e-waste generated is still substantial and may be used as a source for REE. Belgian-based Umicore Group has developed a battery recycling program for nickel-metal-hydride (NiMH) batteries at its new recycling plant in Hoboken (Belgium). After the separation of nickel and iron, the company plans to process the REE-rich material into a high-grade concentrate that would be separated into rare-earth materials at the Rhodia Rare Earth Systems plant at La Rochelle, France (Umicore Group, 2011). Recently Jowitt *et al.* (2018) have provided an overview of the current and future potential of the recycling of the REE, including outlining the significant but currently unrealized potential for increased amounts of REE recycling from end-uses such as permanent magnets, fluorescent lamps, batteries and catalysts. Currently, a number of studies are in progress over the world for the economic recovery of REE from industrial and e-waste (Hammache *et al.*, 2021; Lukowiak *et al.*, 2020).

4.5 INDUSTRIAL APPLICATIONS

In recent years, REE have become crucial to modern life and people's day-to-day activity is highly dependent on them. REE are being extensively utilized in manufacturing a wide variety of defense, aviation, industrial and light-emitting diode (LED) lights. Modern media and communication devices such as mobile phones, iPads, smart television sets and watches, and computers employ REE as magnets for speakers and hard drives, and REE-containing phosphors for optical displays. A few REE are used in oil refining and nuclear power and more specialized uses occur in medicine and manufacturing of fertilizers in agriculture. Much of the demand for REE is driven by many current and emerging alternative low-carbon energy technologies, such as electric vehicles, energy-efficient lighting, wind turbines, rechargeable batteries, radar systems, laser crystals and other advanced defense systems. An overview of the general industrial and defense applications of REE is presented in Table 4.5. As shown in Figure 4.6, catalysts and magnets dominate the REE applications.

REE play an essential role in defense applications. These metals are used in making devices for military applications such as night-vision goggles, precision-guided weapons, communications equipment, GPS equipment, batteries and electronic communication systems which give an enormous military advantage to any country. REE are key ingredients for making the very hard alloys used in armored vehicles and projectiles that shatter upon impact. The samarium-cobalt magnet retains its powerful magnetic properties even when very hot, thus making more powerful radar instruments possible.

REE are chemically very similar to one another, and about 25% of their uses are based on this close similarity. The other 75% usage is, however, based on the unique properties of the individual elements. There are no economic alternatives to these REE. Other substances that can be substituted for REE in their most important uses are usually less effective and more costly. Despite chemical similarities,

Table 4.5 An overview of the general industrial and defense applications of REE.

REE	General Applications	Defense Applications
La	In camera and telescope lenses, compounds containing La are used extensively in carbon lighting applications, such as studio lighting and cinema projection, petroleum refining. Ni–La metal hydride rechargeable batteries for automobiles and even in certain medicines, autocatalysts, polishing powders, glass additives, phosphors, ceramics and optics.	Night-vision goggles.
Ce	In catalytic converters in automobiles (emissions control), petroleum refining, windshields, mirrors and lenses are polished using cerium oxides, battery alloys, metal alloys, glass additives, phosphors and ceramics.	
Pr	To create strong metal alloys for use in aircraft engines, in cigarette lighters and other handheld fire-starters, coloring ceramics.	
Nd	Small and powerful magnets used in loudspeakers and computer hard drives, magnets for wind turbines and hybrid cars, electric motors in a typical electric vehicle, as well as the speakers of its sound system use neodymium-iron-boron permanent magnets, glass coloration and florescent lighting.	Laser range-finders, guidance systems and communication equipment.
Pm	Most promethium is used only for research purposes. It has few applications, including use in nuclear batteries and also used to measure the thickness of materials by evaluating the amount of radiation from a promethium source that passes through the sample, watches and pacemakers.	Possible future uses in portable X-ray sources and as auxiliary heat or power sources for space probes and satellites.
Sm	Magnets, lasers and carbon-arc lighting, as a neutron absorber in nuclear reactors. Samarium's radioactive isotopes are used to treat severe pain associated with bone cancers.	Permanent magnets that are stable at high temperatures, precision-guided weapons, 'white noise' production in stealth technology.
Eu	Televisions and computer screens, in making control rods in nuclear reactors. The element is also used in screening for genetic diseases.	Fluorescents and phosphors in lamps and monitors.

(*Continued*)

Table 4.5 An overview of the general industrial and defense applications of REE (*Continued*).

REE	General Applications	Defense Applications
	Europium is used in the anti-counterfeiting phosphors of bills.	
Gd	In magnetic resonance imaging (MRI) imaging, in television screens. Gadolinium is used in nuclear reactors as both a shield and a secondary emergency shutdown mechanism in certain types of reactors.	
Tb	Phosphors in optical displays contain yttrium, europium and terbium oxides, televisions and computer screens, in solid-state devices such as the crystal stabilizer of fuel cells.	Naval sonar systems.
Dy	In making laser materials, in hard disks, neutron-absorbing control rods in nuclear reactors, and high intensity lighting.	
Ho	For its magnetic strength, Ho is used as a component in powerful magnets, as well as nuclear control rods, since it has the ability to absorb neutrons expelled by nuclear fission, medical devices, such as lasers made with Ho, which are safe to use on human eyes.	Holmium is used in microwave equipment.
Er	Erbium's main uses include combination with Eu isotopes to give them certain florescent properties. It is also used in optical amplifiers, fiber optic cables, medical and dental lasers.	Amplifiers in fiber-optic data transmission.
Tm	Thulium has been used in surgical lasers, portable x-ray devices and high temperature superconductors.	
Yb	In portable x-ray devices, the strengthening of steel, solid state lasers and stress gauges that monitor ground deformation from earthquakes or explosions.	
Lu	In the petroleum industry for petroleum cracking and in certain cancer treatments.	
Y	Y is primarily used in LED televisions, where yttrium oxide is combined with europium to help produce red pixels. Electrical sensors employ yttria-stabilized zirconia to measure and control the oxygen content of the fuel and in preparing colored pigments.	

(*Continued*)

Table 4.5 An overview of the general industrial and defense applications of REE (*Continued*).

REE	General Applications	Defense Applications
Sc	Sc is added to Al, creating an alloy that increases the strength of the metal in heat-affected zones, such as around welds, in high-intensity discharge lamps and light bulbs.	

numerous practical uses of REE often depend on their unusual physical properties (fluorescent, conductive, electrical, magnetic and thermal) which are specific to particular elements, so the challenge of separating them must be overcome. These metals are normally alloyed, or mixed, in small quantities with more common metals such as iron. In addition, our increased understanding of the unique properties of REE has generated their expanded use in contemporary society.

4.5.1 Glass industry

Europium was an essential material for producing color images. Due to their differing spectroscopic properties, europium (as Eu^{3+}, in the solid-state) exhibits red luminescence, while terbium's (as Tb^{3+}) is green. In the early 1950s, europium was used to produce red phosphors for the newly developed color

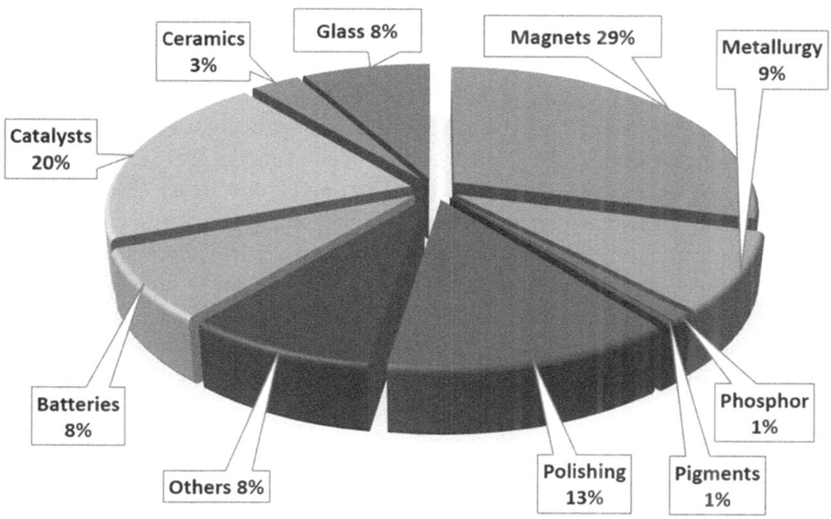

Figure 4.6 Estimated current end uses of REE (https://roskill.com/market-report/rare-earths/).

television technology. The demand for REE saw its first explosion in the mid-1960s when the first color television sets were entering the market. At that time, the glass industry was the leading consumer of REE raw materials, which were used for providing color and as additives that give special optical properties to the glass in addition to their use in glass polishing (Goonan, 2011). REE are used even now in polishing powders for tablet and mobile phone screens. Currently, europium is widely used as a phosphor in LED televisions, computer screens and fluorescent lamps. Cerium oxide is widely used in the production of glass types that require a precision polish, such as flat panel display screens. Cerium is also used to decolorize glass. Lanthanum and lutetium greatly increase the refractive index of optical glass. Lanthanum is widely used in camera lenses, whereas the more expensive lutetium is used in immersion lithography, which requires a high-refractive-index. Erbium, holmium, neodymium, praseodymium, ytterbium and yttrium are used as special colorants and to provide filtering and glare-reduction qualities for glass. Europium is a common dopant (doping agent) for optical fibers. Currently many REE, especially yttrium, cerium, lanthanum, europium and terbium are used individually or in combination to make phosphors for many types of cathode ray tubes and flat panel display screens, and in some incandescent, fluorescent and light-emitting diode lighting (Goonan, 2011; Long, 2011).

4.5.2 Energy-efficient lighting

Though the 2015 Nobel prize winner, Shuji Nakamura used gallium nitride (GaN) for making blue LEDs initially, currently cerium, yttrium and other REE are being increasingly used in making different colored LED lights, because of several advantages such as the possibility for miniaturization and compliance with environmental and energy-efficiency standards (Song *et al.*, 2013). Glass materials containing REE have been central to dramatic light-based innovations like semiconductor lasers, also called laser diodes (Tanabe, 2015). For example, REE are used to create red (Eu and Y) and blue (Eu) phosphors for energy-efficient LEDs, which can provide better energy savings for buildings than incandescent and fluorescent lights. Thus, the LED market is likely to increase REE demand in the future (Ku *et al.*, 2015). Incandescent light bulbs lit the 20th century and the 21st century will be lit by LED lamps. Materials consumption is also diminished as LEDs last up to 100,000 hours, compared to 1000 hours for incandescent bulbs and 10,000 hours for fluorescent lights.

4.5.3 Rechargeable batteries

Battery research in the 1970s and 1980s led to the development of the nickel-metal hydride (NiMH) battery, which used La and Nd. NiMH batteries use anodes made of lanthanum-based alloys. The NiMH batteries in many electric vehicles also contain cerium and lanthanum. These batteries could be recharged repeatedly while holding

Table 4.6 Quantity of REE in some products (modified after Alonso *et al.*, 2012; Ganguli and Cook, 2018).

Item	Amount of REE Used in Manufacturing
Mobile Phone	0.0005 kg
Air conditioner	0.12 kg
Toyota Prius	15 kg per unit
Lockheed-Martin F–35	416 kg
Navy surface ships	1818 kg
Navy submarines	3636 kg
Wind turbine that generates 3.5 MW of electricity	600 kg

a lot of energy relative to their volume (size). Demand for batteries is also driven by demand for portable electronic devices such as cell phones, iPads, laptop computers and video cameras and they are currently widely used in hybrid cars, such as the Toyota Prius, released in 2001. For example, 15 kg of REE are used per unit in the Toyota Prius car, including its battery (Table 4.6). However, some industry analysts expect electric vehicle manufacturers will soon totally transition to lithium-ion batteries, which would reduce REE demand from electric vehicles because lithium-ion batteries do not require REE (Anderson & Patiño-Echeverri, 2009).

4.5.4 Permanent magnets

Permanent magnets are used in demanding applications such as electric motors for hybrid cars, wind turbines and magnetic storage. Permanent magnets also find wide applications in defense. Magnets made with neodymium are among the strongest permanent magnets that are currently known, with the ability to lift around 1000× their own weight. Significant uses include cell phones, electric motors for hybrid vehicles and windmills, actuators in aircraft power windows and doorlocks, windshield wiper motors and electric engine starters (Goonan, 2011; Long, 2011). Magnets made of Nd are used in microphones, professional loudspeakers, headphones and computer hard disks where a magnetic field is required. Neodymium iron boron (NdFeB) magnets accounted for about 90% of the permanent magnets produced in 2018, with samarium cobalt (SmCo) making up the remainder. The largest use for permanent magnets is in synchronous motors, which are found in wind turbines as well as electric and hybrid vehicles. NdFeB magnets have the highest magnetic strength (energy product) among commercially available magnets and enable high energy density and high energy efficiency in energy technologies. These are used when there are restrictions in space and weight. Dysprosium and terbium (HREE) are key critical materials often added to the NdFeB alloy to increase the operating temperature of the magnets. HREE tend to be less abundant and more expensive than LREE. The deployment of energy technologies such as wind turbines and electric vehicles (EVs) could lead to imbalances of supply and demand for these key materials. Unlike smaller, onshore wind turbines that use rotating gearboxes, larger offshore turbines use direct drive technology. This involves a generator composed of a ring of NdFeB. By some estimates, a wind turbine that generates 3.5 megawatts of electricity contains about 600 kg of REE (Table 4.6) (Alonso *et al.*, 2012). Since there is a constant risk to the available supply of REE, research is underway to find substitutes or alternatives such as MnAl magnetic materials for NdFeB magnets (Kontos *et al.*, 2020).

4.5.5 Electronics

Use of REE in electronics expanded through the 1990s and the 2000s. In the early 1990s, Bell Labs developed the erbium-doped fiber amplifier to boost the signal in fiber-optic cables. These small devices made a global network of long fiber-optic cables possible that reduced the price of long-distance telephone calls and now carry internet data around the world. The release of the first iPhone in 2008 showed how far advances in REE metallurgy and applications had developed. Smartphones use lanthanum to reduce distortion in their tiny glass camera lenses, neodymium magnets to improve the sound from tiny speakers, and yttrium and erbium phosphors to make bright colors in an energy-efficient screen. Twenty years ago, very few people owned a mobile phone, but today over 5 billion people own one. The use of REE in computers has grown almost as fast as their use in cell phones.

4.5.6 Catalysts

Catalysts are another major use for REE. The unique catalytic performance of the REE make them a critical resource for industrial applications, such as petroleum refining, the catalytic combustion of fossil fuels, automotive engine emissions control, and the purification of industrial waste, air and solids. In particular, lanthanum and cerium are used in petroleum refining to make gasoline, which constitutes one of the largest end-uses of REE (Figure 4.6). Catalysts enriched in REE are essential to cracking (breaking down) heavy hydrocarbon molecules into smaller molecules, which enables petroleum refineries to obtain significantly more product per barrel of oil processed. These REE ultimately make the fluid catalytic cracking process more efficient, which increases the gasoline yield per unit of catalyst (Sadeghbeigi, 2012). Cerium oxide-based catalysts are used in automotive catalytic converters (Goonan, 2011). Cerium and zirconium oxides are used in autocatalytic converters to create an oxygen-rich environment to help oxidize toxic carbon monoxide into carbon dioxide (Balaram, 2020; Machida *et al.*, 2017). Small amounts of neodymium, praseodymium and yttrium are also used as catalysts in autocatalytic converters to reduce automotive carbon monoxide emissions.

4.5.7 Alloys

REE have unusual fluorescent, conductive and magnetic properties which make them very useful when alloyed, or mixed in small quantities with more common metals such as iron, in combination with other REE and several other metals. The thermal properties of REE lend stability to alloys under high stress and temperature, for instance in jet engines. Cerium, lanthanum, neodymium and praseodymium, commonly in the form of a mixed oxide known as mischmetal, are used in steelmaking to remove impurities, as well as in the production of special steel alloys (Goonan, 2011). REE, along with yttrium, individually or in combination, are also used in various special alloys of chromium, magnesium, molybdenum, tungsten, vanadium and zirconium.

4.5.8 Defense applications

REE are critical particularly for defense applications from aircraft engines and missiles to aircraft carriers and weapons systems. REE also find applications in jet fighter engines, missile guidance systems, missile defense, space-based satellites and communication systems (Machacek & Fold, 2014). Neodymium has the world's mightiest magnetic powers, making it useful for missile guidance systems. Lanthanum enhances the clarity of glass, particularly for high-end camera lenses, such as those used for intelligence, surveillance and reconnaissance. Phosphorescent europium warms the hues in LED lights and plasma displays, while its unique neutron-absorbing properties make it a crucial ingredient in the control rods used in nuclear reactors. Each F-35 jet contains some 416 kg of Y, Tb and other REE, particularly for their advanced targeting systems (Table 4.6).

4.5.9 REE in paints and pigments

Inorganic natural and synthetic pigments produced and marketed as fine powders are integral parts of many decorative and protective coatings. They are used for mass coloration of materials such as plastics, glazes, ceramic bodies and porcelain enamels. The pigments need to be nontoxic, stable, not fading in light with time, or disintegrating with heat. Earlier elements such as Cr, Mn, V and Ni were being used. But recent trends show that REE are being applied in large amounts in producing a variety of colored pigments due to their unique spectroscopic and magnetic properties. Yttrium, lanthanum, cerium, neodymium and praseodymium are used as pigments for ceramics. For example, incorporation of cerium oxide into the praseodymium-zircon lattice has been reported to provide a yellow-orange color (Giable *et al.*, 2005; Sreeram *et al.*, 2008).

Researchers at Oregon State University (USA) led by Mas Subramanian discovered YInMn Blue, the first new blue pigment identified in 200 years while developing special materials for use in electronics (Smith *et al.*, 2009). These researchers have also created a whole range of different colors in the YInMn Blue series, including purples, greens, yellows and oranges, by adding new elements to the mix.

4.5.10 REE in agriculture

REE have been used as a fertilizer to improve crop growth and production in the agriculture of China for about 30 years (Pang *et al.*, 2002). In addition to the use of REE-containing fertilizers, sometimes the soil itself has been enriched with elevated concentrations of REE by the contribution from the mafic and ultramafic lithologies to the soil formation. Millions of tons of REE-containing micro-fertilizers are directly applied on a large-scale to plants in agriculture in China for improving the yield and quality (Diatloff *et al.*, 1995; Guo *et al.*, 1988). Hong *et al.* (1996) reported that the REE can raise the output of wheat by 4–10% every year in a continuous ten-year period after applications of REE at 600 g/ha-yr. However, because of the adverse effects of REE on human health and the ecosystem, in recent times several studies have been initiated to understand the effects of REE in agriculture (Morin-Crini *et al.*, 2021).

4.5.11 REE in medicine

Their unique properties, such as radiation emission or magnetism, allow REE to be used in many different therapeutic and diagnostic applications in modern medicine. Currently, there are a few major applications of REE in medicine, but many more of them are on the horizon. Several studies (Wakabayashi *et al.*, 2016; Zhang *et al.*, 2000a, 2000b) confirmed the antibacterial and antifungal activities of REE which are comparable to those of copper ions, and these elements are beginning to find several pharmaceutical applications. For example, Gd has been used in a chelated form as a contrast agent in magnetic resonance imaging (MRI) measurements (Raju *et al.*, 2010). REE can also be used as nematicide as they can inhibit the formation and germination of fungal spores and thus influence a large number of organisms (Zhang *et al.*, 2000a). New medical applications for these elements are being found at an increasing rate and emerging advancements such as nanotechnology might be used to enhance their use in medicine in the future.

4.5.12 Miscellaneous

REE are also used in making synthetic gems, crystals for lasers, microwave equipment, superconductors, sensors, nuclear control rods and cryo-coolers. Significant potential new uses for REE include their use as nano-filters and in memory devices, power converters, optical clocks, infrared decoy flares and fusion energy. In addition, REE are widely used as environmental tracers for natural systems, such as plant, soil and aqueous systems (Morin-Crini *et al.*,

2021). New developments in medical technology are expected to increase the use of surgical lasers, magnetic resonance imaging and positron emission tomography scintillation detectors. Future research is likely to find many new uses for the REE.

4.6 LOOKING INTO THE FUTURE

Currently, there is an intense search all over the world for new REE reserves, while existing reserves may be made more accessible through technological advances. Carbonatite and peralkaline igneous deposits contain the highest REE concentrations but are exceedingly rare. Abandoned tailing piles from some coal and iron mines may be important resources of REE. Alternate sources of REE, such as coal and its by-products and recycling activities could also change the REE supply chain landscape. Geological and mineralogical research should continue to play an important role in the search for REE ore deposits and their extraction, ensuring that as little damage is done to the environment as possible.

In addition to the opening of new mines and building new processing plants the world over, plans such as deep-sea mining may be a possible alternative as very high concentrations of REE in sea-floor sediments have been reported from Indian and Pacific Oceans. In addition, extraction of REE from the acidic wastewaters of abandoned REE mines, extraction of REE from coal and coal fly ash might become economically viable in the future because of large-scale demand for these metals. From the studies currently underway and the progress made so far, it is expected that the recycling of REE has the potential to be economical and more readily achievable than the exploitation of new mineral deposits.

The variety of high-tech applications of REE has increased very rapidly, a trend which is expected to continue. The ongoing demand for green energies such as solar and wind energy capacity both onshore and expanding offshore and the growing demand for electric vehicles could lead to greater demand for REE in the future especially for neodymium and dysprosium. In fact, electric vehicles will be the biggest driver as we need to meet the goal of limiting the average global temperature rise to 2°C by 2035. They are essential in efforts to reduce greenhouse-gas emissions enough to avoid the most devastating consequences of climate collapse. Thus, REE will continue to remain an important part of our foreseeable future, with demand likely to grow from quantum computing and material sciences to medical applications.

ACKNOWLEDGMENTS

The author thanks the Director, Council of Scientific and Industrial Research-National Geophysical Research Institute (CSIR-NGRI), and the Council of Scientific and Industrial Research (CSIR), New Delhi, India, for support in the form of an Emeritus Scientist Scheme. The author also thanks Dr. P. Krishnamurthy, Former Scientific Officer ‘G’, Atomic Mineral Directorate for

Exploration and Research (AMD), Hyderabad, for going through the earlier version of the manuscript and suggesting some useful corrections.

REFERENCES

Abhilash S. S., Sinha M. K. and Pandey B. D. (2014). Extraction of lanthanum and cerium from Indian red mud. *International Journal of Mineral Processing*, **127**, 70–73.

Akcil A., Akhmadiyeva N., Abdulvaliyev R., Abhilash and Meshram P. (2018). Overview on extraction and separation of rare earth elements from red mud: focus on scandium. *Mineral Processing and Extractive Metallurgy Review*, **39**(3), 145–151.

Alonso E., Sherman A. M., Wallington T. J., Everson M. P., Field F. R., Roth R. and Kirchain R. E. (2012). Evaluating rare earth element availability: a case with revolutionary demand from clean technologies. *Environmental Science & Technology*, **46**(6), 3406–3414.

Althaf S., Babbitt C. W. and Chen R. (2021). The evolution of consumer electronic waste in the U.S. *Journal of Industrial Ecology*, **25**, 693–706.

Anderson D. L. and Patiño-Echeverri D. (2009). An Evaluation of Current and Future Costs for Lithium-Ion Batteries for Use in Electrified Vehicle Powertrains. Duke University Master's thesis, Nicholas School of the Environment, Raleigh-Durham, N.C.

Archambo M. and Kawatra S. K. (2020). Red mud: fundamentals and new avenues for utilization. *Mineral Processing and Extractive Metallurgy Review*, 1–24.

Balaram V. (1996). Recent trends in the instrumental analysis of rare earth elements in geological and industrial materials. *Trends in Analytical Chemistry*, **15**, 475–486.

Balaram V. (2019). Rare earth elements: a review of applications, occurrence, exploration, analysis, recycling, and environmental impact. *Geoscience Frontiers*, **10**, 1285–1303.

Balaram V. (2020). Environmental impact of Pt, Pd and Rh emissions from autocatalytic converters – A brief review of the latest developments. In: Handbook of Environmental Materials Management, C.M. Hussain (ed.), Springer Nature, Switzerland AG.

Balaram V. (2021). New frontiers in analytical techniques – opportunities and challenges in geochemical research. *Journal Geological Society of India*, **97**(4), 330–334.

Balaram V., Banakar V. K., Subramanyam K. S. V., Roy P., Satyanarayanan M., Mohan M. R. and Sawant S. S. (2012). Yttrium and rare earth element contents in seamount cobalt crusts in the Indian ocean. *Current Science*, **110**(11), 1334–1338.

Balaram V., Roy P., Subramanyam K. S. V., Durai L., Mohan R. M., Satyanarayanan M., Sawant S. S., Kamal K. S. and Vani K. (2015). REE Geochemistry of seawater from Afanasy-Nikitin Seamount in the eastern equatorial Indian ocean by high resolution inductively coupled plasma mass spectrometry. *Indian Journal of Geo-Marine Sciences*, **44**(3), 339–347.

Balashov Y. A., Ronov A. B., Migdisov A. A. and Turanskaya N. V. (1964). The effect of climate and facies environment on the fractionation of the rare earth elements during sedimentation. *Geochemical International*, **5**, 951–969.

Banakar V. K. and Borole D. V. (1991). Depth profiles of ^{230}Th excess, transition metals and mineralogy of ferromanganese crusts of the central Indian basin and implications for paleo-oceanographic influence on crust genesis. *Chemical Geology*, **94**, 33–44.

Bao Z. W. and Zhao Z. (2008). Geochemistry of mineralization with exchangeable REY in the weathering crusts of granitic rocks in South China. *Ore Geology Reviews*, **33**, 519–535.

Baron R. (2020). Determination of rare earth elements in power plant wastes. *Mining Machines*, **4**(164), 24–30.

Batapola N. M., Dushyantha N. P., Premasiri H. M. R., Abeysinghe A. M. K. B., Rohitha L. P. S., Ratnayake N. P., Dissanayake D. M. D. O. K., Ilankoon I. M. S. K. and Dharmaratne P. G. R. (2020). A comparison of global rare earth element (REE) resources and their mineralogy with REE prospects in Sri Lanka. *Journal of Asian Earth Sciences*, **200**, 104475.

Bhushan S. K. (2015). Geology of the Kamthai rare earth deposit, District, Rajasthan. *Journal of Geological Society of India*, **85**, 537–546.

Bhushan S. K. and Kumar A. (2013). First carbonatite hosted REE deposit from India. *Journal Geological Society of India*, **81**, 41–60.

Binnemans K., Pontikes Y., Jones P. T., Gerven T. V. and Blanpain B. (2013a). Recovery of Rare Earths from Industrial Waste Residues: A Concise Review. 3rd international Slag Valorization Symposium, Leuven, pp. 19–20.

Binnemans K., Jones P. T., Blanpain B., van Gerven T., Yang Y., Walton A. and Buchert M. (2013b). Recycling of rare earths: a critical review. *Journal of Clean Products*, **51**, 1–22.

Borra C. R., Pontikes Y., Binnemans K. and Van Gerven T. (2015). Leaching of rare earths from bauxite residue (red mud). *Minerals Engineering*, **76**, 20–27.

Borst A. M., Smith M. P., Finch A. A., Estrade G., Villanova-de-Benavent C., Nason P., Marquis E., Horsburgh N. J., Goodenough K. M., Xu C. and Kynický J. (2020). Adsorption of rare earth elements in regolith-hosted clay deposits. *Nature Communications*, **11**(1), 1–15.

Botelho A. B., Costa R. H., Espinosa D. C. R. and Soares T. J. A. (2019). Recovery of scandium by leaching process from Brazilian red mud. In: Rare Metal Technology, G. Azimi, H. Kim, S. Ala, T. Ouchi, N. R. Neelameggham and A. A. Baba (eds), Springer, pp. 73–79.

Brophy J. G. and Basu A. (1990). Europium Anomalies in Mare Basalts as a Consequence of Mafic Cumulate Fractionation from an Initial Lunar Magma. Proc. 20th Lunar and Planetary Science Conference, Lunar and Planetary Institute, pp. 25–30.

Bu W. R., Shi X. F. and Peng J. T. (2003). Geochemical characteristics of seamount ferromanganese nodules from mid-Pacific Ocean. *Chinese Science Bulletin*, **48**, 98–105.

Buccione R., Kechiched R., Mongelli G. and Sinisi R. (2021). REES In the north Africa P-bearing deposits. *Paleoenvironments, and Economic Perspectives: A Review. Minerals*, **11**, 214.

Castor S. B. and Hendrik J. B. (2006). Rare earth elements. In: Industrial Minerals and Rocks: Commodities, Markets, and Uses, J. E. Kogel, N. C. Trivedi, J. M. Barker and S. T. Krukowski (eds), Society for Mining Mineralogy, United States, Vol. 7, pp. 769–792.

Chakhmouradian A. R. and Wall F. (2012). Rare earth elements: minerals, mines, magnets and more. *Elements*, **8**(5), 333–340.

Cockell C. S., Santomartino R. and Finster K. (2020). Space station biomining experiment demonstrates rare earth element extraction in microgravity and Mars gravity. *Nature Communications*, **11**, 5523.

Costis S., Mueller K. K., Coudert L., Neculita C. M., Reynier N. and Blaisa J. N. (2021). Recovery potential of rare earth elements from mining and industrial residues: A review and cases studies. *Journal of Geochemical Exploration*, **221**, 106699. doi: 10.1016/j.gexplo.2020.106699Get rights and content

Cui Y., Liu J., Ren X. and Shi X. (2009). Geochemistry of rare earth elements in cobalt-rich crusts from the Mid-pacific M seamount. *Journal of Rare Earths*, **27**(1), 169–176.

Dai S., Graham I. T. and Ward C. R. (2016). A review of anomalous rare earth elements and yttrium in coal. *International Journal of Coal Geology*, **159**, 82–95.

Deady E. A., Mouchos E., Goodenough K., Williamson B. J. and Wall F. (2016). A review of the potential for rare-earth element resources fromEuropean red muds: examples from Seydişehir, Turkey and Parnassus Giona, Greece. *Mineralogical Magazine*, **80**(1), 43–61.

Diatloff E., Smith F. W. and Asher C. J. (1995). Rare earth elements and plant growth. Third responses of corn and mungbean to low concentrations of cerium in dilution, continuously flowing nutrient solutions. *Journal of Plant Nutrition*, **18**, 1991–2003.

Dostal J. (2017). Rare earth element deposits of alkaline igneous rocks. *Resources*, **6**(34), 1–2.

Dushyantha N., Batapola N., Ilankoon I. M. S. K., Rohitha S., Premasiri R., Abeysinghe B. and Dissanayake K. (2020). The story of rare earth elements (REEs): occurrences, global distribution, genesis, geology, mineralogy and global production. *Ore Geology Reviews*, **122**, 103521.

Emsbo P., McLaughlin P. I., Breit G. N., du Bray E. A. and Koenig A. E. (2015). Rare earth elements in sedimentary phosphate deposits: solution to the global REE crisis? *Gondwana Research*, **27**, 776–785.

Emsley J. (2003). Nature's building blocks: an A-Z guide to the elements. Oxford University Press, New York, 241.

Ercit T. S. (2005). REE-Enriched granitic pegmatites. In Rare-Element Geochemistry and Mineral Deposits, R. L. Linnen and I. M. Samson (eds), Geological Association of Canada, GAC Short Course Notes, Vol. **17**, pp. 175–199.

Francesca P., Mevenkamp L., Pape E., Błażewicz M., Bonifácio P., Riehl T. and Vanreusel A. (2020). A local scale analysis of manganese nodules influences on the Clarion-Clipperton Fracture Zone macrobenthos. *Deep Sea Research Part I: Oceanographic Research Papers*, 103449.

Franus W., Wiatros-Motyka M. M. and Wdowin M. (2015). Coal fly ash as a resource for rare earth elements. *Environmental Science and Pollution Research*, **22**(12), 9464–9474.

Ganguli R. and Cook D. R. (2018). Rare earths: A review of the landscape. MRS energy & sustainability: A review. *Journal of Materials Research Society*, **5**, 1–16.

Ghosal S., Agrahari S., Banerjee S., Chakrabarti R. and Sengupta D. (2020). Geochemistry of the heavy mineral sands from the Garampeta to the Markandi beach, southern coast of Odisha, India: Implications of high contents of REE and Radioelements attributed to Placer Monazite. *Journal of Earth System Science*, **129**, 152.

Giable G., Gisha G., Rao P. P. and Reddy M. L. P. (2005). Synthesis and characterization of environmentally benign nontoxic pigments: RE2Mo_2O_9 (RE ¼ La or Pr). *Chemical Letters*, **34**, 1702.

Gonzalez F. J., Somoza L., Maldonado A., Lunar R., Martinez-Frias J., Martin-Rubi J. A. and Carrion M. C. (2010). High technology elements in Co-rich ferromanganese crusts from the Scotia sea. *Revista de la Sociedad Española de Mineralogía*, **13**, 113–114.

Goonan T. G. (2011). Rare Earth Elements—End Use and Recyclability, U.S. Geological Survey Scientific Investigations Report 2011–5094, p. 15.

K. Gschneidner and A. Cappellen (eds) (1987). 1787–1987 Two Hundred Years of Rare Earths. Rare Earth Information Center, IPRT, North-Holland. IS-RIC 10.

Guo B. S., Zhu W. M., Xiong P. K., Ji Y. J., Liu Z. and Wu Z. M. (1988). Rare Earths in Agriculture. Agricultural scientific technological press, Beijing, China, pp. 23–208.

Hammache Z., Bensaadi S., Berbar Y., Audebrand N., Szymczyk A. and Amara M. (2021). Recovery of rare earth elements from electronic waste by diffusion dialysis. *Separation and Purification Technology*, **254**, 117641.

Haque N., Hughes A., Lim S. and Vernon C. (2014). Rare earth elements: overview of mining, mineralogy, uses, sustainability and environmental impact. *Resources*, **3**, 614–635.

Heggy E., Palmer E. M., Thompson T. W., Thomson B. J. and Patterson G. W. (2020). Bulk composition of regolith fines on lunar crater floors: initial investigation by LRO/Mini-RF. *Earth and Planetary Science Letters*, **541**, 116274.

Hein J. R., Conrad T. A. and Staudigel H. (2010). Seamount mineral deposits, a source of rare-metals for high technology industries. *Oceanography*, **23**, 184–189.

Henderson P. (1983). Rare Earth Element Geochemistry, Volume **2**. 1st Edn. Elsevier, Amsterdam.

Henderson P., Gluyas J., Gunn G., Wall F., Wooley A. and Finlay A. (2011). Rare Earth Elements – A briefing note by the Geological Society of London, pp. 1–13.

Hong W. M., Duan X. B., Gao Z. S., Hu C. P., Zheng W. and Qu H. J. (1996). Long-Term Ocation Test of REE on Agriculture and REE Residual Analysis in Wheat Seeds. Proceeding of the First Sino-Dutch Workshop on the Environmental Behavior and Ecotoxicology of Rare Earth Elements, Beijing, pp. 83–87.

Hower J. C., Granite E. J., Mayfield D. B., Lewis A. S. and Franklin R. B. (2016). Notes on contributions to the science of rare earth element enrichment in coal and coal combustion byproducts. *Minerals*, **6**(32), 1–9.

Hu F. F., Fan H. R., Liu S., Yang K. F. and Chen F. (2009). Samarium-Neodymium and rubidium-strontium isotopic dating of veined REE mineralization for the Bayan Obo REE-Nb-Fe Deposit, Northern China. *Resource Geology*, **59**(4), 407–414.

Jaireth S., Hoatson D. M. and Miezitis Y. (2014). Geological setting and resources of the major rare-earth-element deposits in Australia. *Ore Geology Reviews*, **62**, 72–128.

Jarvis I., Burnett W. C., Nathan Y., Almbaydin F. S. M., Attia A. K. M., Castro L. N., Flicoteaux R., Hilmy M. E., Husain V., Qutawnah A. A., Serjani A. and Zanin Y. N. (1994). Phosphorite geochemistry—state-of-the-art and environmental concerns. *Journal of the Swiss Geological Society*, **87**, 643–700.

Jordens A., Cheng Y. P. and Waters K. E. (2013). A review of the beneficiation of rare earth element bearing minerals. *Minerals Engineering*, **41**, 97–114.

Jowitt S. M., Werner T. T., Weng Z. and Mudd G. M. (2018). Recycling of the rare earth elements. *Current Opinion in Green and Sustainable Chemistry*, **13**, 1–7.

Kato Y., Fujinaga K., Nakamura K., Takaya Y., Kitamura K., Ohta J., Toda R., Nakashima T. and Iwamori H. (2011). Deep-sea mud in the Pacific Ocean as a potential resource for rare-earth elements. *Nature Geoscience*, **4**, 535–539.

Ketris M. P. and Yudovich Y. E. (2009). Estimations of clarkes for carbonaceous biolithes: world averages for trace element contents in black shales and coals. *International Journal of Coal Geology*, **78**, 135–148.

Kontos S., Ibrayeva A., Leijon J., Mörée G., Frost A. E., Schönström L., Gunnarsson K., Svedlindh P., Leijon M. and Eriksson S. (2020). An overview of MnAl permanent magnets with a study on their potential in electrical machines. *Energies*, **13**, 5549.

Krishnamurthy P. (2019). Carbonatites of India. *Journal Geological Society of India*, **94**, 117–138.

Krishnamurthy P. (2020). Rare metal (RM) and rare earth element (REE) resources: world scenario with special reference to India. *Journal Geological Society of India*, **95**, 465–474.

Ku A., Setlur A. and Loudis J. (2015). Impact of light emitting diode adoption on rare earth element use in lighting implications for yttrium, europium, and terbium demand. *Electrochemistry Society Interface*, **24**(4), 45–49.

Lentz D. (1991). Radioelement distribution in U, Th, Mo, and rare-earth-element pegmatites, skarns, and veins in a portion of the Grenville Province, Ontario and Quebec. *Canadian Journal of Earth Sciences*, **28**(1), 1–12.

Liang T., Li K. and Wang L. (2014). State of rare earth elements in different environmental components in mining areas of China. *Environmental Monitoring and Assessment*, **186**(3), 1499–1513.

Lide D. R. (1997). Abundance of elements in the earth's crust and sea. In: CRC Handbook of Physics and Chemistry, 78th Edn. CRC Press, Boca Raton, p. 14.

London D. (2018). Ore-forming processes within granitic pegmatites. *Ore Geology Reviews*, **101**, 349–383.

Long K. R. (2011). The Future of Rare Earth Elements—Will these High-Tech Industry Elements Continue in Short Supply? U.S. Geological Survey Open-File Report 2011–1189, p. 41.

Long K. R., Van Gosen B. S., Foley N. K. and Cordier D. (2012). The principal rare earth elements deposits of the United States: A summary of domestic deposits and a global perspective. *Non-Renewable Resource Issues*, 131–155.

Lukowiak A., Zur L., Tomala R., LamTran T. N., Bouajaj A., Strek W., Righini G. C., Wickleder M. and Ferrari M. (2020). Rare earth elements and urban mines: critical strategies for sustainable development. *Ceramics International*, **46**(16), 26247–26250.

Machacek E. and Fold N. (2014). Alternative value chains for rare earths: The Anglodeposit developers. *Resources Policy*, **42**, 53–64.

Machida M., Ueno M., Omura T., Kurusu S., Hinokuma S., Nanba T., Shinozaki O. and Furutani H. (2017). CeO_2-grafted Mn–Fe oxide composites as alternative oxygen-storage materials for three-way catalysts: laboratory and chassis dynamometer tests. *Industrial Engineering and Chemical Research*, **56**(12), 3184–3193.

Martoyan G. A., Karamyan G. G. and Vardan G. A. (2016). New Technology of Extracting the Amount of Rare Earth Metals from the Red Mud. IOP Conference Series: Materials Science and Engineering, Altay, Russia, p. 112.

Mazumdar A., Banerjee D. M., Schidlowski M. and Balaram V. (1999). Rare-earth elements and stable isotope geochemistry of early Cambrian chert-phosphorite assemblages from the Lower Tal formation of the Krol Belt (Lesser Himalaya, India). *Chemical Geology*, **156**, 275–297.

McLeod C. L. and Krekeler M. P. S. (2017). Sources of extraterrestrial rare earth elements: to the Moon and beyond. *Resources*, **6**(48), 1–28.

Mehmood M. (2018). Rare earth elements – A review. *Journal of Ecology & Natural Resources*, **2**(2), 1–6.

Milinovic J., Rodrigues F. J. L., Barriga F. J. A. S. and Murton B. J. (2021). Ocean-floor sediments as a resource of rare earth elements: An overview of recently studied sites. *Minerals*, **11**, 142.

Morin-Crini N., Lichtfouse E., Liu G., Balaram V., Ribeiro A. R. L., Lu Z., Stock F., Carmona E., Teixeira M. R., Picos-Corrales L. A. and Moreno-Piraján J. C. (2021). Emerging contaminants: policies, analysis, water pollution, fate, toxicity and remediation. In: Emerging Contaminants Vol. 1. Environmental Chemistry for A Sustainable World, vol 65, N. Morin-Crini, E. Lichtfouse and G. Crini (eds), Springer, Cham. doi: 10.1007/978-3-030-69079-3_1

Narayanan R. P., Kazantzis N. K. and Emmert M. H. (2019). Process for scandium recovery from Jamaican bauxite residue: A probabilistic economic assessment. *Materials Today: Proceedings*, **9**, 578–586.

Nath B. N., Balaram V., Sudhakar M. and Pluger W. L. (1992). Rare earth element geochemistry of ferromanganese deposits from the Indian ocean. *Marine Chemistry*, **38**, 185–208.

Nugraheni R. D., Sunjaya D. and Burhannudinnur M. (2020). The enrichment mechanism of rare earth elements in weathered granitoids, tin placer and bauxite laterite. *International Journal of Science and Technology Research*, **9**(3), 1506–1511.

Palaparthi J., Chakrabarti R., Banerjee S., Guin R., Ghosal S., Agrahari S. and Sengupta D. (2017). Economically viable rare earth element deposits along beach placers of Andhra Pradesh, eastern coast of India. *Arabian Journal of Geosciences*, **10**(9), 201.

Pang X., Li D. and Peng A. (2002). Application of rare-earth elements in the agriculture of China and its environmental behavior in soil. *Environmental Science and Pollution Research*, **9**(2), 143–148.

Pazand K. (2015). Rare earth element geochemistry of coals from the mazino coal mine, tabas coalfield. *Iran. Arabian Journal of Geosciences*, **8**(12), 10859–10869.

Piper D. Z. (1974). Rare earth elements in ferromanganese nodules and other marine phases. *Geochimica et Cosmochimica Acta*, **38**, 1007–1022.

Pourmand A., Dauphas N. and Ireland T. J. (2012). A novel extraction chromatography and MC-ICP-MS technique for rapid analysis of REE, Sc and Y: revising CI-chondrite and Post-Archean Australian Shale (PAAS) abundances. *Chemical Geology*, **291**, 38–54.

Raju C. S. K., Cossmer A., Scharf H., Panne U. and Lück D. (2010). Speciation of gadolinium-based MRI contrast agents in environmental water samples using hydrophilic interaction chromatography hyphenated with inductively coupled plasma mass spectrometry. *Journal of Analytical Atomic Spectrometry*, **25**, 55–61.

Ramesh R., Ramanathan R., Ramesh S., Purvaja R. and Subramanyam V. (2000). Distribution of rare earth elements and heavy metals in the surficial sediments of the Himalayan river system. *Geochemical Journal*, **34**, 295–319.

Richardson D. G. and Birkett T. C. (1996). Carbonatite associated deposits. In: Geology of Canadian Mineral Deposit Types: Geological Survey of Canada, O. R. Eckstrand, W. D. Sinclair and R. I. Thorpe (eds), Geology of Canada, Vol. **8**, pp. 541–558.

Sadeghbeigi R. (2012). Fluid Catalytic Cracking Handbook. Elsevier, Waltham, MA.

Sander S. G. and Koschinsky A. (2011). Metal flux from hydrothermal vents increased by organic complexation. *Nature Geoscience*, **4**, 145–150.

Schnabel C., Münker C. and Strub E. (2017). La–Ce isotope measurements by multi-collector-ICP-MS. *Journal of Analytical Atomic Spectrometry*, **32**(12), 2360–2370.

Sengupta D. and Gosen B. S. V. (2016). Placer-type rare earth element deposits. *Reviews in Economic Geology*, **18**, 81–100.

Singh Y. (2020). Chapter 8. Potential Natural Resources Rare Earth Element Resources: Indian Context, Society of Earth Scientists Series. Springer Nature, Switzerland AG, pp. 311–347.

Singh H., Sadiq M. and Sharma B. B. (2014). Exploration for rare earth elements in North East India. *Current Science*, **107**(2), 178–180.

Singh U., Thawrani S. A., Ansari M. S., Puttewar S. P. and Agnihotri A. (2019). Studies on beneficiation and leaching characteristics of rare earth elements in Indian red mud. *Russian Journal of Non-Ferrous Metals*, **60**(4), 335–340.

Singh R., Venkatesh A. S., Sudhakar C., Sethy S. N. and Babu K. P. (2020). Exploration for strategic placer mineral deposits in a fluctuating shoreline: depositional environment and mineralogical characterization of the NE Odisha coast placers, India. *Ore Geology Reviews*, **127**, 103850.

Sinha D. K. (2020). Atomic minerals directorate for exploration and research. *Proc. Indian National Science Academy*, **86**(1), 755–758.

Smith A. E., Mizoguchi H., Delaney K., Spaldin N. A., Sleight A. W. and Subramanian M. A. (2009). Mn^{3+} in trigonal bipyramidal coordination: A new blue chromophore. *Journal of the American Chemical Society*, **131**(47), 17084–17086.

Song X., Chang M. H. and Pecht M. (2013). Rare-earth elements in lighting and optical applications and their recycling. *The Journal of the Minerals, Metals & Materials Society*, **65**(10), 1276–1282.

Sowerbutts L. (2017). https://www.geologyforinvestors.com/rare-earth-element-deposits/ (last accessed October 27, 2021).

Sreeram K. J., Aby C. P., Nair B. U. and Ramasami T. (2008). Colored cool colorants based on rare earth metal ions. *Solar Energy Materials & Solar Cells*, **92**, 1462–1467.

Szucs A. M., Stavropoulou A., O'Donnell C., Davis S. and Rodriguez-Blanco J. D. (2021). Reaction pathways toward the formation of bastnäsite: replacement of calcite by rare earth carbonate. *Crystal Growth & Design*, **21**(1), 512–527.

Tanabe S. (2015). Glass and rare-earth elements: A personal perspective. *International Journal of Applied Glass Science*, **6**(4), 305–328.

Taylor S. R. and McLennan S. M. (1985). The Continental Crust: Its Composition and Evolution. Blackwell Scientific, Boston, US, p. 312.

Temga J. P., Sababa E., Mamdem L. E., Bijeck M. L. N., Azinwi P. T., Tehna N., Zame P. Z. O., Onana V. L., Nguetnkam J. P., Bitom L. D. and Ndjigui P. D. (2021). Rare earth elements in tropical soils, Cameroon soils (Central Africa). *Geoderma Regional*, **25**(1–2), e00369.

Ujaczki E., Zimmerman Y. S., Feigl V. and Lenz M. (2015). Recovery of rare Earth elements from Hungarian red mud with combined acid leaching and liqud-liquid extraction. Proceedings of the Bauxite Residue Valorization and Best Practices Conference, Leuven, Belgium, pp. 1–7.

Ujaczki É, Feigl V., Molnár M., Cusack P., Curtin T., Courtney R., O'Donoghue L., Davris P., Hugi C., Evangelou M. W. and Balomenos E. (2018). Reusing bauxite residues: benefits beyond (critical raw) material recovery. *Journal of Chemical Technology and Biotechnology*, **93**(9), 2498–2510.

Umicore Group. (2011). Umicore and Rhodia develop unique rare earth recycling process for rechargeable batteries: Umicore Group press release CP–2–11–18–R, June 16, accessed October 19, 2015, *at* http://www.umicore.com/en/media/press/20110616REE recyclingEN/

US Geological Survey (2021). Mineral Commodity Summaries. Rare Earths. 2020. Available online: https://pubs.usgs.gov/periodicals/mcs2020/mcs2020-rare-earths.pdf (last accessed October 27, 2021).

Van Gosen B. S., Verplanck P. L., Seal R. R., II, Long K. R. and Gambogi J. (2017). Rare-earth elements, chap. O of In: Critical Mineral Resources of the United States—Economic and Environmental Geology and Prospects for Future Supply, K. J. Schulz, J. H. DeYoung, Jr., R. R. Seal, II and D. C. Bradley (eds), U.S. Geological Survey Professional Paper 1802, pp. O1–O31.

Verplanck P. L., Van Gosen B. S., Seal R. R. and McCafferty A. E. (2014). A Deposit Model for Carbonatite and Peralkaline Intrusion-Related Rare Earth Element Deposits. U.S. Geological Survey Scientific Investigations Report 2010–5070-J, p. 58.

Vind J., Malfliet A., Blanpain B., Tsakiridis P. E., Tkaczyk A. H., Vassiliadou V. and Panias D. (2018). Rare earth element phases in bauxite residue. *Minerals*, **8**, 77.

Voncken J. H. L. (2016). The Rare Earth Elements. Springer Briefs in Earth Sciences. Springer Nature, Switzerland AG.

Wakabayashi T., Ymamoto A., Kazaana A., Nakano Y., Nojiri Y. and Kashiwazaki M. (2016). Antibacterial, antifungal and nematocidal activities of rare earth ions. *Biological Trace Element Research*, **174**(2), 464–470.

Walters A., Lusty P., Chetwyn C. and Hill A. (2010). Rare Earth Elements. Mineral Profile Series. British Geological Survey, United Kingdom, p. 45.

Wang W., Pranolo Y. and Cheng C. Y. (2013). Recovery of scandium from synthetic red mud leach solutions by solvent extraction with D_2EHPA. *Separation and Purification Technology*, **108**, 96–102.

Welsbach V. and Auer C. (1885). Die Zerlegung des Didyms in seine Elemente. *Monatshefte für Chemie und verwandte Teile anderer Wissenschaften*, **6**(1), 477–491.

Wu C., Yuan Z. and Bai G. (1996). Rare element deposits in China. In: Rare Earth Minerals—Chemistry, Origin and ore Deposits, A. P. Jones, F. Wall and C. T. Williams (eds), Chapman and Hall, London, UK, *The Mineralogical Society Series no.* 7, pp. 281–310.

Xu Z., Lim D., Choi J., Yang S. and Jung H. (2009). Rare earth elements in bottom sediments of major rivers around the Yellow Sea: implications for sediment provenance. *Geo-Marine Letters*, **29**(5), 291–300.

Zhang J., Cheng H., Gao Q., Zhang Z. and Liu Q. (2000a). Effect of lanthanum on growth and biochemical property of *Sclerotinia sclerotiorum*. *Chinese Journal of Applied Ecology*, **11**(6), 382–384.

Zhang H., Feng J., Zhu W. F., Liu C. and Gu J. (2000b). Bacteriostatic effect of cerium-humic acid complex: An experimental study. *Biological Trace Element Research*, **73**(1), 29–36.

Zhang W., Noble A., Yang X. and Honaker R. (2020). A comprehensive review of rare earth elements recovery from coal-related materials. *Minerals*, **10**, 451.

Zhou B., Li Z. and Chen C. (2017). Global potential of rare earth resources and rare earth demand from clean technologies. *Minerals*, **7**, 203.

Zhu W., Kennedy M., de Leer E. W. B., Zhou H. and Alaerts G. J. F. R. (1997). Distribution and modelling of rare earth elements in Chinese river sediments. *Science of The Total Environment*, **204**(3), 233–243.

Zou J., Cheng L., Guo Y., Wang W., Tian H. and Li T. (2020). Mineralogical and geochemical characteristics of lithium and rare earth elements in high-Sulfur coal from the donggou mine, chongqing, southwestern China. *Minerals*, **10**, 627.

Part III

Recovery of Rare Earth Elements from Waste Resources

Chapter 5

Rare earth elements recovery from secondary sources

Shuronjit K. Sarker, Shanjida Sultana, Nawshad Haque, Anthony E. Hughes, Warren Bruckard and Biplob K. Pramanik

5.1 INTRODUCTION

Rare earth elements (REE) are a group of chemical elements with unique electrical, optical and magnetic properties. The group contains 17 elements that include 15 lanthanides (Ln), scandium (Sc) and yttrium (Y), as classified by the International Union of Pure and Applied Chemistry (IUPAC). The lanthanides are (in alphabetic order) cerium (Ce), dysprosium (Dy), erbium (Er), europium (Eu), gadolinium (Gd), holmium (Ho), lanthanum (La), lutetium (Lu), neodymium (Nd), praseodymium (Pr), promethium (Pm), samarium (Sm), terbium (Tb), thulium (Tm), and ytterbium (Yb). Commonly, they are classified based on their atomic masses into two main groups (Figure 5.1): heavy rare earth elements (HREE) and light rare earth elements (LREE). The HREE include Tb through Lu with atomic numbers from 65 to 71, and the LREE include La through Gd with atomic numbers from 57 to 64 (Van Gosen *et al.*, 2017). Although, Y, atomic number 39, is light it belongs to the HREE group due to the resemblance of its chemical and physical properties to those of HREE. Sc, atomic number 21, also light but its chemical and physical properties do not call for grouping with either of LREE or HREE. Thus, it makes up a class by itself (Castor & Hedrick, 2006;

doi: 10.2166/9781789062236_0117

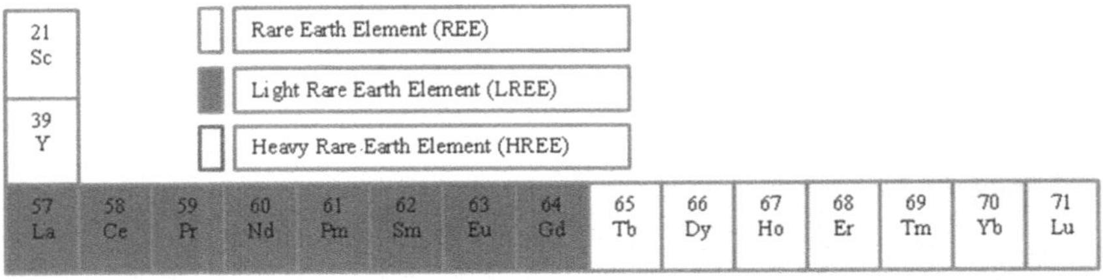

Figure 5.1 Rare earth elements in the periodic table (Öhrlund, 2011).

Dutta *et al.*, 2016; Humphries, 2013; Rydel & Nowak, 2015). Blissett *et al.* (2014) classified the REE into three categories: light rare earth elements (LREE — La, Ce, Pr, Nd and Sm), medium rare earth elements (MREE — Eu, Gd, Tb, Dy and Y) and heavy rare earth elements (HREE — Ho, Er, Tm, Yb and Lu). Seredin and Dai (2012) have classified REE considering their supply and demand relationship. They classified according to recent trends related to future industrial supply and demand. The classification includes **critical** (Dy, Er, Eu, Nd, Tb, and Y), **uncritical** (Gd, La, Pr, and Sm), and **excessive** (Ce, Ho, Lu, Tm, and Yb) (Blissett *et al.*, 2014).

The applications of REE have increased significantly in high technology devices including daily items such as flat-screen televisions, fluorescent and light-emitting-diode lights, smartphones, digital cameras, computer hard disks and computer monitors (see also Chapter 4). In addition, REE are also used in large quantities in defense technologies and renewable energy technologies such as wind turbines. As the demand for wind energy is on the rise, the use of permanent magnets made of these REE is predicted to rise as well. Another emerging application of REE is in electric vehicles, where they are required in large quantities. With 7.3 million electric vehicle (cars and vans) sales reported by the International Energy Agency (IEA) in 2020, requiring up to 14 times more REE production by the industry in 2020 than in 2015 (IEA, 2020). According to a study by Alonso *et al.* (2012), the demand for REE was nearly 100,000 t in 2010. The demand increased steadily throughout the decade and reached roughly 150,000 t at the end of 2020. However, the COVID-19 pandemic has negatively impacted the demand for REE and the demand fell back in 2020 due to disruptions to industrial production. However, a strong recovery in demand is forecasted in 2021, increasing by 10% by the end of the year (Roskill, 2021). The demand for REE will continue to grow throughout this decade and the next. It is estimated that the demand will reach about 400,000 t by the end of 2035 (Alonso *et al.*, 2012).

The REE are one of the most abundant elements in the earth's crust and their deposits have been found in 34 nations (Chen, 2011). However, their extraction from all deposits is not economical and can often cause environmental issues. Therefore, the largest REE producing and exporting nations have decreased their export to preserve their national reserves. For example, China, which once produced and exported approximately 97–98% of the world's REE, has decreased their export to about 40% between 2009 and 2010 as a result of increased taxes on export and regulated export rates (Alonso *et al.*, 2012). Generally, REE is extracted from ores collected from hard rock mines. Recent advancement in mineral extraction technologies and difficulty in finding primary deposits with economically recoverable concentrations has resulted in coal fly ash and brine being considered as potential secondary sources of REE.

Australia has a wealth of identified rare earth deposits and the majority are LREE, although some HREE deposits have also been identified. Australia has developed technologies and a research capability for REE processing. The technology is competent up to the point of producing mixed oxides. Producing single rare earth oxides is limited at the current capability. One of the practical issues is how far down the processing chain REE companies should process. This creates significant opportunities for research and innovation in this field. One of the main requirements is making the REE leaching and extraction process simpler and more environmentally friendly. The separation technologies should have less specific capital and energy intensity. If these constraints are met, it should be possible to develop and establish a niche and a high value rare earth elements production technology in Australia (Haque *et al.*, 2014).

In relation to environmental concerns, as stated by the Chinese Society of Rare Earths, 8.5 kilograms of fluorine and 13 kilograms of dust are created for every ton of rare earth produced. The production of about one ton of REE by high temperature calcination techniques using concentrated sulfuric acid can produce 9600 to 12,000 m^3 of waste gas consisting of hydrofluoric acid, sulfur dioxide and sulfuric acid, dust concentrate, about 75 m^3 of acidic wastewater, and nearly one ton of radioactive aqueous waste residue.

5.2 SOURCES OF REE

The schematic of REE life cycle is shown in Figure 5.2. This chapter aims to discuss the potential for REE recovery from brine and fly coal ash as sources. The discussion will include the REE recovery processes commonly used in commercial practice and their associated challenges due to low REE concentration, REE having similar properties to each other, and environmental issues in the recovery processes.

5.2.1 Brine

Brine is a solution containing a high concentration of salt in water. The salt concentration in brine varies from 3.5 to 26%. Brine with the lowest salt concentration typically represents seawater and brine with the highest salt concentration is a saturated brine solution. Brine occurs in different environments in nature. Brine occurring on the earth's surface is known as a salt lake and when it is found in the earth's crust it is called geothermal brine. It can also be found

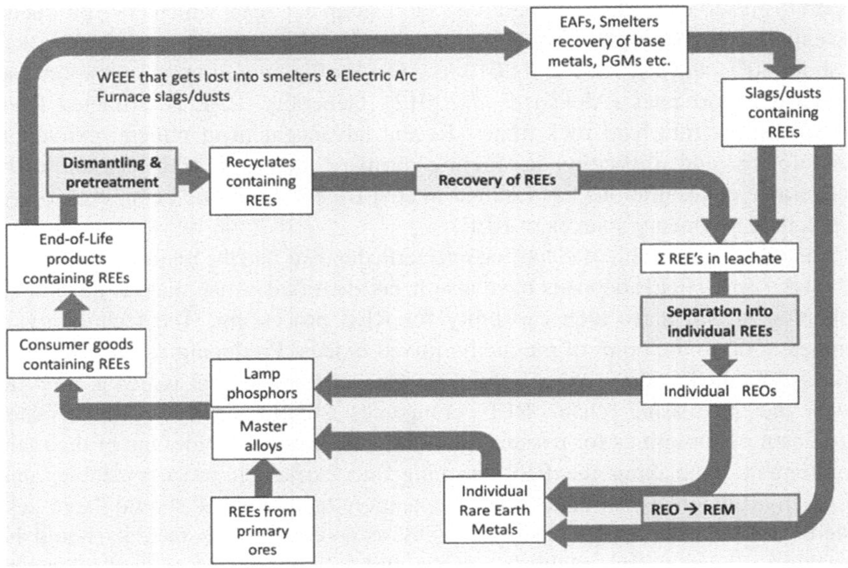

Figure 5.2 Life cycle of REE in major technological applications (Binnemans *et al.*, 2013). WEEE: waste electrical and electronic equipment; REO: rare earth oxide; REM: rare earth metal; PGM: platinum group metals; EAF: electric arc furnace.

on the ocean bottom as a brine pool or brine lake. It can also be a by-product of industrial processes such as desalination. Brine is a known source of common salt and components such as calcium, magnesium, potassium, lithium, gypsum, bromine and iodine. It can also be a source of REE found in dissolved form in brine solutions contained within the hot rocks below the earth's surface. The earth's crust contains relatively more REE than other elements that are usually extracted. However, finding the REE in economic concentrations is rare (Gupta & Krishnamurthy, 1992). Therefore, increased research is being conducted to recover REE from low-grade sources such as brine. Due to improvements in resource recovery technologies and desalination processes in recent decades, brine and seawater are now being considered as important untapped sources of REE (Bardi, 2010).

5.2.2 Coal fly ash

Coal fly ash (CFA) is the coal combustion product generated during production of electricity in coal-fired power plants. The power plants that use coal as the primary biomass for power generation produce over 70% coal ash, known as CFA, with particle sizes of 0.5 to 300 μm (Koukouzas *et al.*, 2006; Yao *et al.*, 2015). With the increased dependency on coal-based power generation worldwide, CFA generation has grown significantly in recent decades. The amount of CFA generated worldwide was 500 million tons in 2005. The amount of CFA increased by 50% within just 10 years to reach 750 million tons in 2015 (Gollakota *et al.*, 2019). CFA contains aluminous and siliceous materials known as a pozzolan. Because of its chemical composition resembling cementing properties, it is widely used as the base material for construction including concrete products, cement, and structural fills. Despite its widespread applications, only a fraction, estimated around 25% of the total CFA produced worldwide, is used in these applications and the rest is considered waste. A large volume of the waste CFA is disposed of into the environment, which causes soil, water and air pollution to the adjacent areas. Due to the growing concerns regarding the environmental impacts from CFA disposal, recent research has been directed towards alternative uses of CFA waste.

The CFA generated at coal fired power plants can contain a significant amount of REE. The concentration of REE in CFA from different power plants is shown in Table 5.1. Notably, in some cases, REE concentration in CFA can be higher than that of the earth's crust, which has an average REE concentration of around

Table 5.1 Concentration of REE in CFA from different coal fired power plants (adapted from Blissett *et al.*, 2014; Franus *et al.*, 2015; Mayfield & Lewis, 2013).

REE	CFA From Polish and UK Power Plants (ppm)	CFA From Power Plants in Europe, Mexico, and USA (ppm)
Cerium (Ce)	30.7–176.5	405–565
Lanthanum (La)	15.5–85.7	206–286
Yttrium (Y)	17.9–73.2	191–259
Neodymium (Nd)	12.7–81.3	183–256
Praseodymium (Pr)	3.3–20.51	49–68.4
Samarium (Sm)	2.8–17	—
Dysprosium (Dy)	2.61–12.18	32.1–50.3
Gadolinium (Gd)	2.85–14.65	—
Erbium (Er)	1.79–7.41	—
Terbium (Tb)	0.45–2.4	4.9–7.3
Europium (Eu)	0.56–3.81	3.9–5.9
Holmium (Ho)	0.59–2.58	—
Ytterbium (Yb)	1.8–6.74	—
Thulium (Tm)	0.27–1.07	—
Scandium (Sc)	7–45	—
Lutetium (Lu)	0.3–1.03	—

130–240 ppm (Zepf, 2013). But the characteristics and nature of CFA have changed due to the recent trend of co-combustion of biomass with coal in coal fired power plants, which often makes it difficult to use in many common applications, such as production of cement and concrete. Hence, there is a need for investigation of alternative uses of CFA. In response to the need, a number of methods for recovering REE from CFA are now being investigated to determine their economic and environmental implications compared to traditional mining and processing of REE (Mayfield & Lewis, 2013). For example, the chemical properties of CFA produced and supplied by a Colorado Springs Utilities power plant were tested. The results showed that the CFA samples contained a significant amount of REE and other metals including 14 rare earth and strategic metals, of which more than 60% were successfully extracted from the CFA. The company calculated that every ton of CFA produced at the Martin Drake plant contained about $600 worth of REE. It was estimated that a single plant, such as Martin Drake in Colorado Springs, United States of America (USA), could produce minerals worth nearly $49 million in a year. Furthermore, the extraction of REE processes use by-products of the off gas scrubbing (sulfuric and nitric acids) to dissolve REE out of CFA, which reduces emissions and provides an additional benefit to the REE recovery from CFA.

5.3 REE RECOVERY FROM BRINE SOLUTIONS

Traditionally, pyrometallurgical or hydrometallurgical methods are used in REE recovery from brine. These methods include solvent extraction, adsorption, precipitation and co-precipitation, ion exchange, electrochemical and membrane processes as well as oxidation and reduction methods. The processes are essential for extracting pure REE from the brine solution obtained after the leaching of the raw materials containing these elements (Kołodyńska *et al.*, 2019). Among them, solvent extraction and ion exchange have been used in practice. In these processes, the ore concentrate containing REE is acid-leached, and then undergoes an ion-exchange or solvent extraction process to separate the REE. For example, the solvent extraction method has been used to separate individual REE from rare earth nitrate solutions using tri-n-butyl phosphate (Figure 5.3). In this process, a feed solution of rare earth nitrate is produced by chemical treatment of monazite ore. This method is proven to be efficient in recovering metals and minerals from low-grade sources.

Solvent extraction is one of the most widely used metal and mineral extraction methods for both primary ores and secondary sources such as brine. In this method an organic solution consisting of an extractant (special reagent) is used to transport the target elements from one aqueous solution to another in order for them to be separated and recovered in purified form. Commercially, a number of solvent extractants are widely used, notably, di-(2-ethylhexyl) phosphoric acid (D_2EHPA) and 2-ethylhexyl phosphoric acid-2-ethylhexyl ester (EHEHPA).

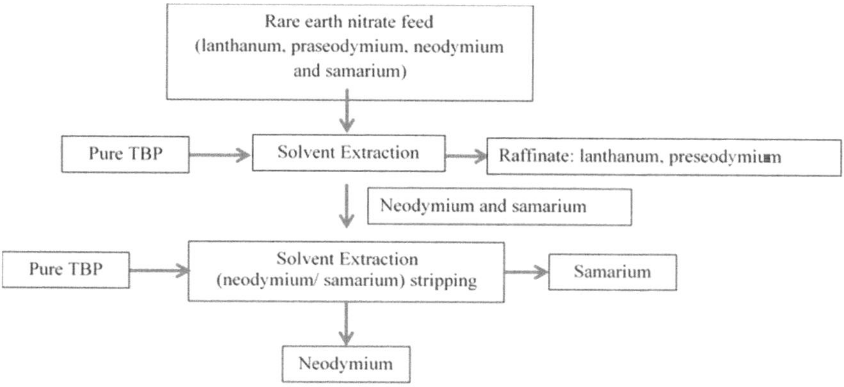

Figure 5.3 Flowsheet of solvent extraction method to separate REE at the Alwaye plant of Indian Rare Earths Ltd, India Pramanik *et al.*, 2020; Gupta and Krishnamurthy, 1992). TBP: tributyl phosphate.

Studies by Jha *et al.* (2016) to recover REE using D_2EHPA have shown that both LREE and HREE can be extracted effectively using this solvent extractant with an efficiency ranging from 77 to 100% (Perez *et al.*, 2019). Other studies have used EHEHPA and found that it can also extract some HREE, such as dysprosium, with 98% purity from concentrated HCl leaching solutions (Larsson *et al.*, 2012; Shimojo *et al.*, 2008; Singh *et al.*, 2008).

The ion-exchange method has been found to work very well for metal and mineral recovery from dilute solutions such as brine. The ion-exchange method has advanced significantly since the discovery of synthetic resins which led to the commercial application of the process. Soon after the development of synthetic resins, the chelating resin was developed. Combined use of these two resins in commercial applications is now common to selectively separate metal ions. A large-scale application of the method was in the Manhattan Project during World War II, in which the ion-exchange method was used to separate individual REE (Rao, 2011). The separation of individual elements of REE in the project was a major success. Ion-exchange with chelating resin is a common practice for selective separation of metal ions. For example, Xiong *et al.* (2011) used the D113-III chelating resin to extract neodymium from brine. However, the high cost in the metal recovery process using the ion-exchange method, combined with the low concentrations of the target element, has remained a major challenge.

5.4 REE RECOVERY FROM COAL FLY ASH

Several processes have been used to recover REE from CFA. The recovery of REE is achieved through the beneficiation of CFA by using physical separation methods.

The most widely used physical separation methods include magnetic, size and density separation, electrostatic separation and froth flotation. However, limited studies have applied these processes for REE recovery from CFA. Studies by Blissett *et al.* (2014), Hower *et al.* (2013) and Lin *et al.* (2017) have found that the conventional flotation method for REE enrichment using fatty acids and octanohydroxamic acid collectors was inefficient. The use of multistage physical separation methods consisting of magnetic separation, froth flotation, and hydrocyclone separation was studied by Blissett *et al.* (2014) for beneficiation of CFA for REE enrichment. They achieved enrichment from 429 to 529 ppm, which was poor.

The resulting upgraded product is then treated for REE recovery after beneficiation of CFA. The recovery of REE process is completed in a few steps consisting of acid leaching and extraction of REE from the leached solution with appropriate reagents. In the acid leaching process, CFA is brought into direct contact with acids or lixiviants to allow the soluble components of the solid phase to dissolve leading to separation of the dissolved metals from the mixture. However, leaching with acid often results in low REE recovery due to a low leaching efficiency. Therefore, alkaline leaching and the combination of acid leaching with alkaline leaching were introduced to maximize the REE recovery. Figure 5.4 shows a flow sheet of the alkaline-acid leaching process for separation of REE from coal and coal refuse.

Acid leaching has been used in the REE recovery processes. A range of acids such as HCl, H_2SO_4 and HNO_3 with varied concentrations is generally used for such applications. A study by Cao *et al.* (2018) showed that the acid leaching

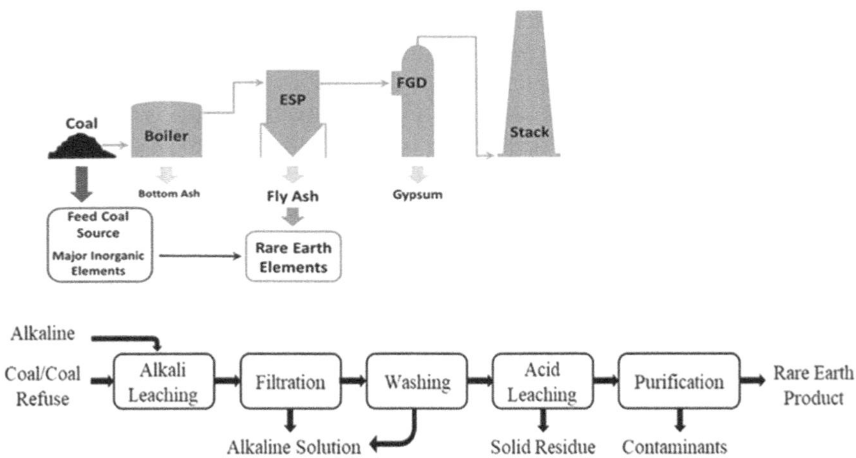

Figure 5.4 Graphical representation of the alkaline-acid leaching process to extract REE from coal and coal refuse (modified from Kuppusamy *et al.*, 2019; Taggart *et al.*, 2016). ESP: electrostatic precipitators; FGD: flue gas desulfurization.

processes could recover 98% of Ce, 75% of Eu and 83% of La from CFA. They noticed in their experiments that the REE recovery efficiency was influenced by the processing temperature (optimum at approximately 80°C), characteristics of the CFA, and type of acid used. Recently, a combination of acid leaching and alkaline leaching methods were used for effective recovery of REE from secondary waste sources, such as REE-containing waste magnets, and CFA. For example, Mutlu *et al.* (2018) used the combined leaching processes and found effective recovery of REE from CFA. However, they noted that the high pH and presence of chemical elements, such as Al, Ca, Fe, Na and Si, at much higher concentration than the REE can make the recovery processes ineffective.

5.5 OTHER WASTE SOURCES

A large quantity of REE is used in fluorescent materials which can be a potential source of REE after end-use. The fluorescent materials include fluorescent lamps, cathode ray tube displays and other devices/applications that require lighting phosphors. More than half of some of the produced REE, such as, Eu, Tb and Y, are used for lighting phosphors to manufacture green light, particularly for fluorescent lamps. It is reported that the waste fluorescent lamps can contain more REE than natural minerals. Since phosphor-containing fluorescent lamps are gradually replacing traditional incandescent lighting equipment due to the latter being energy inefficient, the use of these energy saving fluorescent lamps is increasing rapidly. As a result, recovery potential of REE from these waste sources will increase significantly in future. A flow sheet of recovering REE from the lamp phosphors is shown in Figure 5.5. The efficiency of the processes for REE recovery from lamp phosphors can be influenced by the factors of processing cost, grade of resultant products, and recovery rate. The recovery processes produce two key products: trichromatic phosphor and monochrome phosphor. Phosphors are used in cathode ray tube displays, fluorescent lamps and other applications that require color in the light exhibited. Not all phosphors contain REE, but many do. More than 70% of the production of Eu, Tb and Y is used for phosphors. The number of fluorescent lamps manufactured will increase rapidly with the gradual phasing out of incandescent lighting equipment. Consequently, there is a large potential for the recovery of REE from waste fluorescent lamps as shown in Figure 5.5. The recovery of Y and Eu from red phosphor using hydrometallurgy methods has been a focus recently. There are difficulties in leaching Ce, La, Tb and Eu in green and blue phosphors. The final recovery rate of Y and Eu can reach more than 80%. However, a higher recovery is desirable considering the total value of the REE in fluorescent lamps (Tan *et al.*, 2014).

End-of-life neodymium-iron-boron magnets are also potential sources of REE (Figure 5.6). The magnets containing these elements are found in bigger motors, usually in electric cars, and can contain up to 30 g of dysprosium and 200 g of

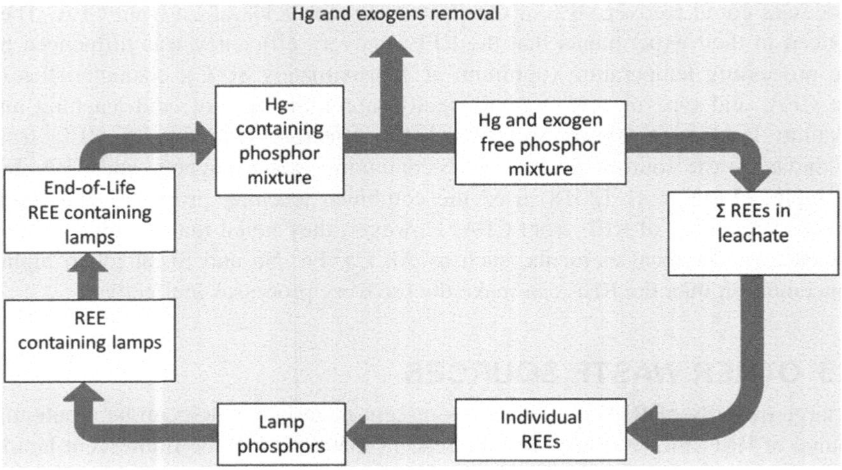

Figure 5.5 Flow sheet (simplified) of REE recycling from lamp phosphors (Binnemans *et al.*, 2013).

neodymium per motor (Keane, 2009). Another good source of REE is spent wind turbine generators which can contain 1140 kg of neodymium and 176 kg of dysprosium in a single motor weighing 4000 kg (Rabe *et al.*, 2017). Although it is likely to be in small amounts, some battery alloys also contain REE that can potentially be recovered by recycling the end-of-life battery waste (see also Chapter 4).

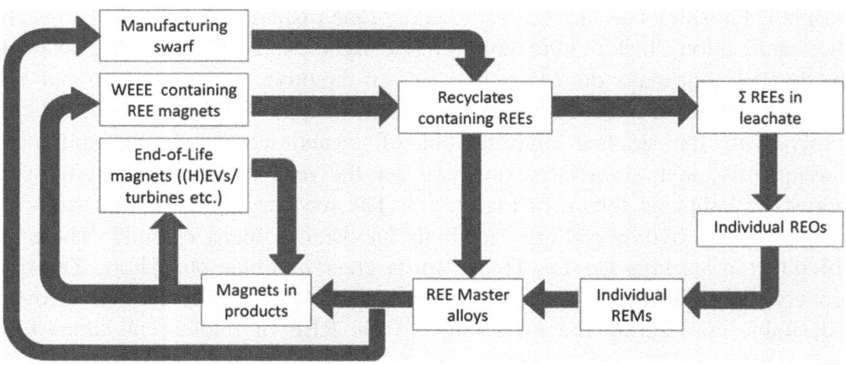

Figure 5.6 Flow sheet for recycling of REE magnets (simplified) (Binnemans *et al.*, 2013). HEV: hybrid and electric vehicles; REM: rare earth metal; REO: rare earth oxide; WEEE: waste electrical and electronic equipment.

5.6 CONCLUSIONS

Recovery of REE from low-grade sources, such as brine and coal fly ash, specifically thermal brine, is technically challenging due to the low concentrations of the REE present. However, methods, notably ion exchange and solvent extraction, have been developed to recover REE from these sources. These methods have only been tested at lab-scale, and therefore, further detailed and larger-scale studies are required to firm up commercial applications. The methods are primarily used for the separation of REE. In order for a separation process to be viable for REE recovery, the value of the total recovered REE must be greater than the total operating cost. The environmental impact and cost must also be taken into consideration.

ACKNOWLEDGEMENT

We would like to acknowledge The Commonwealth Scientific and Industrial Research Organisation and RMIT University for funding this research.

REFERENCES

Alonso E., Sherman A. M., Timothy J. W., Mark P., Everson F. R., Field R. R. and Randolph E. K. (2012). Evaluating rare earth element availability: a case with revolutionary demand from clean technologies. *Environmental Science & Technology*, **46**(6), 3406–3414.

Bardi U. (2010). Extracting minerals from seawater: an energy analysis. *Sustainability*, **2**(4), 980–992.

Binnemans K., Jones P. T., Blanpain B., Gerven T. V., Yang Y., Walton A. and Buchert M. (2013). Recycling of rare earths: a critical review. *Journal of Cleaner Production*, **51**, 1–22.

Blissett R. S., Smalley N. and Rowson N. A. (2014). An investigation into six coal fly ashes from the United Kingdom and Poland to evaluate rare earth element content. *Fuel*, **119**, 236–239.

Cao S., Zhou C., Pan J., Liu C., Tang M., Ji W., Hu T. and Zhang N. (2018). Study on influence factors of leaching of rare earth elements from coal fly ash. *Energy Fuels*, **327**, 8000–8005.

Castor S. B. and Hedrick J. B. (2006). Rare Earth Elements, 7th edn. Society for Mining, Metallurgy, and Exploration, Littleton, Colorado, pp. 769–792.

Chen Z. (2011). Global rare earth resources and scenarios of future rare earth industry. *Journal of Rare Earths*, **29**, 1–6.

Currie A. (2012). Rare earth from fly ash: a method explored. Available at: http://rareearthinvestingnews.com/7284-rare-earth-from-fly-ash-a-method explored.html

Dutta T., Kim K. H., Uchimiya M., Eilhann E. K., Jeon B. H., Deep A. and Yun S. T. (2016). Global demand for rare earth resources and strategies for green mining. *Environmental Research*, **150**, 182–190.

Franus W., Wiatros-Motyka M. M. and Wdowin M. (2015). Coal fly ash as a resource for rare earth elements. *Environmental Science Pollution Research*, **22**, 9464–9474.

Gollakota A. R. K., Vikranth V. and Chi-Min S. (2019). Progressive utilisation prospects of coal fly ash: a review. *Science of The Total Environment*, **672**, 951–989.

Gupta C. K. and Krishnamurthy N. (1992). Extractive metallurgy of rare earths. *International Materials Reviews*, **37**(1), 197–248.

Haque N., Hughes A., Lim S. and Vernon C. (2014). Rare earth elements: overview of mining, mineralogy, uses, sustainability and environmental impact. *Resources*, **3**, 614–635.

Hower J. C., Dai S., Seredin V. V., Zhao L., Kostova I. J. and Silva L. F. O. (2013). A note on the occurrence of yttrium and rare earth elements in coal combustion products. *Coal Combustion Gasification Products*, **5**, 39–47.

Humphries M. (2013) Rare earth elements: the global supply chain. *Congressional Research Service*, **R41347**, 1–31.

IEA (2020). World Energy Outlook 2020. IEA, Paris, 1–464. https://www.iea.org/reports/world-energy-outlook-2020

Jha M. K., Kumari A., Panda R., Kumar J. R., Yoo K. and Lee J. Y. (2016). Review on hydrometallurgical recovery of rare earth metals. *Hydrometallurgy*, **165**, 2–26.

Keane E. (2009). Neodymium magnets provide key to understanding rare earth trends: seeking Alpha, June 23. http://seekingalpha.com/instablog/345817-eamon-keane/9675-neodymium-magnets-provide-key-to-understanding-rare-earth-trends (accessed 10 May 2021).

Kołodyńska D., Fila D., Gajda B., Gęga J. and Hubicki Z. (2019). Rare earth elements-separation methods yesterday and today. In: Applications of Ion Exchange Materials in the Environment, M. Inamuddin Ahamed and A. Asiri (eds), Springer, Cham, 161–185.

Koukouzas N. K., Zeng R., Perdikatsis V., Xu W. and Kakaras E. K. (2006). Mineralogy and geochemistry of Greek and Chinese coal fly ash. *Fuel*, **85**(16), 2301–2309.

Kuppusamy V. K., Kumar A. and Holuszko M. (2019). Simultaneous extraction of clean coal and rare earth elements from coal tailings using alkali-acid leaching process. *Journal Energy Resources Technology*. ASME, **141**, 1–7.

Larsson K., Ekberg C. and Ødegaard-Jensen A. (2012). Using Cyanex 923 for selective extraction in a high concentration chloride medium on nickel metal hydride battery waste. *Hydrometallurgy*, **129**, 35–42.

Lin R., Howard B. H., Roth E. A., Bank T. L., Granite E. J. and Soong Y. (2017). Enrichment of rare earth elements from coal and coal by-products by physical separations. *Fuel*, **200**, 506–520.

Mayfield D. B. and Lewis A. S. (2013). Environmental review of coal ash as a resource for rare earth and strategic elements. World of Coal Ash (WOCA) Conference. Lexington, Kentucky, USA. Available at: http://www.flyash.info/

Mutlu B. K., Cantoni B., Turolla A., Antonelli M., Hsu-Kim H. and Wiesner M. R. (2018). Application of nanofiltration for rare earth elements recovery from coal fly ash leachate: performance and cost evaluation. *Chemical Engineering Journal*, **349**, 309–317.

Öhrlund I. (2011). Future metal demand from photovoltaic cells and wind turbines – investigating the potential risk of disabling a shift to renewable energy systems. doi: 10.13140/RG.2.1.4524.9049

Perez J. P. H., Folens K., Leus K., Vanhaecke F., Van Der Voort P. and Du Laing G. (2019). Progress in hydrometallurgical technologies to recover critical raw materials and precious metals from low-concentrated streams. *Resources Conservation & Recycling*, **142**, 177–188.

Pramanik B. K., Nghiem L. D. and Hai F. I. (2020). Extraction of strategically important elements from brines: constraints and opportunities. *Water Research*, **168**, 115149.

Rabe W., Kostka G. and Smith S. K. (2017). China's supply of critical raw materials: risks for Europe's solar and wind industries? *Energy Policy*, **101**, 692–699.

Rao S. R. (2011). Resource Recovery and Recycling from Metallurgical Wastes. Elsevier, Amsterdam, The Netherlands, Vol. **7**.

Roskill (2021). Rare Earths. Outlook to 2030, 20th edn. Geopolitical Risk and Security-of-Supply. London, United Kingdom. https://roskill.com/market-report/rare-earths/

Rydel P. and Nowak M. (2015). Review of the major minerals of rare earth elements-gold of the 21st century. *Przegląd Geology*, **63**, 348–362.

Seredin V. and Dai S. (2012). Coal deposits as potential alternative sources for lanthanides and yttrium. *International Journal of Coal Geology*, **94**, 67–93.

Shimojo K., Kurahashi K. and Naganawa H. (2008). Extraction behavior of lanthanides using a diglycolamide derivative TODGA in ionic liquids. *Dalton Transaction*, **37**, 5083–5088.

Singh D., Kotekar M. and Singh H. (2008). Development of a solvent extraction process for production of nuclear grade dysprosium oxide from a crude concentrate. *Desalination*, **232**(1), 49–58.

Tan Q., Lia J. and Zenga X. (2014). Rare earth elements recovery from waste fluorescent lamps: a review. *Critical Reviews in Environmental Science and Technology*, **45**, 749–776.

Taggart K. R., Hower C. J., Dwyer S. G. and Hsu-Kim H. (2016). Trends in the rare earth element content of U.S.-based coal combustion fly ashes. *Environmental Science and Technology*, **50**(11), 5919–5926.

Van Gosen B. S., Verplanck P. L., Seal R. R., II, Long K. R. and Gambogi J. (2017). Rare-earth elements. In: Critical Mineral Resources of the United States—Economic and Environmental Geology and Prospects for Future Supply, K. J. Schulz, J. H. DeYoung, Jr, R. R. Seal, II and D. C. Bradley (eds), U.S. Geological Survey Professional Paper, **1802**, O1–O31, doi: 10.3133/pp1802O

Xiong C., Xinyi C. and Caiping Y. (2011). Enhanced adsorption behavior of Nd (III) onto D113-III resin from aqueous solution. *Journal of Rare Earths*, **29**(10), 979–985.

Yao Z. T., Ji X. S., Sarker P. K., Tang J. H., Ge L. Q., Xia M. S. and Xi Y. Q. (2015). A comprehensive review on the applications of coal fly ash. *Earth Science Reviews*, **141**, 105–121.

Zepf V. (2013). Rare Earth Elements: What and Where They are Rare Earth Elements. Springer, Berlin, Heidelberg, pp. 11–39. Springer Theses (Recognizing Outstanding PhD Research) doi: 10.1007/978-3-642-35458-8_2

Chapter 6

Rare earth elements recovery from red mud

Upendra Singh, Sonali A. Thawrani and Anupam Agnihotri

6.1 INTRODUCTION

Aluminum metal is commercially produced from bauxite ore involving two main process steps. In the first step, alumina is obtained by the Bayer's process and in the second step, the alumina is reduced in a Hall Heroult cell to yield metal. Production of alumina from bauxite by the Bayer's process is associated with the generation of red mud, also called bauxite residue, as the major waste material. Depending on the quality of bauxite, the quantity of red mud generated varies from 55–65% of the bauxite processed (Samal *et al.*, 2013). The generation of red mud is huge and the general practice is to dispose of it off the plant site by pumping it into mud ponds, depending on the availability of land space (Evans, 2016). Though substantial development has taken place in the area of dewatering and the disposal of red mud, its treatment and utilization still remain a challenge for aluminum processing industries.

The average utilization rate of red mud in the world is less than 10% (Archambo & Kawatra, 2020). Bauxite residue has several potential applications in the pigment and paint industry, ceramic production, soil amendment, building materials and pollution control. However, these applications have not been exploited at industrial scale (Qu & Lian, 2013) due to various hindrances and barriers such as

doi: 10.2166/9781789062236_0131

the high alkali content in red mud, complex processing techniques, quality, costs and associated risks.

The tailings have a complex mineralogy. They are associated with different gangue minerals, mainly hematite and goethite followed by sodalite, calcites, siliceous, titanous and aluminum bearing oxide gangue materials with sparing amounts of rare earth elements (REEs). Bauxite residue contains a substantial amount of REEs and also the generation volume in India is sizeable (>9 MT). Therefore, the total amounts of REEs locked in them are very large and they can become a significant source for these elements.

Rare earth elements (REEs) seldom occur in concentrated deposits and so there is limited primary production of REEs. On the supply side, they are not scarce but sporadically distributed in the earth's crust as they lack affinity to combine with ore-forming anions. REEs are therefore being produced as a by-product of other metal production. The most abundant REE in the earth's crust is cerium (20–70 mg/kg), followed by neodymium and lanthanum, with heavier REEs being less abundant, and thulium being the least abundant rare earth element (0.2–1 mg/kg). Among the different REEs, scandium is the most strategic one. There has been no record of scandium deposits with concentrations over 100 ppm. Hence, resources such as red mud with a scandium content between 20 and 50 ppm can be considered as an ore (Reid *et al.*, 2017).

Table 6.1 highlights the minerals that are likely to be critical to India's manufacturing sector by 2030 and the reasons why they take on more

Table 6.1 Key determinants of the transition of minerals to the most-critical quadrant.

S. No	Critical Minerals–2030	Key Parameters to Impact Economic Importance	Key Parameters to Impact Supply Risk
1.	Rare earths (heavy)	(a) All the major green technologies depend on heavy rare earths imparting their special properties (b) Extensive applications within the defense industry	India is 100% import dependent, with 94% of global supplies controlled by China. India bears mainly deposits for lighter rare earth elements (in form of monazite).
2.	Rare earths (light)	(a) Major use of light rare earth elements as an alloying material to impart the pyrophoricity property to the steel (b) Modern day applications in electronics manufacturing is yet to begin in India.	India is 100% import dependent, its reserves are associated with coastal beach sands of India, but its mining is not open for private sector to date.

importance. The Council on Energy, Environment and Water (CEEW) published a list of 12 critical minerals that could play an important role in the success of the 'Make in India' program and the sustainable growth of the Indian economy. The critical minerals include beryllium, germanium, rare earths, rhenium and tantalum (Gupta *et al.*, 2016). India is totally import dependent for the identified critical minerals, and the country does not have any declared resources of them, except light rare earths (found along with monazite sands) and beryllium.

China is the world leader in rare earth oxide (REO) supply comprising 94% of the global market (Haque *et al.*, 2014). India produces less than 2% of the total REO produced globally (Singh 2020). It is crucial for India and other countries to explore, expand and exploit available sources including industrial wastes for recovering the valuable mineral wealth to meet the growing domestic demand to a larger extent. The development of processing techniques for residues from alumina extraction processes has been identified as key, not only for getting the minerals to production, but also perpetuated by increasing statutory and regulatory scrutiny with regard to the handling of tailings.

In this chapter, process know-how for effective and economic separation and recovery of rare earth elements (Sc, La and Ce) from the industrial waste-bauxite residue is discussed in detail.

6.2 BAUXITE RESIDUE

6.2.1 Production

Red mud (bauxite residue) is a waste by-product generated from the refinery during processing of bauxite ore. A typical bauxite plant produces 1–2 times as much red mud as alumina. This ratio is dependent on the type of bauxite used in the refining process and the extraction conditions. Global generation of red mud is more than 150 million tons (Evans, 2016) and there exists a global inventory of more than 3 billion (10^9) tons (European Aluminium Document, 2015).

In India, there are three main primary metal producers in the aluminum sector namely, Hindalco (Aditya Birla Group), National Aluminium Company (Nalco) and Vedanta. It is estimated that India generated around 9.17 million tons of red mud in 2017–18 (Rai *et al.*, 2019). Plant-wise generation of red mud is shown in Table 6.2.

6.2.2 Composition

Red mud is composed of a mixture of solids and metallic oxides like Al_2O_3, Fe_2O_3, SiO_2, TiO_2, CaO and Na_2O. In addition to the main components, bauxite residue contains numerous trace elements such as V, P, K, Ga, Zn, Mg and rare earths elements. In Tables 6.3 and 6.4, the typical composition of red mud generated by different Indian alumina plants is presented.

Table 6.2 Red mud generation in Indian bauxite ore refineries (Rai *et al.*, 2019).

Indian Industries	Red Mud Generation (MT*/Annum)
NALCO, Damanjodi, Odisha	3.10
HINDALCO, Utkal, Odisha	2.05
HINDALCO, Belgavi, Karnataka	0.44
HINDALCO, Muri, Jharkhand	0.61
HINDALCO, Renukoot, Uttar Pradesh	0.97
VEDANTA, Langigarh, Odisha	2.00
Total	9.17

*MT-Million ton.

Table 6.3 Chemical composition of Indian Red mud (Singh *et al.*, 2020).

Element	Concentration Range (%)
Al_2O_3	14–23
Fe_2O_3	35–62
SiO_2	4–9
TiO_2	4–18
Na_2O	3–6
CaO	0.5–5
Loss on Ignition	8–14

Table 6.4 Chemical composition of major REEs in Indian refinery red mud (Singh *et al.*, 2020).

Element	Concentration Range (ppm)
Sc	30–65
La	20–65
Ce	45–105
Pr	4.9–6.6
Nd	15–30
Y	4–10
Gd	4–8

6.2.3 Particle size distribution of red mud

The bauxite residue contains very fine particles, which range from 0.186 to 187.6 µm. The particle distribution of red mud is presented in Figure 6.1. Ninety percent of the particles (d_{90}) are below 123 µm and 50% (d_{50}) are smaller than 6.4 µm.

6.3 TECHNOLOGY FOR EXTRACTION OF REES FROM BAUXITE RESIDUE

Various processes have been described for recovering valuable components from red mud. Smelting of bauxite residue followed by slag leaching, reduction roasting of bauxite residue followed by magnetic separation, and water leaching followed by residue leaching and direct acid leaching are a few of the processes for recovery of REEs (Borra *et al.*, 2016a). Recovery of REEs by adopting conventional processing methods has not been a cost-effective exercise as it results in high acid consumption and high impurity co-extraction tailing materials. Further, high amounts of iron are difficult to separate, require a larger number of processing steps and consume larger amounts of chemicals in the subsequent downstream processes as iron and scandium have a few common chemical characteristics. Therefore, physical beneficiation has been employed to remove iron as much as possible from the mud in order to improve the rare earth

Report by Volume Percent

Low Particle Diameter (µm)	Cumulative Volume Finer Percent	Low Particle Diameter (µm)	Cumulative Volume Finer Percent	Low Particle Diameter (µm)	Cumulative Volume Finer Percent	Low Particle Diameter (µm)	Cumulative Volume Finer Percent
187.610	99.0	18.882	70.0	3.564	40.0	0.890	10.0
153.621	95.0	13.830	65.0	2.853	35.0	0.543	5.0
123.431	90.0	10.667	60.0	2.320	30.0	0.186	1.0
68.911	85.0	7.537	55.0	1.889	25.0		
48.177	80.0	6.372	50.0	1.528	20.0		
29.278	75.0	4.455	45.0	1.209	15.0		

Incremental Volume Percent vs. Particle Diameter Graph

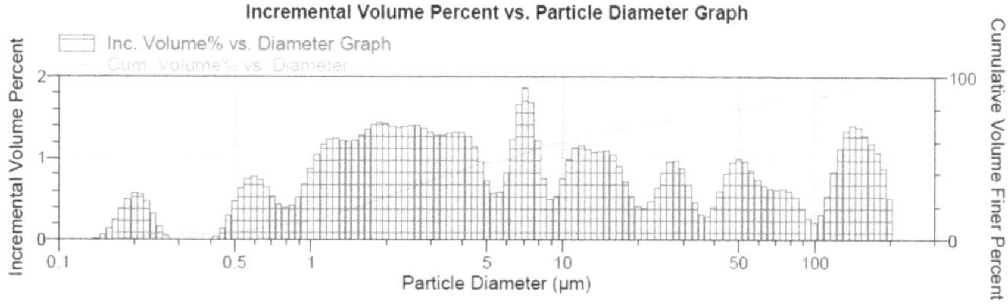

Summary Report

Weighted Statistics (Volume Distribution)

	Std Dev of 1		Std Dev of 1
Mean	29.639	Median	6.372	0.000

Figure 6.1 Particle size distribution of red mud (Singh *et al.*, 2020).

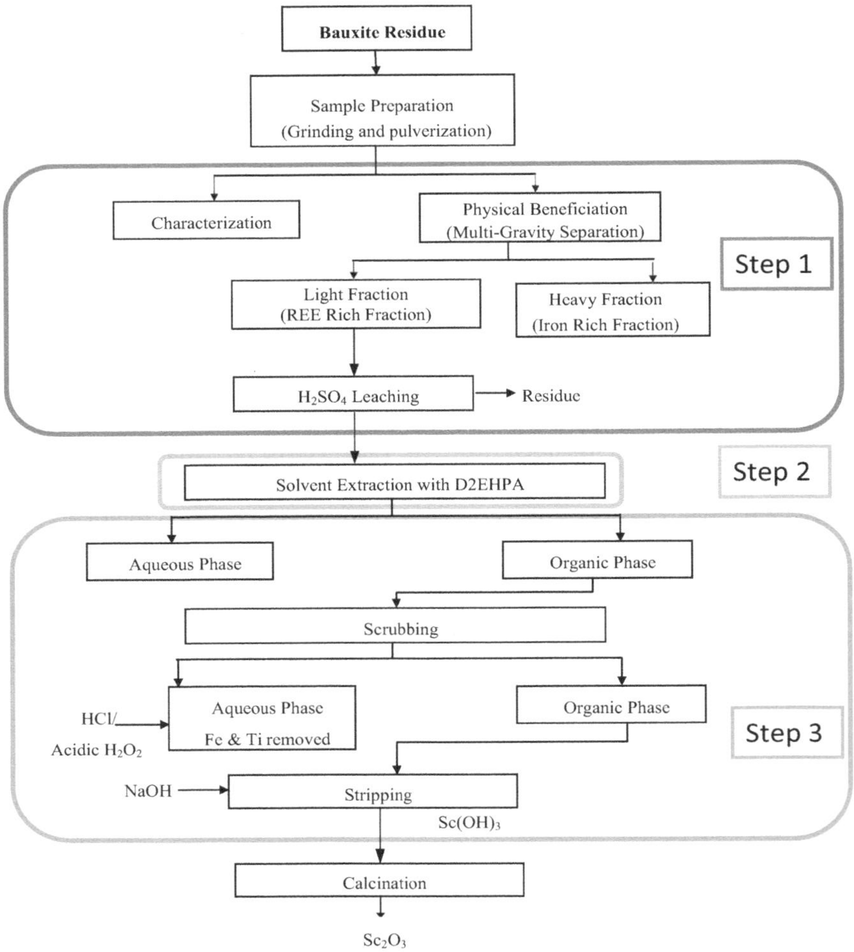

Figure 6.2 Conceptual flow-sheet for recovery of scandium from red mud (Singh *et al.*, 2020).

elements extraction efficiency. The major process steps of this conceptual flow sheet are shown in Figure 6.2.

The key to successful utilization of the beneficiation process is to develop a beneficiation strategy, which is economical and consistent with its mineralogy and the fineness of the particles. It has been found from detailed chemical and mineralogical studies that the iron content in the finer size, below 10 microns, is low. Also, the rare earth content increases with decreasing particle size. It is therefore apparent that a preliminary concentration by hydrocyclone and

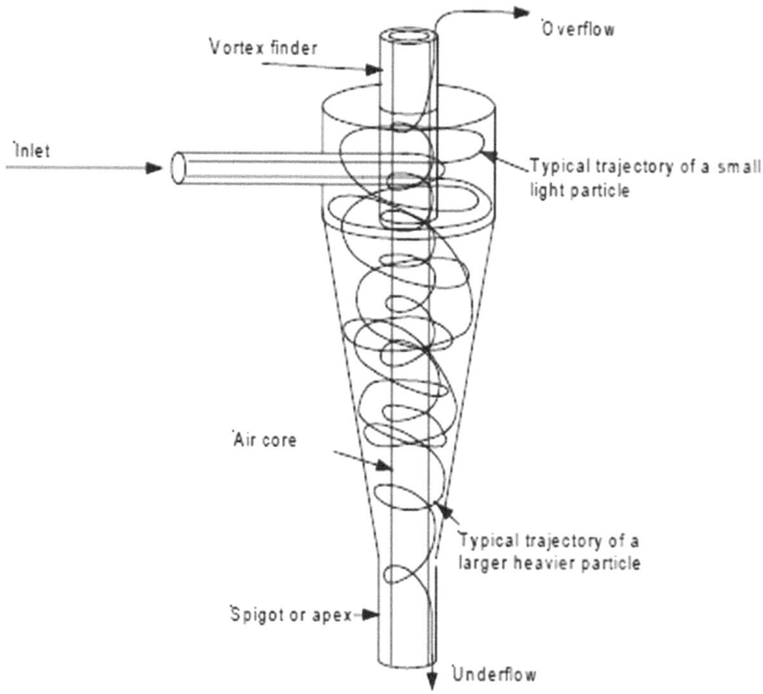

Figure 6.3 Schematic diagram of a hydrocyclone (*Source*: Mineral Technologies International Inc.).

multi-gravity separator (MGS) can be achieved as the rare earth constituent part is distributed at the finer size.

6.3.1 Methods for physical beneficiation

6.3.1.1 General principle of a hydrocyclone operation process

The basic structure of a hydrocyclone is shown in Figure 6.3. The principle of operation of the hydrocyclone is built on the concept of the terminal settling velocity of a solid particle in a centrifugal field. The feed enters the hydrocyclone tangentially and circulates through the circular portion, with a net inward flow of fluid from the outside to the vortex finder on the axis. Since the circulating velocities are very high, large centrifugal fields form within the hydrocyclone. The centrifugal field is usually high enough to generate an air core along the axis that passes from the spigot opening at the bottom of the conical section through the vortex finder to the overflow at the top. In order for this to happen, the centrifugal force field must be many times larger than the gravitational field.

Because of their higher density, particles in this centrifugal field may appear to travel outwards in relation to the carrier fluid. The bigger, heavier particles will quickly migrate to the cylindrical section's outside walls, where they will be forced to travel downward on the inside of the conical wall. The fluid, on the other hand, can drag small, light particles inwards as it travels toward the vortex finder. Any particle's drag force is a complex function of the hydrodynamic conditions within the hydrocyclone, as well as the particle's shape and size.

6.3.1.2 General principle of the multi-gravity separator process

The multi-gravity separator (MGS) is a separation system that uses the combined effects of centrifugal acceleration and an enhanced gravitational field to beneficiate ores with fine particle distribution. The schematic of an MGS is shown in Figure 6.4. The feed is sprayed onto the drum's surface in a slurry shape. A similar launder near the open end of the drum is used to add wash water. Under the increased gravity field, the heavier (denser) particles settle easily to the bottom drum surface and scrape up the drum surface to the outer end, where they are fully discharged. Fines appear to stay suspended and discharged in the reverse end after a period of shaking the drum and continuous washing of the settled material. MGS has a wide range of applications in mineral processing such as recovery of precious metals, pre-concentration of heavy minerals and up-grading concentrate.

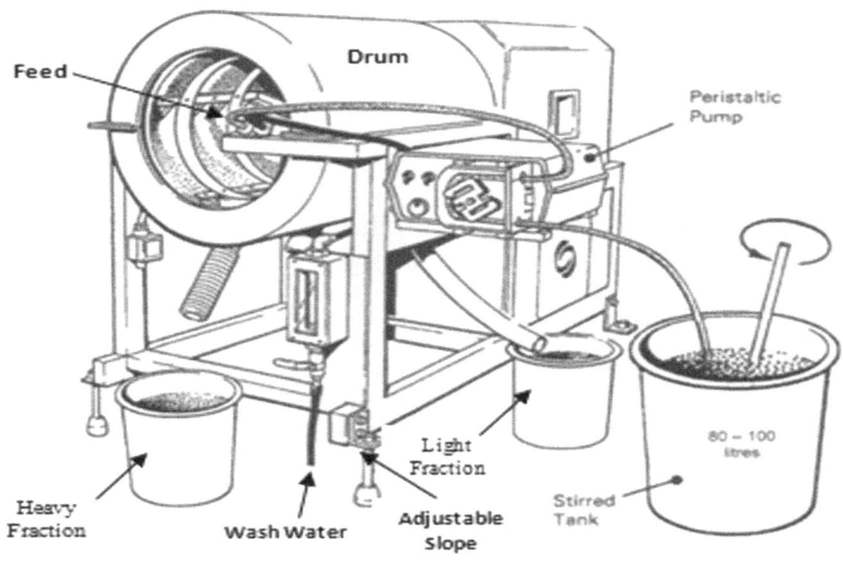

Figure 6.4 Schematic diagram of a multi-gravity separator (Sunil *et al.*, 2012).

6.3.1.3 Hydrocyclone and multi-gravity separator for red mud treatment

Both the hydrocyclone and the multi-gravity separator are useful techniques for beneficiating rare earth elements in bauxite residue and are particularly effective in reducing the iron content. Considering the cost of separation and a better REE grade (about 25% enrichment of Sc, La and Ce) with reasonable recovery (up to 70%) of the lighter fraction of red mud by adopting an MGS route in comparison to a hydrocyclone, a process with MGS has been suggested.

6.3.2 Alkali roasting, smelting and leaching

Reduction roasting/smelting to separate iron from red mud is another appropriate technique for the Bayer process red mud consisting of a high Fe_2O_3 content to recover/remove the iron prior to the leaching step. Borra *et al.* (2017) studied the recovery of rare earths and major metals from bauxite residue (red mud) by alkali roasting, smelting and leaching.

Alkali roasting of bauxite residue at 950°C for 4 h with sodium carbonate followed by water leaching at 80°C for 60 min can remove about 75 wt% of the alumina. The Na_2CO_3 roasted sample is completely molten at 1500°C, which helps achieve a clear slag/metal separation. More than 98% of the iron can be removed (recovered) by smelting. Acid leaching of the slag at 90°C can leach about 80% of the scandium. However, the recovery yields of the other REEs and titanium are very low, especially for the light REEs (<5%). More than 80% of titanium and REEs can be recovered by slag quenching followed by leaching at 25°C. The alkali roasting temperature can be decreased by replacing Na_2CO_3 with NaOH. The process developed is a near-zero-waste process and can recover most of the major metals and REEs.

However, high acid consumption, high temperature requirement during leaching and a still considerable amount of iron dissolution are the major concerns in this route. Furthermore, large amounts of other impurities are also dissolved into the leach solution during direct and slag leaching, which generates large volumes of effluent.

6.3.3 Sulfation, roasting and leaching

Borra *et al.* (2016b) developed a process to selectively leach REEs from red mud (Greece) by a sulfation-roasting-leaching process. Most of the elements in the bauxite residue convert to their respective sulfates during sulfation. Iron and aluminum sulfates decompose to their respective oxides at temperatures around 700°C. The water-soluble rare-earth sulfates, on the other hand, are stable at such roasting temperature ranges and they dissolve easily during the subsequent water leaching step leaving the iron oxides in the residue. Decreasing the roasting temperature increases the iron, aluminum and titanium dissolution. Increasing the roasting time beyond a certain duration decreases the dissolution of iron, titanium and REEs. At a roasting temperature around 700°C for a roasting duration of 1 h and with a sulfuric acid to bauxite residue mass ratio of 1:1, 60 wt% of scandium and >80 wt% of other REEs can be extracted after room temperature leaching. The dissolution of some of the major elements expressed as a fraction of their total availability was found to be less than 1% for Fe and Ti and less than 20% for Al. No silicon dissolution was observed. Sodium was completely soluble and calcium was soluble up to the solubility limit of its sulfate precipitate (gypsum) in water. The pH of the residue generated after leaching is close to neutral, and it is poor in sodium and rich in calcium sulfate content. Therefore, the residue can be considered for applications in cementitious binders.

The main advantages of the sulfation-roasting-leaching process can be summarized as follows: (1) limited leaching of iron, titanium and aluminum; (2) no or very limited silica leaching, meaning filtration problems can be avoided; (3) the possibility of (consumed) acid regeneration and therefore low acid consumption and (4) small volumes of wastewater generation compared to direct leaching. Additionally, because the pH of the residue after water leaching is close to neutral and very low in sodium content, management and utilization of this stream in other applications is easier. The main drawback of this process is that it needs an extra processing (roasting) step.

6.3.4 Direct leaching of mineral acid

Many studies (Ochsenkühn-Petropulu *et al.*, 1996; Wang *et al.*, 2010; Zhang *et al.*, 2005) reported on direct acid leaching of bauxite residue for recovering REEs from red mud. According to Ochsenkühn-Petropulu *et al.* (1996), among all the mineral acids, diluted HNO_3 showed the best recoveries, especially for Y, Yb and Sc and had the best selectivity over Fe. However, the proposed leaching procedure was more effective for the heavy lanthanides (Yb, Er and Y) than for the light ones (La, Ce and Pr). Diluted HCl showed similar recoveries as HNO_3, but it is not so selective, dissolving much more Fe. Therefore, the leaching process with dilute HNO_3 was performed under moderate conditions and without using any preliminary treatment. Their results showed that the best recoveries of scandium (80%) and yttrium (96%) were obtained by using HNO_3 leaching for 24 h at a concentration of 0.5 mol/L, liquid-to-solid (L/S) ratio of 50 and temperature of 25°C. Under this condition, the leaching selectivity of scandium over iron was the highest with only 3% of iron dissolution.

Zhang *et al.* (2005) employed HCl to leach scandium from red mud, and the best leaching conditions were determined to be a HCl concentration of 6 mol/L, L/S ratio of 4, temperature of 50°C, and leaching for 1 hour, resulting in an Sc_2O_3 leaching rate of 82.09%. Wang *et al.* (2010) also recovered scandium from red mud using HCl as a leaching agent, and observed that the L/S ratio was the most critical factor influencing scandium extraction, and that the HCl concentration had a significant impact on iron dissolution. At a concentration of 6 mol/L HCl,

L/S ratio of 5, temperature of 60°C, and a reaction time of 1 hour, the leaching efficiency of scandium was above 85%. Furthermore, the study estimated that the HCl acid consumption was about 21.2 mol for 1 kg red mud.

6.3.5 Pre-concentration-acid leaching

The bauxite residue after beneficiation is subjected to leaching with mineral acids such as HCl, H_2SO_4 and HNO_3 for selective recovery of REEs. The effect of several parameters, like the type of acid, the concentration of the leaching agent, the leaching time, the temperature and the pulp density, (i.e., solid to liquid ratio) were taken into account in order to achieve high recovery efficiencies for the elements of interest under selective and moderate experimental conditions.

The highest leaching efficiency for scandium (Sc) was reported with H_2SO_4, followed by HCl and HNO_3. The maximum leaching efficiencies for lanthanum and cerium were found using HNO_3 and HCl as the leaching agents, respectively. While HNO_3 showed the best selectivity with respect to iron dissolved, followed by HCl and H_2SO_4.

Leaching with H_2SO_4 resulted in 60%, 80% and 89% of Sc, La and Ce, respectively. The optimum conditions during leaching were in the range of 10% pulp density, acid concentration of 2 M at 95–100°C for 2 hours leaching time. Pre-concentration followed by acid leaching shows high leaching efficiencies of REEs in the leachate and better selectivity against iron (Singh *et al.*, 2019).

6.4 REE SEPARATION PROCESSES

Following leaching, the next step is to separate the REEs from the leachate. Processes for separating individual rare earth elements from the solution use small differences in acidity, resulting from the decrease in ionic radius from lanthanum to lutetium. Acidity differences influence the solubility of salts, the hydrolysis of cations and the formation of complex species which forms the basis of rare earth element separation processes.

Fractional crystallization and precipitation are the basic techniques for REE separation (James, 1999) and were widely used in the past but are now considered as uneconomical and have been replaced by ion exchange and liquid-liquid extraction.

6.4.1 Fractional crystallization and precipitation

Fractional crystallization was the original method for separating REEs. In fractional crystallization, a part of a salt in a solution is precipitated either by a temperature change or by evaporation of a saturated solution. The composition of the formed crystals is different compared to the original solution due to different solubilities of the components. The result is a crystal crop with fewer soluble components and a solution enriched with more soluble components. A number of REE salts and double salts have been used in separation with fractional crystallization. For example, double ammonium nitrates have been used in lanthanum removal and to separate praseodymium and neodymium.

Fractional precipitation is a difficult method utilizing the small difference in basicity between the separate REEs. This causes REEs to precipitate at different concentrations and/or pH values when a precipitant is added to the solution. Several compounds have been used in REE fractional precipitation, with hydroxides and double sulfates being the most widely applied (Royen & Fortkamp, 2016).

6.4.2 Ion exchange

The ion exchange technique was acknowledged as the most practical method used for separating REEs just before the booming of the REE industry by solvent extraction in the 1960s (Suli *et al.*, 2017). Although ion exchange is another method of REE separation, which has been employed in attaining >99.9999% purities of REO, this technique is deemed not economical.

Ion exchange involves the adsorption of REE ions from the solution onto an exchanger, which is often selective for specific REEs, followed by a desorption of the REE ions during an elution step. Ion exchange methods are highly selective, but costly and therefore, generally reserved for the production of extremely pure materials. In modern applications, typically sulfonated polystyrene or its Na^+ salt is used as a resin. The cations exchange with H^+ or Na^+ and are then removed from the resin utilizing a complexing agent, such as ethylenediaminetetraacetic acid ($EDTA^{4-}$) (Royen & Fortkamp, 2016).

In an ion exchange resin or an ion exchanger, there are negative or positive ions attached to an insoluble organic matrix. In a cation exchange resin, the ions are positive and in an anion exchange resin, they are negative. When the resin is brought into contact with the solution, the ions in the organic resin can be replaced by the ions in the solution. Typically, the ion with a higher charge displaces the one with a lower charge, and if the ions have the same charge, the ion with the larger radius displaces the ion with a smaller radius. Displacement also occurs according to the law of mass action.

After the adsorption stage, where the ions in the solution are loaded onto the resin, the ions are desorbed from the resin into a solution in an elution stage. If a solution contains several ions that exhibit selectivity in the exchange, ion exchange can be called ion exchange separation.

6.4.3 Solvent extraction

Liquid-liquid extraction, also called solvent extraction, can be used to produce REE compounds of >99.99% purity (Royen & Fortkamp, 2016).

In this technique, two immiscible liquids are mixed and the solutes are separated between the two liquid phases. One of the liquid phases is aqueous and the other organic. The organic phase usually consists of two or more chemicals. The extractant collects the rare earth elements in the organic phase and a suitable solvent is used to dissolve the extractant, which is usually too viscous to be used on its own in practical applications. Kerosene and certain aromatics are commonly used as solvents. Acidic, basic and neutral extractants are the most common industrial extractants and include bis(2-ethylhexyl) phosphoric acid (D_2EHPA), 2-ethylhexyl(2-ethylhexyl) phospho-nate (EHEHPA), tributyl phosphate (TBP), versatic 911, versatic 10 and Aliquat 336. The selectivity of commonly used extractants for scandium extraction under various conditions are summarized in Table 6.5.

Among all the other techniques, solvent extraction has the advantages of processing higher capacity, high loading of rare earths in the organic phase and operational ease at larger scale with lower operational costs, which makes this the most widely used technique. Some of the problems faced during solvent extraction methods, that is, need for multiple stages of all operation steps especially for trace amount element recovery, loss of extractant during operation and formation of emulsions, can be solved using ion exchange resins. Although this method (Section 6.4.2) offers some advantages over solvent extraction, it also has disadvantages which limit its usage. For instance, slow exchange rates which make the process time-consuming, increase the operating costs as well as making it less efficient for highly impure solution treatment. Since, the effectiveness of the resin significantly decreases when it is used for highly impure solutions, these resins are not the optimum material choice to recover Sc from highly impure feeds. In addition, ion exchange resins are relatively expensive to use in trace Sc recovery operations.

Zhang *et al.* (2006) recovered scandium from a HCl leaching solution of red mud using 1 vol% P507+kerosene as extractant. With an organic/aqueous (O/A) ratio of 1 and a 15 minute extraction time, the scandium extraction efficiency was >90%. After solvent extraction, the organic phase was washed twice with 6 mol/L HCl and distilled water, respectively, at an O/A ratio of 3, before being stripped for 15 min with 2 mol/L NaOH at an O/A ratio of 3 and temperature of 50°C. The $Sc(OH)_3$ obtained after filtration was further dissolved in 6 mol/L HCl with ammonium hydroxide, resulting in a pH of about 1.5. Following oxalic acid precipitation and calcination at 850°C, a material-enriched 66.09 wt% of Sc_2O_3 was obtained.

Wang *et al.* (2013) conducted an experiment on the recovery of scandium from a synthetic leach solution of an Australian red mud by using solvent extraction. For the scandium extraction and separation from other metals in the synthetic leach solution, a number of extractants were investigated, including three organophosphorus acids (D_2EHPA, Ionquest 801 and Cyanex 272), a carboxylic acid (Versatic 10), a primary amine (Primene JMT) and two chelating reagents (LIX 984 N and LIX54–100). The results of the extraction efficiencies showed that over 99% scandium was extracted using D_2EHPA, Ionquest 801, Cyanex

Table 6.5 Selectivity of scandium extraction using commonly used extractants under various conditions reported in the literature (Wang & Cheng, 2011).

Extractant	pK_a	Organic Solution	Aqueous feed Solution	Selectivity	Reference
Di-(2-ethylhexyl) phosphoric acid	—	0.75 M HDEHP in n-heptane or cyclohexane	1–11 M HCl, $HClO_4$ or HNO_3	$Sc^{3+} \sim Ti^{4+}, Zr^{4+}, Hf^{4+} > Y^{3+} > La^{3+} > Mn^{2+}$	Qureshi *et al.* (1969)
D_2EHPA	—	Purified HDEHP in *n*-octane	pH 3–10 M HCl	$Sc^{3+} >> Fe^{3+} > Lu^{3+} > Yb^{3+} > Er^{3+} > Y^{3+} > Ho^{3+}$	Xue and Li (1992)
Di-(2-ethylhexyl) phosphoric acid	2.16	20% HDEHP, 15% TBP and in kerosene 0.2 M D2EHPA and 1% TBP in Escaid 110 0.1 M HDEHP in toluene Purified HEHEHP in *n*-heptane Purified HEHEHP in *n*-heptane	2.5 g/L Sc, 25 g/L Mg, Al and Fe and 0.5 M HCl pH 1.5–3.5 H_2SO_4 0.5–11 M $HClO_4$ 0.5–1.5 M H_2SO_4 1.5–5 M H_2SO_4	$Sc^{3+} > Fe^{3+} > Al^{3+}, Mg^{2+}$ $Sc^{3+} \sim Zn^{2+} > Ca^{2+} \sim Al^{3+} > Mn^{2+} > Cr^{3+} \sim Mg^{2+} \sim Ni^{2+} \sim Si$ $Sc^{3+} > Fe^{3+} > Al^{3+} > Mg^{2+}$ $Sc^{3+} \sim Th^{4+} > Ce^{4+} > {}_{Fe}3+$ $Sc^{3+} > Ce^{4+} > Th^{4+} > Fe3+$	Ditze and Kongolo (1997) Haslam and Arnall (1999) Singh and Dhadke (2003) Li *et al.* (1980) Li *et al.* (1980)
2-ethylhexylphosphonic acid mono-2-ethylhexyl ester	3.36	0.2 M Ionquest 801 and 1% TBP 0.1 M PC-88A in toluene	pH 1–5.5 H_2SO_4 0.01–1 M $HClO_4$	$Sc^{3+} > Zn^{2+} > Al^{3+} > Mn^{2+} \sim Cr^{3+} \sim Ca^{2+} \sim Mg^{2+} > Ni^{2+} \sim Si$ $Sc^{3+} > Fe^{3+} > Al^{3+} > Mg^{2+}$	Haslam and Arnall (1999) Singh and Dhadke (2003)

Cyanex 272	5.32	4.8×10^{-2} M Cyanex 272 in *n*-hexane 0.1 M Cyanex 272 and 5% TBP	2×10^{-4}–6×10^{-4} M metals, pH 3–10 M H_2SO_4 pH ~ 1 H_2SO_4	$Sc^{3+} \sim Th^{4+} > Fe^{3+} > Lu^{3+}$ $Sc^{3+} >> Al^{3+} > Ni^{2+} > Si > Mn^{2+} \sim Mg^{2+} \sim Ca^{2+} > Cr^{3+}$	Wang and Li (1994) Haslam and Arnall (1999)
Cyanex 302	4.32	4.8×10^{-2} M Cyanex 302 in *n*-hexane	2×10^{-4}–6×10^{-4} M metals, pH 3–10 M H_2SO_4	$Zr^{4+} > Sc^{3+} > Th^{4+} > Fe^{3+} > Lu^{3+}$	Wang and Li (1995)
Cyanex 301	3.86	4.8×10^{-2} M Cyanex 301 in *n*-hexane	2×10^{-4}–6×10^{-4} M metals, pH 3–10 M H_2SO_4	$Zr^{4+} > Sc^{3+} \sim Fe^{3+} \sim Th^{4+} > Lu^{3+}$	Wang and Li (1995)
Tributyl phosphate	—	100% TBP 100% TBP	7—8 M HCl 4—6 M $HClO_4$	$Sc^{3+} \sim Zr^{4+} > Th^{4+}$ $Sc^{3+} > Zr^{4+}$	Peppard *et al.* (1956) Zhang *et al.* (1997)
P350	—	40% P350 in kerosene	5.8 M HCl	$Sc^{3+} > Ti^{4+}, Y^{3+}, Al^{3+}, Ca^{2+}, Mg^{2+}$	Zhang *et al.* (1997)
Cyanex 923	—	5% Cyanex 923 in kerosene	2.0—7.0 M H_2SO_4 1—5 M HCl	$Zr^{4+} > Sc^{3+} > Ti^{4+} \sim Lu^{3+} > Fe^{3+}$ $Sc^{3+} > Th^{4+} > Lu^{3+}$	Li and Wang (1998)
Cyanex 925	—	5% Cyanex 925 in kerosene	2.0—7.0 M H_2SO_4 0.5—2.5 M HCl 1—5 M HCl	$Zr^{4+} > Sc^{3+} > Lu^{3+} > Ti^{4+} > Fe^{3+}$ $Th^{4+} > Sc^{3+} > Lu^{3+}$ $Sc^{3+} > Th^{4+} > Lu^{3+}$	Li and Wang (1998)

272 and Primene JMT. However, D_2EHPA, Ionquest 801 and Primene JMT had a high iron co-extraction efficiency of over 54%. Because phase separation was poor when D_2EHPA, Ionquest 801 and Cyanex 272 were used individually, TBP was added to these three organophosphorus extractants as a phase modifier. The organic system of 0.05 mol/L D_2EHPA and 0.05 mol/L TBP in Shellsol D70 at an A/O ratio of 5, temperature of 40°C and pH of 0.25 gave the best scandium extraction. Scandium extraction was >99% under these ideal conditions, with co-extraction of 25% Zr, 15% Ti, 2% V, and nearly no Fe, Cr, Ga, Ca and Al. The $Sc(OH)_3$ product was obtained after the loaded organic system was scrubbed twice and subsequently stripped from the D_2EHPA/TBP system with NaOH.

Singh *et al.* (2020) used bis(2-ethylhexyl) phosphoric acid (D_2EHPA) in white kerosene (commercial) to extract scandium from a leached solution. Scandium and impurity elements such as Fe, Ti and Al were co-extracted. The metals were extracted in the order of: Sc(III) >Ti (IV) >Fe(III) >Al(III). More than 95% scandium was extracted into the organic phase from the aqueous leach liquor. The scandium-loaded organic phase was scrubbed with HCl and H_2SO_4/H_2O_2 to separate scandium from other metals. Further, the alkaline solution was used to strip the scandium into an aqueous solution, which was diluted to precipitate the scandium as $Sc(OH)_3$. The $Sc(OH)_3$ precipitate was further purified to obtain a high purity scandium oxide product. The purity of the scandium oxide obtained was >80% with a total recovery of more than 95%.

6.5 CONCLUSION

Red mud generated in the process of industrial production of alumina is a worldwide problem. At current levels of technology and practice, the capacity of consumption and secondary utilization is seriously insufficient. The secure stockpiling of red mud needs to see a reduction of stockpiling costs and an efficiency improvement. So, stockpiling is not a viable solution to resolve the problems of red mud. Only economical and viable comprehensive utilization can effectively resolve the problem in the long term. The present work focuses on finding a lasting solution for productive utilization of red mud.

This book chapter demonstrates that the processing approach routed through pre-concentration- leaching-solvent extraction-precipitation is promising for extraction of REEs, particularly scandium from bauxite residue. The key issue of scandium extraction in the leaching process is the co-extraction of other rare earth elements and other base metals such as iron and aluminum due to their similar chemical behavior. Hence, a physical beneficiation operation for selective recovery of scandium is essential.

REEs are strategic elements crucial for sustainable and high end technologies. China today controls 90% of the global rare earth production posing a vulnerability to manufacturing industries. It is crucial to reduce the import dependence on China for rare earth elements and it is thus essential to extract REEs from locally available sources.

Scandium is one of the strategic elements in the group of REEs. India produces around 10 million tons of red mud annually which may contain about 3000–4000 tons of scandium. Moreover, India has a stockpile of more than 1 billion (10^9) tons of red mud from which it can extract a huge amount of REEs to meet its own domestic demand.

REFERENCES

Archambo M. and Kawatra S. K. (2020). Red mud: fundamentals and new avenues for utilization. *Mineral Processing and Extractive Metallurgy Review*, **42**(7), 1–24. doi: 10.1080/08827508.2020.1781109 (last accessed date June 25, 2020).

Borra C. R., Blanpain B., Pontikes Y., Binnemans K. and Van Gerven T. (2016a). Recovery of rare earths and other valuable metals from bauxite residue (red mud): a review. *Journal of Sustainable Metallurgy*, **2**, 365–386.

Borra C. R., Mermans J., Blanpain B., Pontikes Y., Binnemans K. and Van Gerven T. (2016b). Selective recovery of rare earths from bauxite residue by combination of sulfation, roasting and leaching. *Mineral Engineering*, **92**, 151–159.

Borra C. R., Blanpain B., Pontikes Y., Binnemans K. and Van Gerven T. (2017). Recovery of rare earths and major metals from bauxite residue (red mud) by alkali roasting, smelting, and leaching. *Journal of Sustainable Metallurgy*, **3**(2), 393–404.

Ditze A. and Kongolo K. (1997). Recovery of scandium from magnesium, aluminium and iron scrap. *Hydrometallurgy*, **44**, 179–184.

European Aluminium. (2015). Document, Bauxite Residue Management: Best Practice.

Evans K. (2016). The history, challenges, and new developments in the management and use of bauxite residue. *Journal of Sustainable Metallurgy*, **2**, 316–331.

Gupta V., Biswas T. and Ganesan K. (2016). Report on critical non-fuel mineral resources for India's manufacturing sector a vision for 2030 study, council on energy, environment and water (CEEW) and national science and technology management information system (NSTMIS) division.

Haque N., Hughes A., Lim S. and Vernon C. (2014). Rare earth elements: overview of mining, mineralogy, uses, sustainability and environmental impact. *Resources*, **3**(4), 614–635.

Haslam M. and Arnall B. (1999). An investigation into the feasibility of extracting scandium from nickel laterite ores. Proceeding of ALTA 1999 Nickel/Cobalt Pressure Leaching & Hydrometallurgy Forum, Perth, Australia.

James C. (1999). Separation of Rare Earth Elements. National Historic Chemical Landmarks. American Chemical Society, Durham, New Hampshire, United States. https://www.acs.org/content/acs/en/education/whatischemistry/landmarks/earthelements.html (last accessed date October 29, 1999).

Li D. and Wang C. (1998). Solvent extraction of scandium(III) by Cyanex 923 and Cyanex 925. *Hydrometallurgy*, **48**, 301–312.

Li D. Q., Wan X., Lin D., Xie Y., Lin S., Wang Z., Li Y. and Ji E. (1980). Extraction separation of rare earth elements, scandium and thorium with mono(2-ethyl hexyl)

2-ethyl hexyl phospnonate (HEH(EHP)). Proceedings of the International Solvent Extraction Conference ISEC'80, Liege, Belgium, Paper No. 80–202, Vol. **3**.

Mineral Technologies International, Inc. http://www.mineraltech.com/MODSIM/ModsimTraining/Module2/TechnicalNotes3.pdf (last accessed date May 21, 2000).

Ochsenkühn-Petropulu M., Lyberopulu T., Ochsenkühn K. M. and Parissakis G. (1996). Recovery of lanthanides and yttrium from red mud by selective leaching. *Analytica Chimica Acta*, **319**, 249–254.

Peppard D. F., Mason G. W. and Maier J. L. (1956). Interrelationships in the solvent extraction behaviour of scandium, thorium, and zirconium. *Journal of Inorganic and Nuclear Chemistry*, **3**, 215–228.

Qu Y. and Lian B. (2013). Bioleaching of rare earth and radioactive elements from red mud using Penicillium tricolor RM-10. *Bioresource Technology*, **136**, 16–23.

Qureshi I. H., McClendon L. T. and LaFleur P. D. (1969). Extraction studies of the Group IIIB-VIIB elements and the lanthanides utilizing Bis(2-Ethyl-Hexyl) Oryhophosphoric acid. *Radiochimica Acta*, **12**, 107–111.

Rai S., Bahadure S., Chaddha M. J. and Agnihotri A. (2019). Disposal practices and utilization of red mud (bauxite residue): a review in indian context and abroad. *Journal of Sustainable Metallurgy*, **6**, 1–8.

Reid S., Tam J., Yang M. and Azimi G. (2017). Technospheric mining of rare earth elements from bauxite residue (red mud): process optimization, kinetic investigation, and microwave pretreatment. *Scientific Reports*, **7**(1), 1–9.

Royen H. and Fortkamp U. (2016). Rare Earth Elements – Purification, Separation and Recycling, Technical Report, Report No. C211.

Samal S., Ray A. and Bandopadhyay A. (2013). Proposal for resources, utilization and processes of red mud in India: a review. *International Journal of Mineral Processing*, **118**, 43–55

Singh Y. (2020). Rare Earth Element Resources: Indian Context. Society of Earth Scientists Series. Cham, Switzerland.

Singh R. K. and Dhadke P. M. (2003). Extraction and separation of scandium(III) from perchloratemedia, by D_2EHPA and PC-88A. *Bulletin of the Chemists and Technologists of Macedonia*, **22**, 1–11.

Singh U., Thawrani S. A, Ansari M. S. and Agnihori A. (2019). Studies on beneficiation and leaching characteristics of rare earth elements in indian red mud. *Russian Journal of Non-Ferrous Metals*, **60**(4), 335–340.

Singh U., Thawrani S. and Agnihotri A. (2020). Utilization and development of process for recovery of strategic rare earths from industrial waste – Bauxite Residue at lab scale, Report No. DST/SSTP/Maharashtra/464.

Suli L. M., Ibrahim W. H. W., Aziz B. A., Deraman M. R. and Ismail N. A. (2017). A review of rare earth mineral processing technology. *Chemical Engineering Research Bulletin*, **19**, 20–34.

Sunil K. T., Rama M. Y., Tathavadkar V. and Mark B. D. (2012). Efficacy of multi gravity separator for concentrating ferruginous chromite fines. *Journal of mining and Metallurgy*, **48**, 39–49.

Wang W. and Cheng C. Y. (2011). Separation and purification of scandium by solvent extraction and related technologies: a review. *Journal of Chemical Technology & Biotechnology*, **86**, 1237–1246.

Wang C. and Li D. Q. (1994). Extraction mechanism of Sc(III) and separation from Th(IV), Fe (III), and Lu(III) with bis(2,4,4–trimethylpentyl)phosphinic acid in n-hexane from sulphuric acid solutions. *Solvent Extraction and Ion Exchange*, **12**, 615–631.
Wang C. and Li D. Q. (1995). Solvent extraction of Sc(III), Zr(IV), Th(IV), Fe(III), and Lu (III) with thiosubstituted organophosphinic acid extractants. *Solvent Extraction and Ion Exchange*, **13**, 503–523.
Wang K., Yu Y., Wang H. and Chen J. (2010). Experimental investigation on leaching scandium from red mud by hydrochloric acid. *Chinese Rare Earths*, **31**(1), 95–98.
Wang W., Pranolo Y. and Cheng C. Y. (2013). Recovery of scandium from synthetic red mud leach solutions by solvent extraction with D2EHPA. *Separation and Purification Technology*, **108**, 96–102.
Xue L. Z. and Li D. Q. (1992). Extraction of scandium(III), yttrium(III), lanthanides(III) and iron(III) fromhydrochloric acid solutionswith di- (2-ethylhexyl)phosphinic acid. *Chinese Journal of Applied Chemistry*, **9**, 21–25.
Zhang P., You S., Zhang L, Feng S. and Hou S. (1997). A solvent extraction process for the preparation of ultrahigh purity scandium oxide. *Hydrometallurgy*, **47**, 47–56.
Zhang J., Deng Z. and Xu T. (2005). Experimental investigation on leaching metals from red mud. *Light Metals*, **2**, 13–15.
Zhang J., Deng Z. and Xu T. (2006). Recovery scandium from leaching liquor of red mud. *Light Metals*, **7**, 16.

Part IV

Technologies to Recover Rare Earth Elements

Chapter 7

Adsorptive recovery of rare earth elements

Arul N. Manikandan and Piet N. L. Lens

7.1 INTRODUCTION

Extensive application of rare earth elements (REEs) in emerging technologies accompanied by their high supply chain risk has projected REEs to be critical materials having international importance (Ballinger *et al.*, 2020; Keilhacker & Minner, 2017; Pinto *et al.*, 2020). There are a total of 17 elements grouped under the category of rare earth elements, out of which 15 are from the lanthanide series, and the other two are frequently alluded to as pseudo lanthanides (Tyler, 2004; Vukojevič *et al.*, 2019). Based on the atomic number of the REEs, they are further classified as light and heavy REEs. In the modern periodic table, lanthanum (La) through gadolinium (Gd) ($Z = 57$ to 64) are classified as light rare elements (LREEs), and terbium (Tb) through lutetium ($Z = 65$ to 71) are classified as heavy rare earth elements (HREEs) (Dushyantha *et al.*, 2020; Pinto *et al.*, 2020). Nevertheless, scandium (Sc), due to its unique properties which contrast with those of other REEs, is referred to as the lightest rare earth element (Huskič *et al.*, 2020). All the REEs along with their density and atomic number are listed in Figure 7.1.

REEs have revolutionized various industrial sectors such as the electrical, electronic, metallurgical and petrochemical industries and even the medical field (Iftekhar *et al.*, 2020; Işıldar *et al.*, 2019). The unique magnetic, optical, catalytic and chemical properties of REEs have posed them to be inevitable and

doi: 10.2166/9781789062236_0153

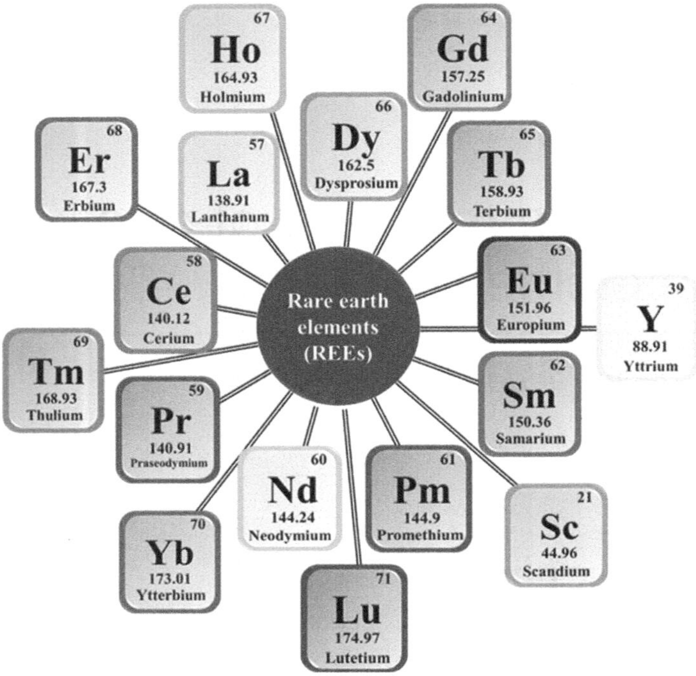

Figure 7.1 Overview of rare earth elements (REEs) along with their salient properties (i.e., molar mass and atomic number) from the modern periodic table.

irreplaceable materials for establishing almost all electronic goods (Dushyantha *et al.*, 2020; Işıldar *et al.*, 2019). Some of the well-known REEs, along with their uses, are portrayed in Figure 7.2. The scattered availability of REEs in any geographical region and the tedious process mandated for the separation of the REEs has led to these elements being named 'rare earths' (Tavares *et al.*, 2004). More specifically, 90–95% of the ores for REE production are sourced from China. After China, Australia and the USA are the leading REE producers having shares below 5% (Iftekhar *et al.*, 2020; Paulick and Machacek, 2017). However, it is worthwhile to note that REEs are more abundantly available than some other precious metals like gold (Au), silver (Ag) and platinum (Pt). The dwindling nature of REEs and reduction in the export quota offered by China has induced a worldwide search for rare earth mines since 2009 (Smith Stegen, 2015).

On the other hand, disposed materials containing REEs are continuously heaped worldwide as electronic waste, metallurgical waste, mineral processing waste and coal combustion waste. Further, improper disposal of REEs has devastating effects on the environment, and their bioaccumulation in lifeforms exhibits deleterious effects on aquatic life and eventually on the human body (Pagano *et al.*, 2019). The exhaustible nature of REEs and their ill effects on the

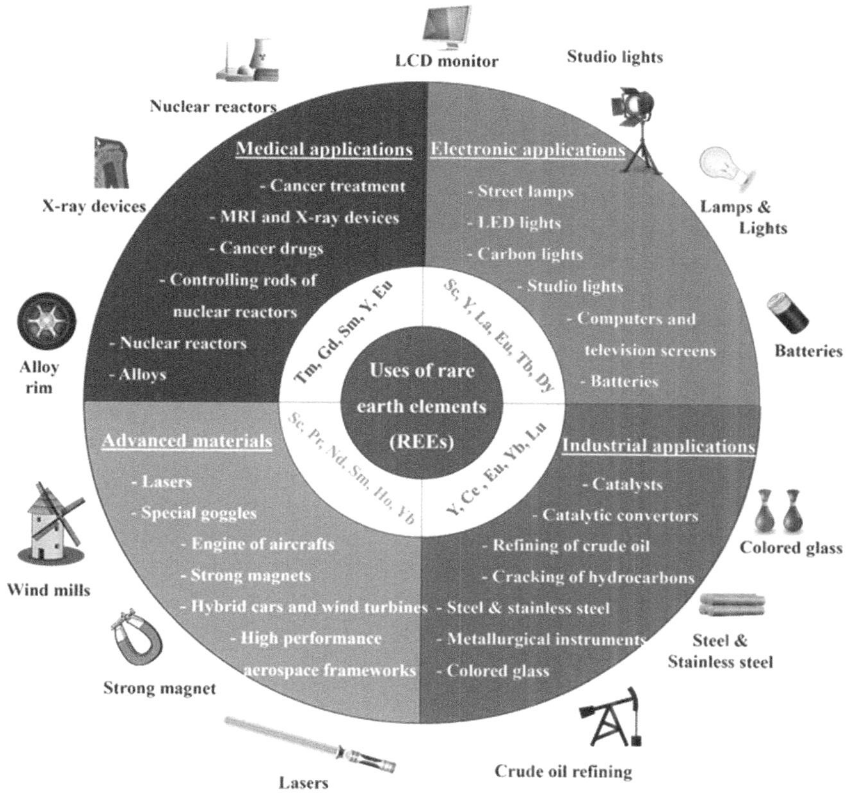

Figure 7.2 Major applications of rare earth elements (REEs). Upon end-of-life, these devices are potential resources for the recovery of REEs, thus avoiding primary mining and preserving the environment.

environment due to improper disposal have accelerated the need for recovery of REEs from used electronic goods and other materials constructed using REEs (Islam *et al.*, 2020). It is expected that identification of a safe practice for recovery of REEs from waste resources will help to avoid REE price volatility and supply augmentation. Basically, almost all the materials listed in Figure 7.2 can be used as a source for REE recovery.

Up until 2011, only 1% of the mined REEs are accounted to have been recovered and reused (Iftekhar *et al.*, 2018a). However, increased demand for REEs representing a global demand of 119 650 metric tons per year has invoked the search for alternative REE sources, such as recycling of e-waste. Several well-known techniques, like ion exchange, separation, chemical precipitation and solvent extraction, are established and commercially operational for the recovery of various heavy metals (da Costa *et al.*, 2020). Similar technologies have been optimized and extended to realize their application in the area of REEs recovery. However, all the techniques mentioned above used for recovering REEs from the liquid phase are inevitably preceded by leaching or bio-leaching (Rasoulnia *et al.*, 2021). While bioleaching is not essential for the recovery of REEs from wastewater, recovery of REEs from solid waste inevitably demands a leaching/bio-leaching process to solubilize REEs from the solid to the liquid phase. Once the REEs are phase transferred from the solid to liquid phase, they are enriched and refined for their potential use. REEs can be enriched and modified by one of the following techniques, viz. ion exchange, separation, chemical precipitation, membrane separation, supercritical extraction, solvent extraction, electro-winning and electro-refining (Ponou *et al.*, 2014).

Among the aforementioned recovery techniques, adsorption is frequently used on the grounds that the adsorptive recovery enables the extraction of even a low concentration of REEs (Huang *et al.*, 2017). Besides scalability, cost-efficiency associated with adsorption makes it an interesting and affordable technique (Barcelos *et al.*, 2020). Adsorption is often referred to as solid-phase extraction, as it involves solid materials for the recovery of REEs suspended in the leachate/liquid phase. Further, adsorption enables continuous recovery of REEs by adsorption followed by desorption in packed bed columns (Arul Manikandan *et al.*, 2016; Barcelos *et al.*, 2020). Typically two types of adsorbents are used for the adsorption of REEs: chemisorbents (Figure 7.3) and biosorbents (Figure 7.4).

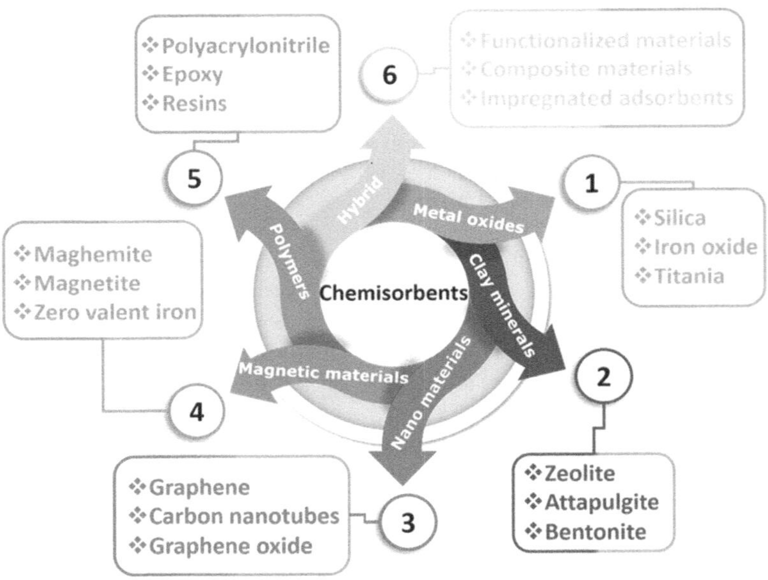

Figure 7.3 Classification of chemisorbents for the adsorptive recovery of REE based on their chemical structure.

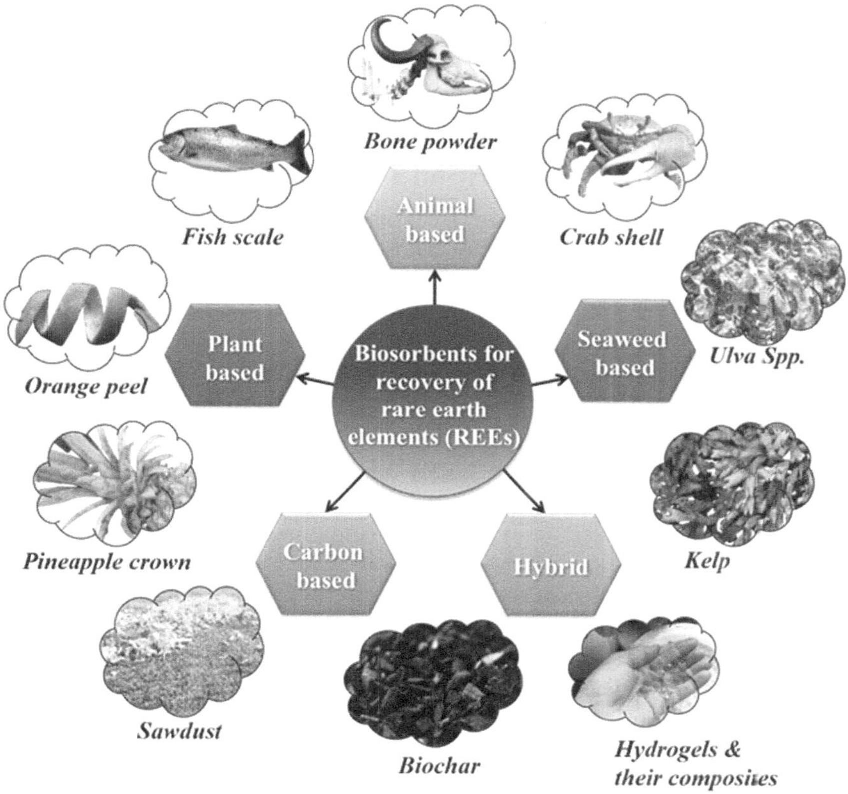

Figure 7.4 Classification of biosorbents for the adsorptive recovery of REEs based on their origin.

Chemical-based adsorbents such as clay minerals, polymers and magnetic materials are chemisorbents, and bio-based materials derived from plants, animals and seaweeds are classified as biosorbents. Both offer promising potential for the recovery of REEs (Keshtkar *et al.*, 2019).

Chemisorbents are synthetic adsorbents derived from organic and inorganic materials, whereas biosorbents are natural adsorbents originating from biological sources. Chemisorbents were the earlier one to hit the commercial market and, therefore, are well-known and widely used to adsorb and extract valuable resources (Beaver *et al.*, 2009). At the same time, biosorption acquired prevalence after the green technology initiatives and is therefore continuously researched and is gaining a stronghold in industrial operations. Due to technological advancements, nanomaterials in chemisorption and composite materials in biosorption are gaining more attention. This chapter discusses various chemisorbents and biosorbents used for the recovery of REEs. Finally, an account of desorption and future directions is provided that needs to be considered to promote the adsorptive recovery of REEs for industrial-scale applications.

7.2 REE REMOVAL BY CHEMISORBENTS

In the adsorptive removal of REEs, the adsorbents play a vital role in the recovery of REEs. Hence, it becomes pivotal to choose the right adsorbents to make the adsorption process an economically viable and efficient one. The properties of the adsorbent, like surface area and porosity, are the preliminary screening options for selection of an efficient adsorbent for REEs recovery. Some of the well-known chemisorbents or synthetic adsorbents are listed in Figure 7.3.

7.2.1 Silica based adsorbents

Silica is one of the successful and widely used chemisorbents employed for the adsorptive recovery of REEs. The large surface area, high porosity, long-term use and eco-friendly nature add an exceptional adsorption capacity to the silica (Gao *et al.*, 2017). Further, the adsorption capacity can be enhanced by imparting binding sites through constant modification and functionalization of silica using chemical agents. Especially, mesoporous silica (Santa Barbara Amorphous-15 or SBA-15) prepared under acidic conditions using a triblock copolymer surfactant as a template is widely used to remove and recover various REEs (Dolatyari *et al.*, 2016), including La, Sc, Y, Eu, Th, Sm, Dy, Gd and Lu. However removal of Eu and Th by silica has been repeatedly studied because these elements are homologous and act as representative elements for different radioactive elements like neptunium, uranium and plutonium (Dolatyari *et al.*, 2016). Studies on removing these radioactive elements are of prime importance as they are present in the depleted fuels refused by nuclear power plants. In that regard, SBA-15 functionalized through N-propyl salicylaldimine (SBA/SA) has exhibited an adsorption capacity of 4.7 and 18.5 mg/g for Eu and Th, respectively (Table 7.1).

Further SBA-15 functionalization through ethylenediamine propylesalicylaldimine (SBA/EnSA) has improved the adsorption capacity of Eu and Th to 15 and 81 mg/g, respectively. Lysine, one of the accessible and non-toxic biomolecules packed with electro attracting groups like carbonyl and amino groups, was used for SBA-15 functionalization (Ma *et al.*, 2014). Consequent application of this lysine-modified SBA-15 for selective adsorption of Sc from a mixture of rare earth elements resulted in a maximum adsorption capacity of 35.29 mg/g (Ma *et al.*, 2014). Gao *et al.* (2017) modified SBA-15 with phosphorous acid to attain ultrafast and high adsorption of Gd(III) ions. Indeed the results are fascinating, wherein the authors could achieve a maximum adsorption capacity of 1.3 mmol/g within 2 minutes. A composite material made of nano-silica and activated carbon was prepared by grafting polyacrylonitrile

Table 7.1 Predominant chemisorbents and their combination with biobased materials for adsorptive recovery of REEs.

Classification	REEs of Interest	Adsorption Capacity	Reference
Clay minerals			
Carboxylic acid functionalized diatomite	La	139.5 mg/g	Zhou *et al.* (2016)
Alumina-silica based composite nanomaterials	Dy Lu	125.44 mg/g 129.77 mg/g	Awual *et al.* (2017)
Silica with N-propyl salicylaldimine functionalization	Eu Th	4.7 mg/g 18.5 mg/g	Dolatyari *et al.* (2016)
Silica with Ethylenediaminepropyl salicylaldimine functionalization	Eu Th	15 mg/g 81 mg/g	
Mesoporous silicas (SBA-15) with phosphorous acid modification	Gd	1.3 mmol/g	Gao *et al.* (2017)
Bifunctional modification of BSA-15 with EDTA/phosphonic groups	Nd Eu	238.1 mg/g 243.9 mg/g	Dudarko *et al.* (2021)
Zr modified mesoporous silica SBA-15	Sm Dy	1.736 mg/g 1.889 mg/g	Aghayan *et al.* (2013)
Sodium rich heulandite (Na-HEU) zeolite	Eu Th	0.66 mequiv./g 0.62 mequiv./g	Sharma *et al.* (2013)
Lysine-functionalized mesoporous silica (SBA-15)	Sc	35.29 mg/g	Ma *et al.* (2014)
PAN functionalized -APTES nanosilica	La Sc Y	85.72 mg/g 75.5 mg/g 62.92 mg/g	Ramasamy *et al.* (2018)
Nanomaterials			
Oxidized MWCNTs	La Dy	99.01 mg/g 78.12 mg/g	Koochaki-Mohammadpour *et al.* (2014)
Polyethylenimine-cross-linked cellulose nanocrystals	La Eu Er	0.611 mmol/g 0.670 mmol/g 0.719 mmol/g	Zhao *et al.* (2017)

(*Continued*)

Table 7.1 Predominant chemisorbents and their combination with biobased materials for adsorptive recovery of REEs (*Continued*).

Classification	REEs of Interest	Adsorption Capacity	Reference
Graphene oxide magnetic nanoparticles	La Ce	1001 mg/g 982 mg/g	Ghobadi *et al.* (2018)
Polymer			
Poly(methylacrylate) grafted cellulose	La Ce Pr Gd Nd Eu Sm	260 mg/g 245 mg/g 235 mg/g 220 mg/g 210 mg/g 195 mg/g 192 mg/g	Rahman *et al.* (2017)
Polyacrylonitrile grafted onto algal biomass	Sc	66.81 mg/g	Ramasamy *et al.* (2019)
Polysulfone immobilized *Turbinaria conoides*	Tm	157.9 mg/g	Rangabhashiyam and Vijayaraghavan (2019)
Clay minerals			
Carboxylic acid functionalized diatomite	La	139.5 mg/g	Zhou *et al.* (2016)
Alumina-silica based composite nanomaterials	Dy Lu	125.44 mg/g 129.77 mg/g	Awual *et al.* (2017)
Silica with			
N-propyl salicylaldimine functionalization	Eu Th	4.7 mg/g 18.5 mg/g	Dolatyari *et al.* (2016)
Silica with			
Ethylenediaminepropyl salicylaldimine functionalization	Eu Th	15 mg/g 81 mg/g	
Mesoporous silicas (SBA-15) with phosphorous acid modification	Gd	1.3 mmol/g	Gao *et al.* (2017)
Bifunctional modification of BSA-15 with EDTA/phosphonic groups	Nd Eu	238.1 mg/g 243.9 mg/g	Dudarko *et al.* (2021)
Zr modified mesoporous silica SBA-15	Sm Dy	1.736 mg/g 1.889 mg/g	Aghayan *et al.* (2013)

(*Continued*)

Table 7.1 Predominant chemisorbents and their combination with biobased materials for adsorptive recovery of REEs (*Continued*).

Classification	REEs of Interest	Adsorption Capacity	Reference
Sodium rich heulandite (Na-HEU) zeolite	Eu Th	0.66 mequiv./g 0.62 mequiv./g	Sharma *et al.* (2013)
Lysine-functionalized mesoporous silica (SBA-15)	Sc	35.29 mg/g	Ma *et al.* (2014)
PAN functionalized -APTES nanosilica	La Sc Y	85.72 mg/g 75.5 mg/g 62.92 mg/g	Ramasamy *et al.* (2018)

(PAN), and the resulting adsorbent yielded a selective recovery of 75.5 mg/g for Sc, 85.72 mg/g for La and 62.92 mg/g for Y from a multicomponent solution having high salinity, water hardness and presence of oil and organic compounds (Ramasamy *et al.*, 2018). Dudarko *et al.* (2021) have recently imparted a bifunctional modification with EDTA/phosphonic groups onto BSA-15 to achieve a maximum adsorption capacity of 238.1 and 243.9 mg/g for Nd and Eu, respectively. A selective recovery of Th(IV) from a mixture of U(VI) and Ln(III) was achieved using phosphorodiamidate functionalized silica as the adsorbent (Zhang *et al.*, 2020).

7.2.2 Nanomaterials

Nanotechnology, with its continuously extending varieties of nanomaterials, offers a strong promise of novel adsorbents. Nanomaterials with one of its dimensions less than or equivalent to 100 nm offers several interesting properties, like high surface area and high adsorption efficiency (Ramasamy *et al.*, 2018). These intrinsic properties cause them to be a ready alternative as nanoadsorbents for resource retrieval and, more specifically, to recover REEs (Kegl *et al.*, 2020). Nanomaterials come with various geometries, including a sheet-like structure called nanosheets, a tube-like design called nanotubes or a particulate geometry referred to as nanoparticles. Like nanomaterials, magnetic materials were also continuously investigated in the adsorption process (Ashour *et al.*, 2017b). The use of magnetic materials is exciting on the grounds that the use of an external magnetic field guides the process of separation for its consecutive desorption and reuse. Thus, nanomaterials with magnetic properties are frequently studied for REEs adsorption (Kegl *et al.*, 2020).

Some of the well-known magnetic nanomaterials include magnetite, maghemite and zero-valent iron (Figure 7.3) (Kegl *et al.*, 2020). A commonly noticed fact is that magnetic nanomaterials exhibit relatively low adsorption efficiency for REEs. Thus, composite materials with nanoparticles and magnetic materials are formed to exploit the ease of separation associated with magnetic materials. Carbon nanotubes (CNTs), one of the superior nanomaterials discovered by Iijima – a Japanese physicist (Koochaki-Mohammadpour *et al.*, 2014), have opened the gateway for several interdisciplinary investigations, especially CNTs, for the adsorption of heavy metals. Over time, multiwalled carbon nanotubes (MWCNTs) have also been applied for the adsorption of REEs. For instance, oxidized MWCNTs displayed a maximum adsorption capacity of 99.01 and 78.12 mg/g for La and Dy, respectively (Table 7.1) (Koochaki-Mohammadpour *et al.*, 2014). Graphene oxide, a nanomaterial possessing sheet-like geometry, was co-synthesized with $MnFe_2O_4$ to obtain a magnetic nanomaterial exhibiting a high adsorption capacity of 1001 and 982 mg/g for La and Ce, respectively (Ghobadi *et al.*, 2018). As mentioned before in the functionalization of mesoporous silica with L-cystine, functionalization of magnetite nanoparticles was also carried out using different chemical agents like L-cystine and citric acid. To be specific, functionalization of magnetite materials with citric acid resulted in a maximum adsorption capacity of 32.5, 41, 52 and 35.8 mg/g for La, Nd, Gd and Y, respectively (Table 7.2) (Ashour *et al.*, 2017b). Likewise, functionalization of magnetite nanoparticles with L-cysteine yielded a maximum adsorption capacity of 57.2, 85.5, 98 and 73 mg/g for La, Nd, Gd and Y, respectively (Ashour *et al.*, 2017b). Over time, several other organic nanomaterials of biological and polymeric origin have also entered into the list of nanomaterials, and one such material is cellulose nanocrystals cross-linked with polyethyleneimine. This polyethyleneimine cross-linked cellulose nanocrystal exhibited a maximum adsorption capacity of 0.611, 0.670 and 0.719 mmol/g for La, Eu and Er, respectively (Zhao *et al.*, 2017). Other organic materials like calcium alginate beads and chitosan alginate beads loaded with magnetite were tested for the adsorption of La and depicted a maximum adsorption capacity of 123.5 and 97.1 mg/g, respectively (Wu *et al.*, 2010).

7.2.3 Surface modification

While functionalization and modification of the adsorbent surface are attractive, the prospective method of grafting polymers on organic materials and biomass is even more fascinating. This grafting of polymers involves the induction of particular monomers onto the polymer, which is then used for grafting onto other organic materials via covalent binding (Rahman *et al.*, 2017). For example, pure cellulose was grafted by induction of the methacrylate monomer to form a polymethylmethacrylate grafted cellulose biomass. This adsorbent was used to remove and concentrate a range of rare earth elements, including La, Ce, Pr, Gd, Nd, Eu and Sm, with a maximum adsorption capacity of 260, 245, 235, 220, 210, 195 and 192 mg/g, respectively (Rahman *et al.*, 2017). Likewise, polyacrylonitrile and polysulfone were grafted onto algal biomass and *Turbinaria Conoides* to

Table 7.2 Composite and magnetic materials along with their conjugation with biological materials for adsorptive recovery of REEs.

Classification	REEs of Interest	Adsorption Capacity	Reference
Composite materials			
Glutaraldehyde cross-linked chitosan with poly (aminocarboxymethylation)	Er	145 mg/g	Abd El-Magied *et al.* (2017)
Chitosan acryloylthiourea (CATU) derivative	La	2.1 mmol/g	Khalil *et al.* (2018)
Silica-based urea–formaldehyde composite	Eu Nd	3.1 meq/g 2.8 meq/g	Naser *et al.* (2015)
Cellulose-silica nanocomposite	La Sc	29.48 mg/g 23.76 mg/g	Iftekhar *et al.* (2017a)
Gum Arabic grafted polyacrylamide based silica nanocomposites	La Sc	7.90 mg/g 11.05 mg/g	Iftekhar *et al.* (2018b)
O-carboxymethyl chitosan entrapped by silica	Nd	53.04 mg/g	Wang *et al.* (2014)
EDTA-chitosan–silica	Nd	0.42 mmol/g	Roosen *et al.* (2014)
DTPA-chitosan–silica		0.74 mmol/g	
Magnetic materials			
Magnetic silica nanocomposites	La	55.9 mg/g	Wu *et al.* (2013)
Magnetite nanoparticles functionalized with citric acid	La Nd Gd Y	32.5 mg/g 41 mg/g 52 mg/g 35.8 mg/g	Ashour *et al.* (2017b)
Magnetite nanoparticles functionalized with L-cysteine	La Nd Gd Y	57.2 mg/g 85.5 mg/g 98 mg/g 73 mg/g	
Chitosan magnetic nano-based particles	Nd Dy Yb	51 mg/g 52 mg/g 52 mg/g	Galhoum *et al.* (2015)
Magnetite loaded calcium alginate beads	La	123.5 mg/g	Wu *et al.* (2010)
Magnetic alginate–chitosan gel beads	La	97.1 mg/g	Wu *et al.* (2011)
Carboxymethyl cellulose – modified magnetite composite	Eu	0.28 mmol/g	Cai *et al.* (2017)

adsorb and remove Sc and Th with a maximum adsorption capacity of 66.81 and 157.9 mg/g, respectively (Rangabhashiyam & Vijayaraghavan, 2019).

In addition to surface functionalization, modification and grafting of adsorbents, other techniques like cross-linking and impregnation can be used to prepare composite materials exhibiting superior properties. Composite materials are the combination of two or more constituents, often two polymers with different properties. These polymers are then blended to prepare composite materials with tailor-made properties beneficial for the adsorption process. Naser *et al.* (2015) used a solvent impregnation technique to impregnate organophosphorus extractant into the silica-based urea-formaldehyde composite material. In this solvent-impregnated composite material, the adsorbent acts as a selective ion exchanger for the adsorption of rare earth elements like Eu and Nd, exhibiting a maximum adsorption capacity of 3.1 and 2.8 meq/g, respectively. Only a few studies have reported the involvement of two or more surface modification techniques, like cross-linking, grafting or functionalizing, being used together to develop an adsorbent having a maximum adsorption efficiency. As an illustration, a composite material of chitosan was prepared by cross-linking chitosan with glutaraldehyde, and thereafter, the cross-linked chitosan was functionalized by poly(amino carboxymethylation) to result in an adsorbent having a maximum adsorption capacity of 145 mg/g for Er (Abd El-Magied *et al.*, 2017). Often, as discussed before, silica, one of the most exploited chemisorbents for the recovery of REEs, is blended with Arabic gum or chitosan or urea-formaldehyde to prepare composite materials. As an illustration, adsorption of La and Sc was carried out using cellulose-silica nanocomposite with a maximum adsorption capacity of 29.48 and 23.76 mg/g, respectively (Iftekhar *et al.*, 2017a). Similarly, gum Arabic grafted polyacrylamide based silica nanocomposites were used for the adsorption of La and Sc with a maximum adsorption capacity of 7.9 and 11.05 mg/g, respectively (Iftekhar *et al.*, 2018b).

7.3 BIOSORBENTS FOR THE RECOVERY OF REE

7.3.1 Advantages of biosorbents

Biosorbents are often considered as an eco-friendly alternative to chemisorbents. Further, the simplistic and budgetary solutions offered by the biosorbents make them a promising solution for the separation and concentration of rare earth elements from ores and waste solutions (Gupta *et al.*, 2019). In recent decades, enormous literature reports were gathered on developing various biosorbents, spanning both the energy and environmental fields. Biosorption refers to the recovery of adsorbate (more specifically rare earth elements) from a mixture of solution using living or dead microorganisms (Figure 7.4). While the recovery of rare earth elements using live microbes is elaborated in Chapter 8 of this book, this chapter strives to explore various types of dead biomass used for the biosorption of rare earth elements. Indeed, it is interesting to use dead biomass rather than active biomass, because biosorption using dead biomass allows working with a solution having a relatively higher (and potentially toxic) concentration of rare earth elements.

Further, the cost incurred due to supplementing various nutrient sources in active biomass is absent in biosorption with dead biomass (Arul Manikandan *et al.*, 2015). Other advantages of biosorption using dead biomass include utilization of waste materials as adsorbents in a closed circular approach. In addition, the options offered by the chemisorbents like magnetic separation and reusability potential can still be achieved in biosorbents by forming a hybrid material linking both biosorbents and chemisorbents (Giese, 2020). Unlike chemisorbents, biosorbents are often utilized without any surface modification such as functionalization or cross-linking. The inherent presence of various functional groups in biosorbents, including carboxyl, hydroxyl, amino and sulfhydryl groups, eliminates the need for other binding sites introduced through chemical modification techniques such as surface functionalization or cross-linking. Hence, unlike chemisorbents, biosorbents are often utilized without any surface modification. However, acid or alkali treatment of biomass is carried out occasionally to enhance the adsorption efficiency (Barcelos *et al.*, 2020).

7.3.2 Algae based biosorbents

Algae, both micro and macroalgae (Figure 7.4), are often considered suitable biomass for biosorption and have been repeatedly studied for their use in the adsorption of rare earth elements (Cao *et al.*, 2021). The use of algae as a biosorbent for removing REEs is not a recent technique, with biosorption using algae dating back to 1997. In spite of the interesting adsorption performance demonstrated by algae, research on algae as a biosorbent is still pursued with the goal of producing algal biosorbent exhibiting higher reusability potential for industrial applications. Brown algae like *Cystoseira indica* and *Trochomorpha conoides* have been tested as a biosorbent to remove rare earth elements (Cao *et al.*, 2021). For instance, *Cystoseira indica* was used for the adsorption of La and Ce and exhibited a maximum adsorption capacity of 185.44 and 172.33 mg/g, respectively (Table 7.3). Similarly, *Trochomorpha conoides* portrayed a maximum adsorption capacity of 154.7, 152.8, 138.2 and 121.2 mg/g for La, Ce, Eu and Yb, respectively. This adsorption capacity depicted by algae is much higher than the adsorption performance of most of the chemisorbents discussed in Section 7.2. The adsorption performance of algae can be improved by chemical modification techniques. For example, the remarkable adsorption performance of *Cystoseira indica* previously mentioned (maximum adsorption capacity of 185.44 and 172.33 mg/g for La and Ce, respectively) was achieved by xanthation of the biomass of this marine brown algae (Keshtkar *et al.*, 2019).

Table 7.3 Algal based biosorbents for adsorptive recovery of REEs.

Classification	REEs of Interest	Adsorption Capacity	Reference
Algae based			
Brown alga *Cystoseira indica*	La Ce	185.44 mg/g 172.33 mg/g	Cao *et al.* (2021)
Brown alga *Trochomorpha conoides*	La Ce Eu Yb	154.7 mg/g 152.8 mg/g 138.2 mg/g 121.2 mg/g	Cao *et al.* (2021)
Stichococcus bacillaris	La	51.02 mg/g	Birungi and Chirwa (2014)
Desmodesmus multivariabilis		100 mg/g	
Chlorella vulgaris		74.60 mg/g	
Scenedesmus acuminutus		111.1 mg/g	
Chloroidium saccharophilum		129.87 mg/g	
Chlamydomonas reinhardtii		142.86 mg/g	
Turbinaria conoides	Tm	200.5 mg/g	Rangabhashiyam and Vijayaraghavan (2019)
Chemically modified *Cystoseira indica*	La Ce	185.44 mg/g 172.33 mg/g	Keshtkar *et al.* (2019)

7.3.3 Agrowaste

7.3.3.1 Animal waste

Agro-based waste resulting from the agro-industries is often landfilled or improperly disposed of by ignoring the untapped potential lounging behind this enormous resource. Both plant and animal-based debris are categorized under agro-industrial waste. Often animal waste like animal bones, fish scales, crab and eggshells are studied for their application in adsorption (Abdollahi *et al.*, 2021). Such animal-based waste, such as fish scales and prawn carapace, has shown an unprecedented adsorption performance (Table 7.4). For instance, fish scales exhibited a maximum adsorption capacity of 200 and 250 mg/g for Ce and La, respectively (Abdollahi *et al.*, 2021). Likewise, a remarkable adsorption capacity of 1000 mg/g was demonstrated for the adsorption of Ce using prawn carapace. In addition to the excellent adsorption performance offered by this animal-based waste, they are also known for their high reuse potential. For instance, fish scales

Table 7.4 Animal- and plant-based biosorbents for adsorptive recovery of REEs.

Classification	REEs of Interest	Adsorption Capacity	Reference
Fish scales	Ce	200.0 mg/g	Abdollahi *et al.* (2021)
	La	250 mg/g	
Prawn carapace	Ce	1000 mg/g	
	La	200 mg/g	
Crab shell	Ce	90.9 mg/g	
	La	90.9 mg/g	
Crab shell	La	140.1 mg/g	Vijayaraghavan *et al.* (2009)
Egg shell	Ce	166.6 mg/g	Abdollahi *et al.* (2021)
	La	100 mg/g	
Neem saw dust	La	160.2 mg/g	Das *et al.* (2014)
Platanus orientalis leaf powder	La	28.65 mg/g	Sert *et al.* (2008)
	Ce	32.05 mg/g	
Pinus brutia leaf powder	La	22.94 mg/g	Kütahyali *et al.* (2010)
	Ce	17.24 mg/g	
Corn style	Ce	180.2 mg/g	Varsihini *et al.* (2014)
Malt spent rootlets	Eu	156 mg/g	Anagnostopoulos and Symeopoulos (2013)
Grapefruit peel	La	171.20 mg/g	Torab-Mostaedi *et al.* (2015)
	Ce	159.30 mg/g	
Orange peel	Pr	49.9 mg/g	Varshini *et al.* (2015)
Non-treated cactus fibers	Eu	0.16 mol/kg	Prodromou and Pashalidis (2016)
Phosphorylated cactus fibers		0.045 mol/kg	
MnO2-coated cactus fibers		0.46 mol/kg	
Thiourea functionalized cellulose	Eu	27 mg/g	Negrea *et al.* (2018)
	Nd	73 mg/g	

can be reused without any noticeable difference in the adsorption performance, even up to seven cycles of adsorption and desorption (Das *et al.*, 2014). The use of eggshells demonstrated a maximum adsorption capacity of 166.6 mg/g and 100 mg/g for Ce and La, respectively (Abdollahi *et al.*, 2021).

7.3.3.2 Plant-based waste

Like animal waste, plant waste is also a waste generated by agro-based industries. The abundant nature, low economic value and least processing requirement associated with the plant-based waste makes it an industrially viable feedstock for the adsorption of rare earth elements. A variety of plant-based wastes like sawdust, dried leaves, rootlets, plant fibers and fruit peels are studied as biosorbents to remove and recover various rare earth elements such as La, Ce, Eu, Pr and Nd (Das *et al.*, 2014; Kütahyali *et al.*, 2010; Torab-Mostaedi *et al.*, 2015). Specifically, the use of sawdust from neem, a herbaceous tree grown in the tropical and semi-tropical region, has demonstrated a maximum La adsorption capacity of 160.2 mg/g (Das *et al.*, 2014). Grapefruit peel, a typical waste generated from the fruit juice industry, contains several carboxylated groups which are beneficial to the adsorption process (Torab-Mostaedi *et al.*, 2015). Thus, the presence of numerous carboxylated groups in grapefruit peel resulted in a maximum adsorption capacity of 171.2 and 159.3 mg/g for La and Ce, respectively. Leaf powder from *Pinus brutia* demonstrated a maximum adsorption capacity of 22.94 and 17.24 mg/g for La and Ce, respectively (Kütahyali *et al.*, 2010). Further, the leaf powder of *Platanus orientalis* resulted in a maximum adsorption capacity of 28.65 and 32.05 mg/g for La and Ce, respectively (Sert *et al.*, 2008). Chemically treated and un-treated fibers from cactus, a spiny plant, were investigated for their Eu adsorption performance. A MnO_2 coating on the cactus fibers brought about an almost three-fold increase in the Eu adsorption on the biosorbent (Prodromou & Pashalidis, 2016).

7.3.4 Activated carbon

Carbon-based adsorbents are efficient for removing various organic and inorganic pollutants present in the aqueous solution. Moreover, carbon-based adsorbents derived from biomass are abundantly accessible, and the use of waste charcoal generated from the combustion process promotes upcycling of waste. For instance, it was reported that 50% of the industrially available activated carbon is sourced from materials of biological origin (Awwad *et al.*, 2010). Thus, abundant availability linked with the enormous adsorption efficiency led to activated carbon being the most applied adsorbent for several decades for treatment of industrial wastewater. Further, activated carbon is fascinating in its adsorption application as it possesses a vast surface area ranging from 500 to 1500 m^2/g and the well-developed microporous structure present in these adsorbents helps to remove pollutants from the wastewater (Yang *et al.*, 2007).

Research on finding a more suitable carbon-based adsorbent showing affinity to specific pollutants like REEs is still underway. Rice husk, one of the commodity wastes generated from agro-based industries, was used to prepare activated carbon (Awwad *et al.*, 2010). Using rice husk derived carbon is attractive because it tackles waste disposal while adding value to the commodity waste. Further, the chemical modification of this activated carbon often yielded a better adsorption capacity. For instance, phosphoric acid modified activated carbon from rice husk resulted in a maximum adsorption capacity of 175.4 and 250 mg/g for La and Eb, respectively (Table 7.5) (Awwad *et al.*, 2010). A much more exciting adsorption capacity of 350 mg/g for Sm was noticed with activated biochar derived from *Opuntia Ficus Indica* (Hadjittofi *et al.*, 2016). As in the case of silica-based chemisorbents, composite materials made of activated carbon and magnetite were prepared to exploit the advantage of ease of removal associated with magnetic materials. Recently, carbon black magnetite nanocomposites were used for the adsorption of REEs, and a remarkable adsorption capacity of 400, 400 and 384.6 mg/g was noticed for Ce, La and Nd, respectively (Abdollahi *et al.*, 2021).

7.3.5 Hydrogels

Hydrogels are three-dimensional cross-linked polymeric networks and are hydrophilic. High hydrophilicity, porosity, and ample availability of -COOH and -OH binding sites associated with these hydrogels make them an excellent adsorbent with high water holding capacity (Wahlström *et al.*, 2020). Hydrogels prepared from polysaccharides such as cellulose, chitosan, pectin and starch make them cost-effective and eco-friendly adsorbents for various organic and inorganic pollutants (Wahlström *et al.*, 2020). Hydrogels tested for REEs adsorption are often grafted with other polymers or blended with inorganic materials like silica. For instance, acrylic acid grafted onto hydroxypropyl cellulose with attapulgite clay exhibited a maximum adsorption capacity of 270 and 200 mg/g for La and Ce, respectively (Zhu *et al.*, 2015). Likewise, polyacrylic acid and silica-based hydrogel nanofibers exhibited a maximum adsorption capacity of 232.6, 268.8 and 250 mg/g for La, Eu and Tb, respectively (Wang *et al.*, 2016). Hydrogel prepared from carboxymethyl having a honeycomb structure with an open cellular network displayed a maximum adsorption capacity of 384.62 and 333.33 mg/g for La and Ce, respectively (Zhu *et al.*, 2015). Thus, porous hydrogel as an upcoming biosorbent promises a vast potential for the concentration and recovery of REEs from industrial waste.

Table 7.5 Carbon and hydrogel based biosorbents for adsorptive recovery of REEs.

Classification	REEs of Interest	Adsorption Capacity	Reference
Carbon based			
Chemically modified activated carbon from rice husk	La Eb	175.4 mg/g 250 mg/g	Awwad *et al.* (2010)
Active carbon	Eu	86 mg/g	Anagnostopoulos and Symeopoulos (2013)
Activated biochar from *Opuntia Ficus Indica*	Sm	350 mg/g	Hadjittofi *et al.* (2016)
Activated carbon obtained from apricot stone treated with phosphoric acid	Eu	46.5 mg/g	Gad and Awwad (2007)
Bamboo charcoal	La	120 mg/g	Chen (2010)
PAN functionalized Activated carbon-APTES nanosilica	La Sc Y	103.5 mg/g 112.74 mg/g 84.1 mg/g	Ramasamy *et al.* (2018)
Carbon black magnetite nanocomposites	Ce La Nd	400 mg/g 400 mg/g 384.6 mg/g	Abdollahi *et al.* (2021)
Hydrogel			
Acrylic acid grafted onto hydroxypropyl cellulose with attapulgite	La Ce	270 mg/g 200 mg/g	Zhu *et al.*, 2015
Carboxymethylcellulose based hydrogels	La Ce	384.62 mg/g 333.33 mg/g	Zhu *et al.* (2015)
Calcium alginate and γ-poly glutamic acid based gel	Nd	1.65 mmol/g	Wang *et al.* (2014)
Poly(vinyl pyrilidone) & Silica hydrogel	La Ce Nd Eu	116 mg/g 103 mg/g 92 mg/g 76 mg/g	Borai *et al.* (2015)
Poly(acrylic acid)–silica hydrogel nanofibers	La Eu Tb	232.6 mg/g 268.8 mg/g 250.0 mg/g	Wang *et al.* (2016)

7.4 DESORPTION FOR THE RECOVERY OF ADSORBED REE

Desorption followed by the adsorption of REEs plays a crucial role in defining the successful scale-up of the adsorption process. The sustainability and affordability of the adsorption process depend on the reuse of adsorbents with negligible loss in the adsorption efficiency. Hence, the eluents selected for the desorption of REEs from the adsorbents govern the success and mechanism of adsorption in the consecutive cycle. Various solvents like hydrochloric acid (HCl), nitric acid (HNO_3), sodium chloride (NaCl), calcium chloride ($CaCl_2$) and thiourea are used as an eluent for the desorption of REEs. Some of the critical criteria for selecting eluents include: effective in desorbing the REEs, eco-friendly and cost-effective for industrial-scale applications (Barcelos *et al.*, 2020).

While most of the studies conducted on the adsorptive recovery of REEs revolve around selecting the best suitable adsorbents and optimizing process parameters, only a few studies consider the final recovery of REEs through desorption. For instance, elution of REEs-loaded Co-Al layered double hydroxide with 0.1 M HCl resulted in a REEs recovery of more than 91%, 92% and 80%, even up to five adsorption and desorption cycles for La, Ce and Y, respectively (Iftekhar *et al.*, 2017b). Likewise, recovery of Ce from Ce-loaded kenaf cellulose using 2 M HCl resulted in more than 92.5% recovery even up to 10 adsorption and desorption cycles (Rahman *et al.*, 2017). Use of 0.5 M thiourea acidified with sulfuric acid as an eluent on biosorbent made of cellulose and chitosan derivatives resulted in more than 92% Er recovery even up to five cycles of adsorption and desorption (Abd El-Magied *et al.*, 2017). More than 90% recovery of Nd was reported from ethylenediaminetriacetic acid functionalized activated carbon using 1 M HCl as the eluent, even up to five cycles of adsorption and desorption (Babu *et al.*, 2018). A few studies reported 99% recovery of REEs, but in these only a single cycle of adsorption and desorption was studied. For example, 99% recovery of La and Nd from graphene oxide nanosheets was reported using 0.1 M HNO_3 as eluent, but as mentioned, only a single cycle of reuse was considered (Ashour *et al.*, 2017a). Therefore, recently developed materials like layered double hydroxide and plant-based cellulose showed better reuse potential than that of the industrially available activated carbon or silica-based adsorbents. However, future studies on adsorptive recovery demand equal importance to be given to desorption as to adsorption.

7.5 FUTURE PERSPECTIVE

This chapter highlighted the potential of adsorptive recovery of REEs. A couple of simplicities offered by chemisorbents, such as simple magnetic separation and surface modification for selective recovery of REEs, suggest adsorption to be a fascinating technique having industrial relevance. Further, the remarkable adsorption capacity exhibited by several biosorbents along with their reuse potential makes adsorption an intriguing technique that can be applied at industrial level. Development of tailor-made composite materials or hybrid adsorbents unifying the properties of chemisorbents and biosorbents has gained a great deal of attention from both academicians and industrialists. However, to realize the actual potential of the adsorption process, pilot-plant studies and industrial level scale-up are inevitable (Iftekhar *et al.*, 2018a). Though industrial scale units for adsorption in petroleum industries have been established and are fully operational, no studies on pilot-scale processes for the adsorptive recovery of REEs have been available until now. The studies on adsorptive recovery of REEs are mostly limited to batch studies, and thus dynamic studies on packed columns are required. Further, such continuous studies on packed bed columns should focus on developing mathematical and kinetic models, which will help the effluent treatment plant operator to run the process without any major perturbation.

Till now, supposedly, only a handful of studies reported on the adsorption of REEs from real effluents. These studies deal with the recovery of lanthanum from ceramic industrial effluent using powdered fish scales as the adsorbent (Das *et al.*, 2014). Most of the studies on REEs adsorption focus on a mono or multi-element system comprising very few rare earth elements. Thus, studies on selective adsorption of REEs from real industrial effluents are essential to evaluate the actual performance of adsorption. Since adsorption is an unsteady state operation (Chen *et al.*, 1968), its performance in a large scale set-up and techno-economic assessment is necessary to evaluate the actual potential of this recovery technique for commercial REEs adsorption.

7.6 CONCLUSION

Increasing supply chain risk for REEs with rare earth mines scarcely available worldwide demands the recovery and reuse of REEs. The deleterious effect of REEs on the environment further catalyzes the need for the recovery of REEs from discarded materials like batteries, magnets and several other electronic goods. As a solution to these pressing issues, adsorptive recovery of REEs presents an excellent potential for large-scale applications. While chemisorbents like graphene oxide and layered double hydroxide promise good adsorption performance with relatively high reuse potential, the fate of these adsorbents with risk for secondary pollution in the environment limits them from acquiring commercial potential. Biosorbents like kenaf cellulose and prawn carapace are promising, as they nullify the risk of secondary pollution and promote upcycling of the waste generated from agro-based industries. Adsorption and desorption of REEs using biosorbents help in establishing a circular bioeconomy with minimal waste generation. More emphasis on column studies using real effluents and kinetic and mathematical modelling are required to realize the true potential of adsorptive recovery of REEs.

REFERENCES

Abd El-Magied M. O., Galhoum A. A., Atia A. A., Tolba A. A., Maize M. S., Vincent T. and Guibal E. (2017). Cellulose and chitosan derivatives for enhanced sorption of erbium (III). *Colloids and Surfaces A: Physicochemical and Engineering Aspects*, **529**, 580–593.

Abdollahi H., Maleki S., Sayahi H., Gharabaghi M., Darvanjooghi M. H. K., Magdouli S. and Brar S. K. (2021). Superadsorbent Fe_3O_4-coated carbon black nanocomposite for separation of light rare earth elements from aqueous solution: GMDH-based Neural Network and sensitivity analysis. *Journal of Hazardous Materials*, **416**, 125655.

Aghayan H., Mahjoub A. R. and Khanchi A. R. (2013). Samarium and dysprosium removal using 11-molybdo-vanadophosphoric acid supported on Zr modified mesoporous silica SBA-15. *Chemical Engineering Journal*, **225**, 509–519.

Anagnostopoulos V. A. and Symeopoulos B. D. (2013). Sorption of europium by malt spent rootlets, a low cost biosorbent: effect of pH, kinetics and equilibrium studies. *Journal of Radioanalytical and Nuclear Chemistry*, **295**, 7–13.

Arul Manikandan N., Alemu A. K., Goswami L., Pakshirajan K. and Pugazhenthi G. (2016). Waste litchi peels for Cr(VI) removal from synthetic wastewater in batch and continuous systems: sorbent characterization, regeneration and reuse study. *Journal of Environmental Engineering*, **142**(9), C4016001.

Ashour R. M., Abdelhamid H. N., Abdel-Magied A. F., Abdel-Khalek A. A., Ali M. M., Uheida A., Muhammed M., Zou X. and Dutta J. (2017a). Rare earth ions adsorption onto graphene oxide nanosheets. *Solvent Extraction and Ion Exchange*, **35**, 91–103.

Ashour R. M., El-sayed R., Abdel-Magied A. F., Abdel-khalek A. A., Ali M. M., Forsberg K., Uheida A., Muhammed M. and Dutta J. (2017b). Selective separation of rare earth ions from aqueous solution using functionalized magnetite nanoparticles: kinetic and thermodynamic studies. *Chemical Engineering Journal*, **327**, 286–296.

Awual M. R., Alharthi N. H., Okamoto Y., Karim M. R., Halim M. E., Hasan M. M., Rahman M. M., Islam M. M., Khaleque M. A. and Sheikh M. C. (2017). Ligand field effect for Dysprosium(III) and Lutetium(III) adsorption and EXAFS coordination with novel composite nanomaterials. *Chemical Engineering Journal*, **320**, 427–435.

Awwad N. S., Gad H. M. H., Ahmad M. I. and Aly H. F. (2010). Sorption of lanthanum and erbium from aqueous solution by activated carbon prepared from rice husk. *Colloids and Surfaces B: Biointerfaces*, **81**, 593–599.

Babu C. M., Binnemans K. and Roosen J. (2018). Ethylenediaminetriacetic Acid-Functionalized activated carbon for the adsorption of rare earths from aqueous solutions. *Industrial & Engineering Chemistry Research*, **57**, 1487–1497.

Ballinger B., Schmeda-Lopez D., Kefford B., Parkinson B., Stringer M., Greig C. and Smart S. (2020). The vulnerability of electric-vehicle and wind-turbine supply chains to the supply of rare-earth elements in a 2-degree scenario. *Sustainable Production and Consumption*, **22**, 68–76.

Beaver M. G., Caram H. S. and Sircar S. (2009). Selection of CO_2 chemisorbent for fuel-cell grade H_2 production by sorption-enhanced water gas shift reaction. *International Journal of Hydrogen Energy*, **34**, 2972–2978.

Birungi Z. S. and Chirwa E. M. N. (2014). The kinetics of uptake and recovery of lanthanum using freshwater algae as biosorbents: comparative analysis. *Bioresource Technology*, **160**, 43–51.

Borai E. H., Hamed M. G., El-kamash A. M., Siyam T. and El-Sayed G. O. (2015). Template polymerization synthesis of hydrogel and silica composite for sorption of some rare earth elements. *Journal of Colloid and Interface Science*, **456**, 228–240.

Cai Y., Yuan F., Wang X., Sun Z., Chen Y., Liu Z., Wang X., Yang S. and Wang S. (2017). Synthesis of core–shell structured Fe_3O_4@carboxymethyl cellulose magnetic composite for highly efficient removal of Eu(III). *Cellulose*, **24**, 175–190.

Cao Y., Shao P., Chen Y., Zhou X., Yang L., Shi H., Yu K., Luo X. and Luo X. (2021). A critical review of the recovery of rare earth elements from wastewater by algae for resources recycling technologies. *Resources, Conservation and Recycling*, **169**, 105519.

Chen Q. (2010). Study on the adsorption of lanthanum(III) from aqueous solution by bamboo charcoal. *Journal of Rare Earths*, **28**, 125–131.

Chen J. W., Buege J. A., Cunningham F. L. and Northam J. I. (1968). Scale-up of a column adsorption process by computer simulation. *Industrial & Engineering Chemistry Process Design and Development*, **7**, 26–31.

da Costa T. B., da Silva M. G. C. and Vieira M. G. A. (2020). Recovery of rare-earth metals from aqueous solutions by bio/adsorption using non-conventional materials: a review with recent studies and promising approaches in column applications. *Journal of Rare Earths*, **38**, 339–355.

Das D., Varshini C. J. S. and Das N. (2014). Recovery of lanthanum(III) from aqueous solution using biosorbents of plant and animal origin: batch and column studies. *Minerals Engineering*, **69**, 40–56.

Dolatyari L., Yaftian M. R. and Rostamnia S. (2016). Adsorption characteristics of Eu(III) and Th(IV) ions onto modified mesoporous silica SBA-15 materials. *Journal of the Taiwan Institute of Chemical Engineers*, **60**, 174–184.

Dudarko O., Kobylinska N., Mishra B., Kessler V. G., Tripathi B. P. and Seisenbaeva G. A. (2021). Facile strategies for synthesis of functionalized mesoporous silicas for the removal of rare-earth elements and heavy metals from aqueous systems. *Microporous and Mesoporous Materials*, **315**, 110919.

Dushyantha N., Batapola N., Ilankoon I. M. S. K., Rohitha S., Premasiri R., Abeysinghe B., Ratnayake N. and Dissanayake K. (2020). The story of rare earth elements (REEs): occurrences, global distribution, genesis, geology, mineralogy and global production. *Ore Geology Reviews*, **122**, 103521.

Gad H. M. H. and Awwad N. S. (2007). Factors affecting on the sorption/desorption of Eu (III) using activated carbon. *Separation Science and Technology*, **42**, 3657–3680.

Galhoum A. A., Mahfouz M. G., Abdel-Rehem S. T., Gomaa N. A., Atia A. A., Vincent T. and Guibal E. (2015). Diethylenetriamine-functionalized chitosan magnetic nano-based particles for the sorption of rare earth metal ions [Nd(III), Dy(III) and Yb(III)]. *Cellulose*, **22**, 2589–2605.

Gao Q., Xie J. F., Shao Y. T., Chen C., Han B., Xia K. S. and Zhou C. G. (2017). Ultrafast and high-capacity adsorption of Gd(III) onto inorganic phosphorous acid modified mesoporous SBA-15. *Chemical Engineering Journal*, **313**, 197–206.

Ghobadi M., Gharabaghi M., Abdollahi H., Boroumand Z. and Moradian M. (2018). MnFe2O4-graphene oxide magnetic nanoparticles as a high-performance adsorbent for rare earth elements: synthesis, isotherms, kinetics, thermodynamics and desorption. *Journal of Hazardous Materials*, **351**, 308–316.

Giese E. C. (2020). Biosorption as green technology for the recovery and separation of rare earth elements. *World Journal of Microbiology and Biotechnology*, **36**, 1–11.

Gupta N. K., Gupta A., Ramteke P., Sahoo H. and Sengupta A. (2019). Biosorption-a green method for the preconcentration of rare earth elements (REEs) from waste solutions: a review. *Journal of Molecular Liquids*, **274**, 148–164.

Hadjittofi L., Charalambous S. and Pashalidis I. (2016). Removal of trivalent samarium from aqueous solutions by activated biochar derived from cactus fibres. *Journal of Rare Earths*, **34**, 99–104.

Huang X., Dong J., Wang L., Feng Z., Xue Q. and Meng X. (2017). Selective recovery of rare earth elements from ion-adsorption rare earth element ores by stepwise extraction with HEH(EHP) and HDEHP. *Green Chemistry*, **19**, 1345–1352.

Huskič I., Arhangelskis M. and Friščič T. (2020). Solvent-free ageing reactions of rare earth element oxides: from geomimetic synthesis of new metal-organic materials towards a simple, environmentally friendly separation of scandium. *Green Chemistry*, **22**, 4364–4375.

Iftekhar S., Srivastava V. and Sillanpää M. (2017a). Enrichment of lanthanides in aqueous system by cellulose based silica nanocomposite. *Chemical Engineering Journal*, **320**, 151–159.

Iftekhar S., Srivastava V. and Sillanpää M. (2017b). Synthesis and application of LDH intercalated cellulose nanocomposite for separation of rare earth elements (REEs). *Chemical Engineering Journal*, **309**, 130–139.

Iftekhar S., Ramasamy D. L., Srivastava V., Asif M. B. and Sillanpää M. (2018a). Understanding the factors affecting the adsorption of Lanthanum using different adsorbents: a critical review. *Chemosphere*, **204**, 413–430.

Iftekhar S., Srivastava V., Ramasamy D. L., Naseer W. A. and Sillanpää M. (2018b). A novel approach for synthesis of exfoliated biopolymeric-LDH hybrid nanocomposites via in-stiu coprecipitation with gum Arabic: application towards REEs recovery. *Chemical Engineering Journal*, **347**, 398–406.

Iftekhar S., Srivastava V. and Sillanpää M. (2020). Synthesis of hybrid bionanocomposites and their application for the removal of rare-earth elements from synthetic wastewater. In: Advanced Water Treatment Adsorption, Mika Sillanpaa (ed.), United States, Elsevier, pp. 505–564.

Işıldar A., van Hullebusch E. D., Lenz M., Du Laing G., Marra A., Cesaro A., Panda S., Akcil A., Kucuker M. A. and Kuchta K. (2019). Biotechnological strategies for the recovery of valuable and critical raw materials from waste electrical and electronic equipment (WEEE) – a review. *Journal of Hazardous Materials*, **362**, 467–481.

Islam A., Ahmed T., Awual M. R., Rahman A., Sultana M., Aziz A. A., Monir M. U., Teo S. H. and Hasan M. (2020). Advances in sustainable approaches to recover metals from e-waste-a review. *Journal of Cleaner Production*, **244**, 118815.

Kegl T., Košak A., Lobnik A., Novak Z., Kralj A. K. and Ban I. (2020). Adsorption of rare earth metals from wastewater by nanomaterials: a review. *Journal of Hazardous Materials*, **386**, 121632.

Keilhacker M. L. and Minner S. (2017). Supply chain risk management for critical commodities: a system dynamics model for the case of the rare earth elements. *Resources, Conservation & Recycling*, **125**, 349–362.

Keshtkar A. R., Moosavian M. A., Sohbatzadeh H. and Mofras M. (2019). La(III) and Ce(III) biosorption on sulfur functionalized marine brown algae *Cystoseira indica* by xanthation method: response surface methodology, isotherm and kinetic study. *Groundwater for Sustainable Development*, **8**, 144–155.

Khalil M. M. H., Atrees M. S., Abd El Fatah A. I. L., Salem H. and Roshdi R. (2018). Synthesis and application studies of chitosan acryloylthiourea derivative for the separation of rare earth elements. *Journal of Dispersion Science and Technology*, **39**, 605–613.

Koochaki-Mohammadpour S. M. A., Torab-Mostaedi M., Talebizadeh-Rafsanjani A. and Naderi-Behdani F. (2014). Adsorption isotherm, kinetic, thermodynamic, and desorption studies of lanthanum and dysprosium on oxidized multiwalled carbon nanotubes. *Journal of Dispersion Science and Technology*, **35**, 244–254.

Kütahyali C., Şert S., Çetinkaya B., Inan S. and Eral M. (2010). Factors affecting lanthanum and cerium biosorption on *Pinus brutia* leaf powder. *Separation Science and Technology*, **45**, 1456–1462.

Ma J., Wang Z., Shi Y. and Li Q. (2014). Synthesis and characterization of lysine-modified SBA-15 and its selective adsorption of scandium from a solution of rare earth elements. *RSC Advances*, **40**, 41597–41604.

Manikandan N. A., Pakshirajan K. and Syiem M. B. (2015). Cu(II) removal by biosorption using chemically modified biomass of *Nostoc muscorum* – a cyanobacterium isolated from a coal mining site. *International Journal of ChemTech Research*, **7**, 80–92.

Naser A. A., El-deen G. E. S., Bhran A. A., Metwally S. S. and El-Kamash A. M. (2015). Elaboration of impregnated composite for sorption of europium and neodymium ions from aqueous solutions. *Journal of Industrial and Engineering Chemistry*, **32**, 264–272.

Negrea A., Gabor A., Davidescu C. M., Ciopec M., Negrea P., Duteanu N. and Barbulescu A. (2018). Rare earth elements removal from water using natural polymers. *Scientific Reports*, **8**, 1–11.

Pagano G., Thomas P. J., Di Nunzio A. and Trifuoggi M. (2019). Human exposures to rare earth elements: present knowledge and research prospects. *Environmental Research*, **171**, 493–500.

Paulick H. and Machacek E. (2017). The global rare earth element exploration boom: an analysis of resources outside of China and discussion of development perspectives. *Resources Policy*, **52**, 134–153.

Pinto J., Henriques B., Soares J., Costa M., Dias M., Fabre E., Lopes C. B., Vale C., Pinheiro-Torres J. and Pereira E. (2020). A green method based on living macroalgae for the removal of rare-earth elements from contaminated waters. *Journal of Environmental Management*, **263**, 110376.

Ponou J., Wang L. P., Dodbiba G., Okaya K., Fujita T., Mitsuhashi K., Atarashi T., Satoh G. and Noda M. (2014). Recovery of rare earth elements from aqueous solution obtained from Vietnamese clay minerals using dried and carbonized parachlorella. *Journal of Environmental Chemical Engineering*, **2**, 1070–1081.

Prodromou M. and Pashalidis I. (2016). Europium adsorption by non-treated and chemically modified opuntia ficus indica cactus fibres in aqueous solutions. *Desalination and Water Treatment*, **57**, 5079–5088.

Rahman M. L., Biswas T. K., Sarkar S. M., Yusoff M. M., Sarjadi M. S., Arshad S. E. and Musta B. (2017). Adsorption of rare earth metals from water using a kenaf cellulose-based poly(hydroxamic acid) ligand. *Journal of Molecular Liquids*, **243**, 616–623.

Ramasamy D. L., Puhakka V., Repo E., Ben Hammouda S. and Sillanpää M. (2018). Two-stage selective recovery process of scandium from the group of rare earth elements in aqueous systems using activated carbon and silica composites: dual applications by tailoring the ligand grafting approach. *Chemical Engineering Journal*, **341**, 351–360.

Ramasamy D. L., Porada S. and Sillanpää M. (2019). Marine algae: a promising resource for the selective recovery of scandium and rare earth elements from aqueous systems. *Chemical Engineering Journal*, **371**, 759–768.

Rangabhashiyam S. and Vijayaraghavan K. (2019). Biosorption of Tm(III) by free and polysulfone-immobilized *Turbinaria conoides* biomass. *Journal of Industrial and Engineering Chemistry*, **80**, 318–324.

Rasoulnia P., Barthen R. and Lakaniemi A. M. (2021). A critical review of bioleaching of rare earth elements: the mechanisms and effect of process parameters. *Critical Reviews in Environmental Science and Technology*, **51**, 378–427.

Roosen J., Spooren J. and Binnemans K. (2014). Adsorption performance of functionalized chitosan-silica hybrid materials toward rare earths. *Journal of Materials Chemistry A*, **2**, 19415–19426.

Sert Ş., Kütahyali C., Inan S., Talip Z., Çetinkaya B. and Eral M. (2008). Biosorption of lanthanum and cerium from aqueous solutions by *Platanus orientalis* leaf powder. *Hydrometallurgy*, **90**, 13–18.

Sharma P., Sharma M. and Tomar R. (2013). Na-HEU zeolite synthesis for the removal of Th (IV) and Eu(III) from aqueous waste by batch process. *Journal of the Taiwan Institute of Chemical Engineers*, **44**, 480–488.

Stegen K. S. (2015). Heavy rare earths, permanent magnets, and renewable energies: an imminent crisis. *Energy Policy*, **79**, 1–8.

Tavares L. Z., Da Silva E. S. and da Cruz Pradella J. G. (2004). Production of poly (3-hydroxybutyrate) in an airlift bioreactor by *Ralstonia eutropha*. *Biochemical Engineering Journal*, **18**, 21–31.

Torab-Mostaedi M., Asadollahzadeh M., Hemmati A. and Khosravi A. (2015). Biosorption of lanthanum and cerium from aqueous solutions by grapefruit peel: equilibrium, kinetic and thermodynamic studies. *Research on Chemical Intermediates*, **41**, 559–573.

Tyler G. (2004). Rare earth elements in soil and plant systems – a review. *Plant and Soil*, **267**, 191–206.

Varsihini C. J. S., Das D. and Das N. (2014). Optimization of parameters for cerium(III) biosorption onto biowaste materials of animal and plant origin using 5-level Box-Behnken design: equilibrium, kinetic, thermodynamic and regeneration studies. *Journal of Rare Earths*, **32**, 745–758.

Varshini C. J. S., Das D. and Das N. (2015). Optimization of parameters for praseodymium (III) biosorption onto biowaste materials using response surface methodology: equilibrium, kinetic and regeneration studies. *Ecological Engineering*, **81**, 321–327.

Vijayaraghavan K., Mahadevan A., Joshi U. M. and Balasubramanian R. (2009). An examination of the uptake of lanthanum from aqueous solution by crab shell particles. *Chemical Engineering Journal*, **152**, 116–121.

Vukojevič V., Đurđič S., Stefanovič V., Trifkovič J., Čakmak D., Perovič V. and Mutič J. (2019). Scandium, yttrium, and lanthanide contents in soil from Serbia and their accumulation in the mushroom *Macrolepiota procera* (Scop.) Singer. *Environmental Science and Pollution Research*, **26**, 5422–5434.

Wahlström N., Steinhagen S., Toth G., Pavia H. and Edlund U. (2020). Ulvan dialdehyde-gelatin hydrogels for removal of heavy metals and methylene blue from aqueous solution. *Carbohydrate Polymers*, **249**, 116841.

Wang F., Zhao J., Wei X., Huo F., Li W., Hu Q. and Liu H. (2014). Adsorption of rare earths (III) by calcium alginate-poly glutamic acid hybrid gels. *Journal of Chemical Technology & Biotechnology*, **89**, 969–977.

Wang M., Li X., Hua W., Shen L., Yu X. and Wang X. (2016). Electrospun poly(acrylic acid)/silica hydrogel nanofibers scaffold for highly efficient adsorption of lanthanide ions and its photoluminescence performance. *ACS Applied Materials & Interfaces*, **8**, 23995–24007.

Wu D., Zhao J., Zhang L., Wu Q. and Yang Y. (2010). Lanthanum adsorption using iron oxide loaded calcium alginate beads. *Hydrometallurgy*, **101**, 76–83.

Wu D., Zhang L., Wang L., Zhu B. and Fan L. (2011). Adsorption of lanthanum by magnetic alginate-chitosan gel beads. *Journal of Chemical Technology & Biotechnology*, **86**, 345–352.

Wu D., Sun Y. and Wang Q. (2013). Adsorption of lanthanum (III) from aqueous solution using 2-ethylhexyl phosphonic acid mono-2-ethylhexyl ester-grafted magnetic silica nanocomposites. *Journal of Hazardous Materials*, **260**, 409–419.

Yang C., Kheireddine M., Mohd W. and Wan A. (2007). Review of modifications of activated carbon for enhancing contaminant uptakes from aqueous solutions. *Separation and Purification Technology*, **52**(3), 403–415.

Zhang F., Ma K. Q., Li Y., Ran Q., Yao C. Y., Yang C. T., Yu H. Z., Hu S. and Peng S. M. (2020). Selective separation of thorium from rare earths and uranium in acidic solutions by phosphorodiamidate-functionalized silica. *Chemical Engineering Journal*, **392**, 123717.

Zhao F., Repo E., Song Y., Yin D., Hammouda S. B., Chen L., Kalliola S., Tang J., Tam K. C. and Sillanpää M. (2017). Polyethylenimine-cross-linked cellulose nanocrystals for highly efficient recovery of rare earth elements from water and a mechanism study. *Green Chemistry*, **19**, 4816–4828.

Zhou Q., Yang H., Yan C., Luo W., Li X. and Zhao J. (2016). Synthesis of carboxylic acid functionalized diatomite with a micro-villous surface via UV-induced graft polymerization and its adsorption properties for Lanthanum(III) ions. *Colloids and Surfaces A: Physicochemical and Engineering Aspects*, **501**, 9–16.

Zhu Y., Zheng Y. and Wang A. (2015). A simple approach to fabricate granular adsorbent for adsorption of rare elements. *International Journal of Biological Macromolecules*, **72**, 410–420.

Chapter 8

Microbial recovery of rare earth elements

J. Castillo, M. Maleke, J. Unuofin, S. Cebekhulu and A. Gómez-Arias

8.1 INTRODUCTION

Microbial-metal interactions for the recovery of precious metals have been studied extensively in the last decade thanks to interdisciplinary research areas such as geomicrobiology and biogeochemistry that have allowed a better understanding of metal-microbe processes such as bioleaching, biosorption, bioaccumulation, biomineralization, and bioreduction (Andrès *et al.*, 2003; Kumar *et al.*, 2019; Lumpe *et al.*, 2018; Moriwaki & Yamamoto, 2013). The interaction between microorganisms and rare earth elements (REE) has not been an exception: the scarcity of economic REE deposits boosted the search for alternative technologies, such as biohydrometallurgy, to extract REE from primary deposits as well as from secondary resources. Secondary REE-resource investigations are focused on industrial waste such as phosphogypsum stacks (fertilizer industry), red mud (aluminum industry), coal ash (thermal power plants), E-waste, wastewater streams and mining waste (slags, tailings and rock dumps) (Humsa & Srivastava, 2015; Jowitt *et al.*, 2018; Zhang *et al.*, 2014). For instance, according to the United States Geological Survey (USGS), there are 652 Mt of tailings in the Phalaborwa Complex (South Africa) that contain 1.5 kg of REE per ton (Orris & Grauch, 2002). According to Apple, the potential recovery of REE from iPhones is 0.8 kg per ton (Balaram, 2019).

doi: 10.2166/9781789062236_0179

Bioleaching has been broadly utilized in the recovery of metals, such as Zn, Ni, Pb, Cu and Au, amongst others, from sulfidic low-grade ores, industrial solid wastes and printed circuit boards (Ayangbenro *et al.*, 2018; Barmettler *et al.*, 2016; Mowafy, 2020; Yavari *et al.*, 2019). This biological process has also been used with varying success to leach REE from mine waste and E-waste (Barmettler *et al.*, 2016; Jin *et al.*, 2019; Rasoulnia *et al.*, 2020; Reed *et al.*, 2016). The efficiency of bioleaching is closely associated with the mineralogical structure of the waste (Park & Liang, 2019; Somasundaran *et al.*, 2018). Therefore, the selection of a suitable microorganism relies on the type of REE-minerals present in the waste. For example, the extraction of Sc from iron-bearing minerals (i.e., sulfide or oxides) is efficiently executed by siderophore-producing chemoautotrophic bacteria, whereas phosphate-rich and carbonate minerals are dissolved by organic acids secreted from chemoheterotrophic bacteria (Dev *et al.*, 2020; Kazak *et al.*, 2018; Keekan *et al.*, 2017; Reed *et al.*, 2016).

Once REE are in solution, their immobilization is induced by various microbial species that can efficiently perform biosorption processes comparable to the binding capacities of commercial ion exchangers (Moriwaki & Yamamoto, 2013; Vijayaraghavan & Yun, 2008). The microbial sorption capacity of metals is approximately in the order of 10^{-5} to 10^{-3} mol metal/g (dry wt) microbe (Gadd, 2004). The binding capacity of these elements is related to several physicochemical parameters (e.g., pH, ionic strength and temperature), which determine the biochemical strategy to extract those REE from the solution. The REE-microbe interaction modifies the parameters of the leachate by organic ligand production, which promotes the sorption of those elements (Ozaki *et al.*, 2006).

In general, the drawback with biological processes such as bioleaching, biosorption, biomineralization, bioaccumulation and bioreduction is their lack of selectivity. However, recent studies have indicated that there is a dependency of a novel thermoacidophilic methanotroph (*Methylacidiphilum fumariolicum* SolV) on lanthanides (Cotruvo, 2019; Ramos *et al.*, 2016). This dependency has led to the discovery of selective binding proteins to REE such as pyrroloquinoline quinone (PQQ)-dependent methanol dehydrogenases (MDH) (Cotruvo, 2019; Lumpe *et al.*, 2018) and Lanmodulin (LanM) (Deblonde *et al.*, 2020), as well as suggesting the presence of a specific chelator named 'lanthanophore' produced by *M. fumariolicum* that assists in the REE uptake (Daumann, 2019). These studies have shed light on the mobilization and the role that REE play in the metabolism of extremophiles, thereby demonstrating that REE are not inert elements. At the same time, these findings open room for developing technologies for REE extraction using highly selective proteins. This chapter describes the role of the microorganisms in rare earth biometallurgy from the bioleaching process to the selective recovery of REE, including the limitations and the environmental implications of the technology.

8.2 MICROBIAL RECOVERY OF RARE EARTH ELEMENTS

8.2.1 Bioleaching

The initial step in REE recovery from mine waste and even industrial waste (including E-waste) starts with the release of metals into solution. Remarkably, microbes can solubilize REE through bioleaching. The bioleaching process involves the direct or indirect microbial mobilization of metals to a soluble state (Jeremic *et al.*, 2016; Qu & Lian, 2013; Shin *et al.*, 2015).

Direct bioleaching of REE might take place because of the scarcity of essential elements in the solution. Microbes selectively solubilize essential metals, such as Zn^{2+}, Mn^{2+}, Fe^{2+} and Cu^{2+} by producing complexing agents named siderophores (Figure 8.1) (Rajkumar *et al.*, 2010). Likewise, REE have been shown to be essential for other microorganisms, therefore, the solubilization of REE-bearing minerals through siderophore production might be a form of a survival mechanism (Cotruvo, 2019). For instance, *Micrococcus* sp. and *Streptomyces* sp. can directly promote the bioleaching of REE through the action

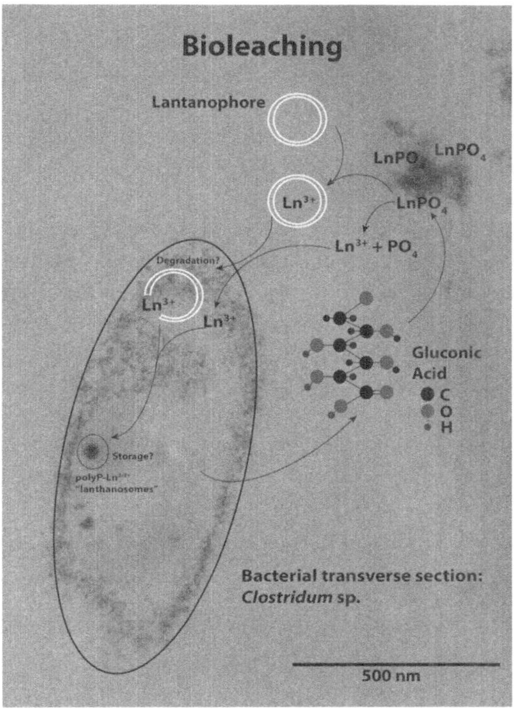

Figure 8.1 Schematic representation of direct bioleaching of rare earth elements from $LnPO_4$ using a *Clostridium* sp. based on a scanning electron microscope (SEM) image obtained from an REE biorecovery experiment.

of siderophores (Christenson & Schijf, 2011; Zhang *et al.*, 2018). The production of these complexing agents has also been demonstrated for other bacterial species, such as *Pseudomonas* sp. and *Burkholderia* sp. (Bau *et al.*, 2013; Brisson *et al.*, 2016; Desouky *et al.*, 2016; Yoshida *et al.*, 2004). Fungal species such as *Aspergillus* sp. could also leach REE from phosphorites/monazite (up to 1.37 mg/L of REE) using siderophores (Castro *et al.*, 2020; Osman *et al.*, 2019). It is worth mentioning that exuded compounds contribute to the bioleaching process (Brisson *et al.*, 2016). Similar to siderophores, the production of organic acids such as aliphatic (citrate and oxalate) and aromatic (gluconic, malic, phthalate and salicylate) acids by microbes has been used as a method to leach metals of economic interest such as REE from diverse sources (Figure 8.1) (Barnett *et al.*, 2018; Brisson *et al.*, 2020). For instance, Brisson *et al.* (2020) demonstrated that the fungus *Paecilomyces* could bioleach up to 42 mg/L of REE from monazite ore using organic acids (i.e., citric and citramalic acids). Similarly, gluconic acids produced by the bacterium *Gluconobacter oxydans* effectively leached REE (in the order of Y > La > Ce > Eu > Tb) from waste phosphor (Reed *et al.*, 2016). The authors reported a direct correlation between the gluconic acid concentration and amount of REE leached. In another study, the gluconic acid-producing *Gluconobacter* strain was able to leach on average 24.7% of the total REE from industrial waste materials, that is, spent fluid catalytic cracking (FCC) catalysts (Jin *et al.*, 2019).

Typical examples of indirect mobilization include the oxidation of REE-bearing minerals (Villar & Garcia, 2006; Wong *et al.*, 2002). For instance, *Acidithiobacillus ferrooxidans* promotes the oxidation of Fe^{2+} to Fe^{3+} at low pH (Ohmura *et al.*, 2002), which indirectly results in the oxidation of metallic sulfide minerals releasing elements of interest such as REE (Vardanyan *et al.*, 2018). In a comparative study of autotrophic *A. ferrooxidans* and heterotrophic *Acetobacter methanolicus* leaching of a zircon mineral, the REE bioleaching efficiency was almost 80% (1.1 mg/h) and 67% (1.4 mg/h), respectively (Glombitza *et al.*, 1987). The study demonstrated distinctive mechanisms between the respective microorganisms. *Acidithiobacillus ferrooxidans* cell biomass adsorbed most of the leached rare earth elements, while *A. methanolicus* did not adsorb the leached rare earth elements.

8.2.2 Rare earth elements microbial interactions as a biorecovery option

8.2.2.1 Microbial cell wall interaction with rare earth elements

In their microenvironment, the fungal/bacterial cell wall is the first barrier against non-toxic and toxic compounds (Zhou *et al.*, 2015). That is, the cell wall plays a major role in facilitating the uptake of nutrients and protecting the cells from metal toxicity (Wagner *et al.*, 2006). The cell wall of Gram-positive bacteria is a heterogeneous insoluble macromolecular polymeric matrix primarily composed of two types of polymers, namely peptidoglycan and teichoic acids, separated by a small periplasmic space (Nygaard *et al.*, 2015; Sarvas *et al.*, 2004). The Gram-negative bacterial cell wall is more complex than the Gram-positive cell wall. Typically, a bacterial cell wall is made up of an outer membrane and a cytoplasmic/inner membrane (Silhavy *et al.*, 2010) separated by a gel-like periplasmic space (Beveridge, 1999; Hobot *et al.*, 1984), which contains a thin layer of peptidoglycan (Maldonado *et al.*, 2016). The outer membrane is made up of phospholipids, proteins and lipopolysaccharides (LPSs) (Koraha *et al.*, 2005; Maldonado *et al.*, 2016; Okuda *et al.*, 2016). Comparatively, the fungal cell wall primarily consists of chitin, glucans and glycoproteins (Pinto *et al.*, 2008).

The presence of negatively charged functional groups (e.g., amines, hydroxyl, carboxyl and phosphoryl) allows sorption of metal cations on the cell surface of microbes (i.e., fungal, as well as both Gram-positive and Gram-negative bacterial) (Boyanov *et al.*, 2003; Hosomomi *et al.*, 2013; MacHalová *et al.*, 2015; Oves *et al.*, 2013). Even though functional groups play an important role in metal sorption, the metal sorption capacity seems to be more dependent on the structural differences of the cell wall (Hosomomi *et al.*, 2013; Morozzi *et al.*, 1986; Tsuruta, 2007). For instance, a comparative investigation between Gram-positive and Gram-negative bacteria showed that *Bacillus subtilis* (Gram-positive) is able to bind 10 times more Sc ($>$10 mM scandium/g (dry weight)) than *Escherichia coli* (Gram-negative). Moriwaki *et al.* (2016) demonstrated that the adsorption of REE onto a wild-type strain was higher than that of a teichoic acid-defective strain, indicating that teichoic acid plays a pivotal role in the adsorption of rare earth elements in Gram-positive bacteria. These studies demonstrate that REE sorption varies with both the REE and bacterial surface characteristics (Andrès *et al.*, 2003).

8.2.2.2 Microbial resistant mechanisms for the recovery of REE

Several studies have reported that Gram-positive bacteria are more resistant to rare earth elements toxicity than Gram-negative bacteria, probably due to their thick peptidoglycan layer. For example, Peng *et al.* (2004) reported that *E. coli* cell wall permeability increases when the cells are exposed to La. Exposure to La resulted in the collapse of some LPS patches on the cell surface, which altered the permeability and functionality of the outer cell membrane. Additionally, Chen *et al.* (2010, 2012) showed that bacterial cells exposed to Ce^{3+} started to dissociate at Ce^{3+} concentrations as low as 0.25 mM. At higher Ce^{3+} concentrations (2.5 mM) the bacterial cell wall was completely dissociated and only the cytoplasmic membrane remained. The authors concluded that Ce^{3+} impeded the physiological activities of bacteria due to a change in permeability because of the disruption of the microstructure of the cell membrane (outer- and inner membrane).

Nevertheless, microbes have developed metal-resistant mechanisms to thrive in the presence of heavy and/or rare earth elements, which can be used for biorecovery

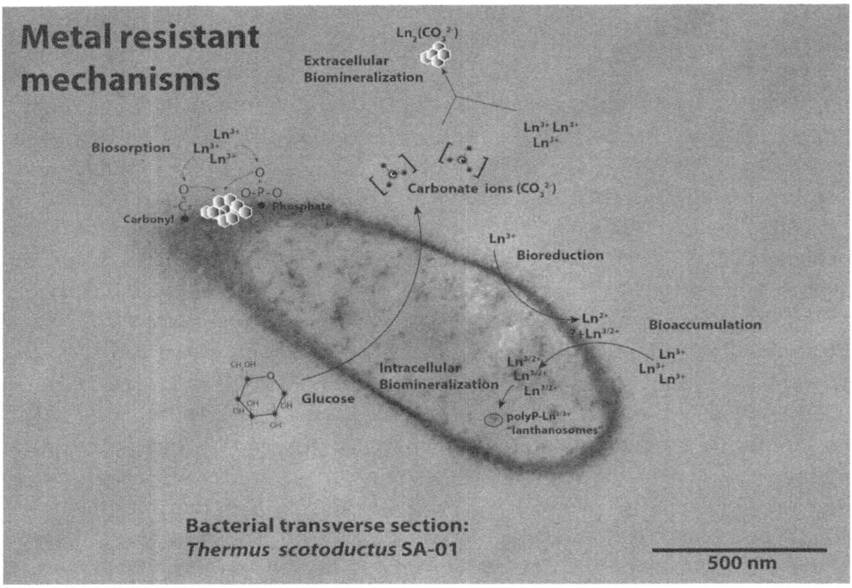

Figure 8.2 Schematic representation of metal resistant mechanisms associated with the mobility of REE based on an SEM image obtained from an REE biorecovery experiment.

processes. These include biosorption, bioaccumulation, biomineralization and bioreduction mechanisms (Figure 8.2).

8.2.2.2.1 Biosorption

Biosorption is defined as a physicochemical process of metals adhering to the microbial cell surface (Andrès *et al.*, 2003; Takahashi *et al.*, 2005). Biosorption of REE is a passive mechanism that involves the sorption of those metals to bacterial functional groups (Hosomomi *et al.*, 2013; Takahashi *et al.*, 2005) by means of electrostatic interactions, covalent bonding and van der Waals forces (Murthy *et al.*, 2012). Gram-positive and negative bacteria as well as fungi have been investigated for the sorption behavior of REE (Table 8.1). The sorption capacity of a biosorbent depends on the polarity and its surface area (Adewuyi, 2020; Ambaye *et al.*, 2021). In other words, due to the larger surface area of a fungal cell wall, which increases the available binding sites (i.e., functional groups), the metal immobilization ability will be higher for fungal cells compared to the bacterial cell wall (Dhankhar & Hooda, 2011).

The sorption capacity also relies on the functional groups. For instance, a study conducted with Eu demonstrated that this metal forms fairly stable complexes with phosphate and carboxyl surface groups (Konishi & Noda, 2001). The binding

Table 8.1 REE biosorption studies using bacterial and fungal biomass.

	Biosorbent	Rare Earth Metals	Remarks	Reference
Gram (+)	*Bacillus cereus*	Ce, Nd	Phosphate and carboxyl groups were responsible for Ce adsorption	Emmanuel *et al.* (2011)
	B. subtilis	La, Sm	Sorption capacity of La and Sm on the cell surface was homogeneous	Giese and Jordão (2019)
		Sc, Yb, Lu, La	Adsorption capacity could be attributed to the strong Lewis acidity ions	Moriwaki *et al.* (2016)
	B. thuringiensis	Eu	The adsorption capacity of Eu^{3+} can be as high as 160 mg/g	Pan *et al.* (2017)
	Mycobacterium smegmatis	Eu, La, Yb	Selective adsorption of rare earth elements	Andres *et al.* (1993)
Gram (−)	*P. aeruginosa*	Eu, Nd, La	Eu^{3+} adsorbs on bacterial cells in the presence of organic ligands with low chelating ability	Ozaki *et al.* (2005)
		La, Pr, Nd, Eu, Dy	Adsorption of rare earth elements can be attributed to functional groups	Texier *et al.* (1999)
		Dy	Adsorption capacity of Dy (1 mM Dy/g) demonstrated at pH 5	Philip *et al.* (2000)
	Myxococcus xanthus	Eu, La	Accumulated 0.9 mM of La/g of wt biomass and 0.99 mM/g of dry biomass	Merroun *et al.* (2003)
	Pseudomonas sp.	La	Adsorption of La in the form of $LaPO_4$ probably due to phosphate groups	Kazy *et al.* (2006)
	E. coli	Tm, Yb, Lu	Preferential adsorption of Tm, Yb and Lu carboxylate and phosphate binding sites	Takahashi *et al.* (2005)

(Continued)

Table 8.1 REE biosorption studies using bacterial and fungal biomass (*Continued*).

	Biosorbent	Rare Earth Metals	Remarks	Reference
	Roseobacter sp.	Tm, Yb, Lu	Preferential adsorption for lanthanides through phosphate groups	Bonificio and Clarke (2016)
	Thermus scotoductus SA-01	Eu	Sorption of Eu is associated with phosphates, carboxyl, carbonyl and amide groups	Maleke *et al.* (2019a, 2019b)
Archaea	*Halobacterium noricense*	Eu	Biosorption of Eu through complexation with phosphate groups	Bader *et al.* (2019)
Fungi	*Botryosphaeria rhodina*	La, Sm	Rare earth sorption capacity increased at pH greater than 5	Giese *et al.* (2019)
	Fusarium sp.	Nd, Dy, Nd, Lu	Preferential sorption of Nd, Dy, Nd and Lu at low pH	Breuker *et al.* (2020)
	Penicillium sp. ZD28	Y	Adsorption of Y $>$0.46 mM/g in conditions of 0.47 mM $<$[Y]$<$ 6.38 mM	Wang *et al.* (2020)
Yeast	*S. cerevisiae*	Gd, Dy, Yb, Nd	Selective biosorption of rare earth elements	Breuker *et al.* (2020)
		Nd, Yb	Phosphorylated cells selectively adsorbed the Nd and Yb ions from solution containing heavy metals and rare earth ions	Ojima *et al.* (2019)
Algae	*Chlorella vulgaris*	Nd	The sorption affinity for Nd was greater than activated carbon in algae species	Kucuker *et al.* (2017)

capacity of phosphate groups to Eu seems to be greater than that of carboxyl groups (Takahashi *et al.*, 2005). Furthermore, the availability of the binding sites is dependent on the pH of the solution (Takenaka *et al.*, 2004). The cell envelope of the bacterium *Halomonas elongate* showed a high affinity towards Eu^{3+} in high ionic strength solution (Takenaka *et al.*, 2004). Interestingly, bacteria have shown preferential sorption of Eu^{3+}. In the presence of organic ligands, Ozaki *et al.* (2005) demonstrated that adsorption of Eu^{3+} on the *P. fluorescens* cell envelope was due to the covalent linkage process. Likewise, Texier *et al.* (1999) speculated that the preferential affinity of Eu^{3+} to the phosphate and carboxyl sites was due to the metal's ionic radius. In addition, the carboxyl and phosphate functional groups present on the fungal cell wall have also been shown to effectively sorb metal cations, including REE (Michalak *et al.*, 2013).

8.2.2.2.2 Bioaccumulation

Bioaccumulation is an active mechanism employed by bacterial cells (Abbas *et al.*, 2014; Bae *et al.*, 2000; Hrynkiewicz *et al.*, 2014) to adsorb metals in a two-step process. The first step (metabolism-independent) involves the rapid sorption of metal ions to the cell wall (MacHalová *et al.*, 2015). For example, *E. coli* accumulates La in the periplasmic space (Bayer & Bayer, 1991). The periplasmic content of REE was thought to be due to the diffusion of the REE into the cell. The authors interpreted the accumulation of La as a result of the phosphate groups of phospholipids and lipopolysaccharides present on the bacterial cell wall (Merroun *et al.*, 2003). The second step (metabolism-dependent) involves the active uptake of the REE within the cytoplasmic space (Abbas *et al.*, 2014). For instance, the intracellular cytoplasmic bioaccumulation of REE has been reported for *Myxococcus xanthus* (Merroun *et al.*, 2003).

Although the bioaccumulation mechanism of REE described for *E. coli* and *M. xanthus* remains unclear, the superfamily adenosine triphosphate (ATP) binding cassette (ABC) transporters are known to participate in the uptake of a variety of metals (Erasmus *et al.*, 2014). For instance, ABC protein transporters were involved in the periplasmic accumulation of Au and U in *Thermus scotoductus* SA-01 (Cason *et al.*, 2012; Erasmus *et al.*, 2014) and the cytoplasmic accumulation of antimony (Sb) in *Leishmaniasis* (Brochu *et al.*, 2003). These studies demonstrate the role that membrane proteins play in the bioaccumulation of metals by bacterial cells. Recently, the ABC transporter protein was suspected to play a role in the bioaccumulation of the rare earth Eu (Maleke *et al.*, 2019b). The homeostasis of Eu within the bacterium could also be related to the presence of inorganic polyphosphates (PolyP). The authors speculated that the intracellular accumulation of Eu may be mediated through the PolyP metabolism. It has been suggested that the bacteria would take advantage of the low solubility of REE polyphosphate complexes for lanthanide storage in 'lanthanosomes' (Cotruvo, 2019). Once bioaccumulated, the REE ions have been reported to act as essential metabolic co-factors for several microbial metabolic processes. For instance,

PyrR (from *B. subtilis*) is a hexameric protein under normal conditions that becomes a dimer in the presence of REE (i.e., La) (Tomchick *et al.*, 1998). These ions bind to groups that would otherwise form the inter-subunit contacts at the three-fold axis of the hexamer (Pidcock & Moore, 2001). Furthermore, *Methylacidiphilum fumariolicum* SolV's methanol dehydrogenase demonstrated dependency towards REE, that is, Eu, Gd, La, Pr, Nd and Sm (Pol *et al.*, 2014).

8.2.2.2.3 Biomineralization

Biomineralization involves the precipitation or crystallization of REE as insoluble inorganic or organic compounds by bacterial excreted ligands (e.g., carbonates, phosphates and/or hydroxides) concentrated on the cell surface or intracellularly (e.g., inorganic polyphosphates) (Choudhary & Sar, 2015; Newsome *et al.*, 2014; 2015). The process occurs at the organic-inorganic interface where a molecular recognition system is involved in the control of crystal nucleation and growth (Mann *et al.*, 1989).

To the best of our knowledge, only a few studies have been documented for biomineralization involving REE and microbes. For example, Jiang *et al.* (2018) reported the biomineralization of Sm using both yeast and bacteria (Jiang *et al.*, 2018). The authors demonstrated that Sm-phosphate was mainly formed on the cell surface with some metal patches accumulating intracellularly. Furthermore, a recent study (Maleke *et al.*, 2019b) demonstrated that the biomineralization of Eu ions as carbonate minerals was achieved using the thermophilic bacterium *T. scotoductus* SA-01 through two different mechanisms: (1) by modifying the conditions of its surrounding microenvironments and/or (2) by acting as nucleation sites.

8.2.2.2.4 Bioreduction

Bioreduction involves the direct or indirect reduction of a metal from a high oxidation state to a lower oxidation state through enzymatic or electron shuttle processes, respectively (Fredrickson *et al.*, 2000a, 2000b; Johnson *et al.*, 2017; Lloyd, 2003; Pfeffer *et al.*, 2012). Direct enzymatic metal reduction is a well-documented process and normally requires direct contact with the metal. The ability of *Desulfovibrio vulgaris* to reduce U^{6+} via cytochrome C_3 is an example of direct enzymatic metal reduction (Lovley *et al.*, 1993). In the process, the reduction of U^{6+} is as a result of electron transfer from hydrogen oxidation to cytochrome C_3 through the cytoplasmic membrane. The electron shuttle process is carried out by the production of cellular electron shuttles, for example, humic acid, flavins or extracellular insoluble metals, which can be reversibly oxidized and reduced, thereby conferring the capacity to serve as electron carriers among multiple redox reactions (Johnson *et al.*, 2017; Kappler *et al.*, 2004; Kotloski & Gralnick, 2013). In other words, electrons can be carried between bacterial cells and the insoluble electron acceptors, enabling indirect electron transfer (Cheng *et al.*, 2013; Lies *et al.*, 2005). This phenomenon has been reported for

Shewanella oneidensis, which can transfer electrons to Fe^{3+} oxide located $\geq$50 μm from the cell surface (Lies *et al.*, 2005).

The reduction of most REE was thought to be thermodynamically unfavorable, as they remain in the +3 oxidation state under different environmental conditions (Deplanche & Macaskie, 2008; Deplanche *et al.*, 2011). Remarkably, only one study has been documented on the bioreduction of REE in bacteria using glucose as an electron donor (Maleke *et al.*, 2019a). The authors described for the first time the bioreduction of Eu^{3+} by a site-specific *Clostridium* strain, which could be used as a biorecovery strategy.

8.2.3 Selectivity of enzymes as REE recovery strategy

8.2.3.1 The role of REE in microbial metabolism

Divalent d-block metal cations (DDMCs), such as Zn^{2+}, Mn^{2+}, Fe^{2+} and Cu^{2+}, are essential for bacterial metabolism (Barber-Zucker *et al.*, 2017). For example, Fe^{2+} and Cu^{2+} readily donate and accept electrons and are important cofactors for many enzymes (Barber-Zucker *et al.*, 2017). Contrary to the prior assumptions of their biological inertia, REE have recently emerged as a biological component (Cotruvo, 2019; Daumann, 2019; Deblonde *et al.*, 2020; Lumpe *et al.*, 2018; Ramos *et al.*, 2016). Investigations made in the early 21st century suggest and support the hypothesis of bacterial cells being able to incorporate certain REE into their biological machinery (Ruming *et al.*, 2003; Wenhua *et al.*, 2003; Zhao *et al.*, 2002). These investigations demonstrated that (1) certain REE induce both endogenic and ectogenic metabolism in *E. coli* (Wenhua *et al.*, 2003), (2) REE facilitate the secretion of antibiotics in *Streptomyces* sp. (Kawai *et al.*, 2007), the secretion of rhamnan-oriented EPS in *Bradyrhizobium* sp. (Fitriyanto *et al.*, 2011a) and enable a late idiophase overproduction of amylase, with a corresponding protease and bacilysin secretion in *Bacillus subtilis* (Inaoka & Ochi, 2011). It is hereby ratiocinated that REE might be housekeeping elements for the regulation of physiological and biochemical processes of extremophilic bacterial species.

The discovery and description of a lanthanoid-affixed enzyme amalgam, methanol dehydrogenase (MDH), confirms the suspicion that lanthanoids (Ln^{3+}) play a role in bacterial metabolism (Fitriyanto *et al.*, 2011b; Hibi *et al.*, 2011; Nakagawa *et al.*, 2012). This protein is a key periplasmic enzyme responsible for the oxidative transformation of methanol to formaldehyde. The role of REE has been interpreted as a cofactor for MDH, which derives energy from a variety of C_1–C_2 substrates in both methylotrophic and non-methylotrophic bacteria (Chistoserdova & Kalyuzhnaya, 2018; Huang *et al.*, 2019). So far, the role of lanthanoids as possible cofactors has been demonstrated in XoxFs, where micromolar quantities of La^{3+} and Cs^{3+} (Fitriyanto *et al.*, 2011b; Hibi *et al.*, 2011; Nakagawa *et al.*, 2012) and La^{3+}–Gd^{3+} (Pol *et al.*, 2014; Wang *et al.*, 2020) were detected. Similarly, REE have been detected in ExaF and PedH, where they facilitate the metabolism of multi-carbon substrates, such as aliphatic, aromatic, primary and secondary alcohols as well as aldehydes (Good *et al.*, 2016; Wehrmann *et al.*, 2017) and in lanmodulin, where they stimulate its maturation and structural reorientation to a functional protein (Cotruvo *et al.*, 2018).

REE dependence is associated with specific mechanisms for their uptake and metabolism. Maleke *et al.* (2019a) suggested that before the intracellular uptake of Eu, *Clostridium* sp. use a pyruvate flavodoxin oxidoreductase located in the periplasm and/or cytoplasm for Eu^{3+} reduction to Eu^{2+}. This study proposed that the reduction of this rare earth element is either to conserve energy or to be used as a co-factor for different enzymes. It is well known that reduced metal species are easily transported intracellularly (Maleke *et al.*, 2019b; Su *et al.*, 2021). However, in the case of REE, most are trivalent and they are rarely found in solution, i.e. they have a solubility of ~0.1–1 pM at pH 7 (Daumann, 2019; Featherston *et al.*, 2019). Therefore, other mechanisms have been explored, until recently where a lanthanide-uptake system was proposed (Figure 8.3) (Cotruvo *et al.*, 2018; Cotruvo, 2019; Featherston *et al.*, 2019; Ochsner *et al.*, 2019; Roszczenko-Jasińka *et al.*, 2020). Here, Ln^{3+} (either La, Ce, Pr or Nd) are

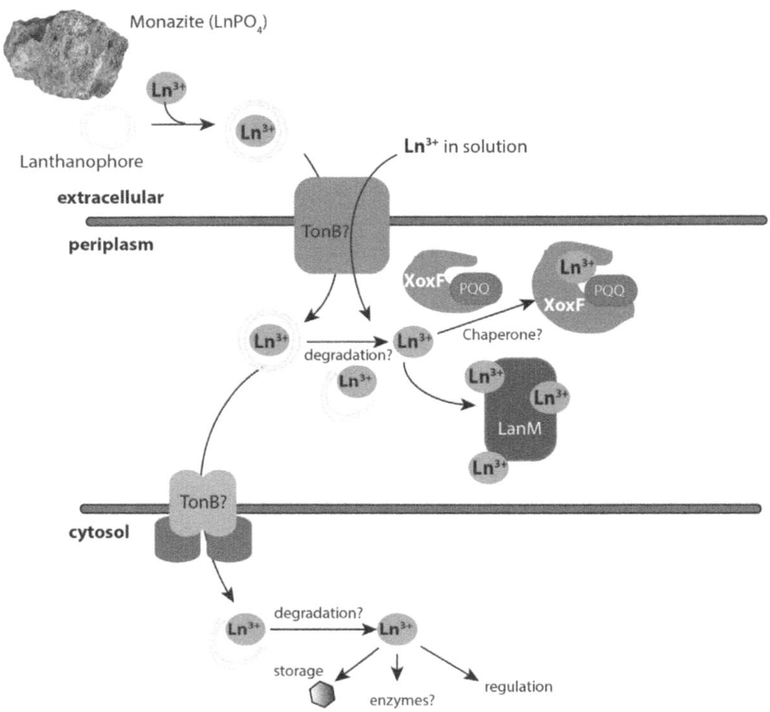

Figure 8.3 Model for Ln^{3+}-uptake system and utilization in *M. extorquens* (modified from Cotruvo, 2019). PQQ: pyrroloquinoline quinone.

presumed to be picked up through a lanthanophore (polydentate lanthanoid-chelating ligand) and transported into the cell through a TonB-dependent system in a similar manner as observed in the siderophore and chalcophore-assisted iron and copper uptake (Daumann, 2019). Once in the cell, Ln^{3+} is incorporated as cofactor in the XoxF active site, thereby facilitating the catalysis of methanol oxidation through some electron transport processes, usually involving the C_L and C_H cytochromes (Ochsner *et al.*, 2019; Roszczenko-Jasińka *et al.*, 2020). Regarding methanol metabolism, certain microorganisms might possess the machinery for easy adaptability in the absence or presence of different Ln^{3+} species with varying concentrations (Chu *et al.*, 2016; Nakagawa *et al.*, 2012; Skovran *et al.*, 2019; Vu *et al.*, 2016; Wehrmann *et al.*, 2018) through a mechanism known as the 'Ln^{3+} switch'. For example, in the absence of Ln^{3+}, the Ca^{2+}-dependent MDH MxaFI of *M. extorquens* AM1 is stimulated; whereas the same strain favorably secretes the XoxF type of MDH in the presence of low Ln^{3+} concentrations (Nakagawa *et al.*, 2012; Vu *et al.*, 2016). The striking similarity of Ln^{3+} to Ca^{2+} ions in size, coordination environment and ligand preferences, along with strong Lewis acidity, suggests that this crossover seems plausible (Lim & Franklin, 2004).

The implications of Ln^{3+} bacterial metabolism in different biotechnology domains are still in their infancy, as new theories regarding this complex phenomenon keep unfolding. Nevertheless, the likely occurrence of a lanthanophore that induces the mobility and uptake of REE and the selectivity of proteins to utilize these elements as cofactors necessitates further investigations regarding the interaction of extremophiles and REE. Certainly, these proteins have potential applications for biohydrometallurgy of REE, from bioleaching to selective bioaccumulation.

8.2.3.2 Selectivity of enzymes for REE

In order to understand the affinity of metals to proteins, it is necessary to be familiar with the Irving–Williams series, also known as the natural order of stability for divalent transitional metals (Frausto da Silva & Williams, 2001; Irving & Williams, 1948). As proteins offer imperfect steric selection between metals, the affinities for metals have a tendency to follow a universal order of preference, which is the Irving–Williams series for essential divalent metals: Mg^{2+} and $Ca^{2+} < Mn^{2+} < Fe^{2+} < Co^{2+} < Ni^{2+} < Cu^{2+} > Zn^{2+}$.

However, the cell can escape from limitations imposed by this universal order of metal preference by using selective metal binding mechanisms. Otherwise, metals such as Ni and Cu with high affinity to proteins would impede the binding of metals with low affinity. The homeostatic mechanism is comprised of storage proteins and metallochaperones that contribute to regulate and deliver specific metals, respectively (Waldron & Robinson, 2009). For instance, vanabins localized in the cytoplasm of vanadocytes of *A. sydneiensis samea* can selectively bind to V^{4+} (Hamada *et al.*, 2005; Pessoa *et al.*, 2015). Vanabins might function as metal chaperone proteins in the cytoplasm rather than proteins for metal storage or detoxification (Ueki *et al.*, 2003). Calmodulin is another selective protein found in the cytoplasm of all eukaryotic cells (Edington *et al.*, 2018; Kawasaki *et al.*, 2019). This protein can selectively bind a total of four Ca^{2+} ions. Other selective proteins include transferrin (Boradia *et al.*, 2014; Holland *et al.*, 2012), ferritin (Honarmand Ebrahimi *et al.*, 2012) and siderocalin (Fischbach *et al.*, 2006; Harris, 1996), which selectively sense, uptake and store Fe ions.

As previously mentioned, several studies have explored the pyrroloquinoline quinone (PQQ)-dependent MDH (XoxF) of methylotrophs because this enzyme is able to bind Ln^{3+}. Lumpe *et al.* (2020) reported the isolated PQQ-lanthanide complexes and determined their molecular structures for the first time in the absence of the enzyme matrix. The results demonstrate that PQQ can selectively uptake Ln^{3+} such as Eu^{3+} and Nd^{3+} from aqueous solutions containing mixtures of salts, without the need for potentially hazardous organic solvents or other additives. Thereupon, this mechanism is of particular interest for sustainable technologies for REE recovery.

Another protein recently discovered that selectively binds to REE is lanmodulin (LanM) (Cotruvo, 2019). The structural stability and metal-binding properties exhibited by this protein outperform many commercial chelators (including synthetics) (Deblonde *et al.*, 2020). LanM is thermostable (up to 95°C) and the REE binding in LanM can take place at low pH of up to 2.5. In addition, the LanM-REE complexes can be produced below molar amounts of competing non-REE metal ions (including Mg, Ca, Zn and Cu) (Deblonde *et al.*, 2020). Since REE must be released from minerals or an artificial matrix, they remain initially in acid leachates with high concentrations of non-REE metal ions, thus the stability of the LanM-REE complexes and selectivity of LanM looms largely. The bacterial mechanisms once again provide new insight into more sustainable strategies to extract critical raw materials.

8.3 CHALLENGES AND FUTURE PERSPECTIVES OF REE BIORECOVERY

The production of REE requires large amounts of material and energy, which generates large volumes of solid waste, industrial wastewater and gas emissions. Therefore, conventional REE mining is not considered an environmentally sustainable industry (Navarro & Zhao, 2014). Despite the environmental footprint of REE production, the current trend of REE consumption indicates that the global demand for REE will keep growing due to technology advancements, which increases the risk of supply disruption at a global scale (Ballinger *et al.*, 2020; Keilhacker & Minner, 2017; Schmid, 2019). Therefore, secondary sources of REE (i.e., mine wastes and E-wastes) are broadly investigated at this stage, not only to mitigate the supply risk, but also to reduce the environmental challenges associated with REE mining (Binnemans *et al.*, 2013, 2015). There are three reasons to further investigate the production of REE from secondary sources: (1) the increase in the demand is catching up with the current supply from the mining sector, (2) the centralization of REE production in China is counterproductive for the rest of the world and implies a geopolitical risk on the supply chain and (3) the REE industry should move towards environmental sustainability and embrace a circular economy approach to minimize the environmental impact of REE production. Therefore, the use of biological processes, which have been demonstrated to have a positive impact on the environmental footprint of the industry, are attractive strategies to extract REE from secondary resources.

One environmentally and economically favorable approach to recover REE is bioleaching. Bioleaching of metals from mineral ores is a well-established industrial process. For instance, techno-economic analysis and life cycle analysis for bioleaching of REE indicated that a bioleaching plant based on *G. oxydans* lixiviant production could be profitable and have a lower environmental impact than a chemical leaching plant (Jin *et al.*, 2019; Thompson *et al.*, 2018). The carbon source used to feed the microorganisms contributes up to 44% of the total cost of the bioleaching process. The use of organic waste, such as agricultural and food waste products, to replace commercial carbon sources is common practice to reduce the operating expenses (OPEX) of bioleaching plants. While the standard bioleaching process extracts REE successfully with a profit of 10% of the total costs (OPEX and extrapolated capital expenditure (CAPEX)), the use of agricultural waste products as an alternative substrate for microbial metabolism maintains the leaching efficiency and increases the profit (60% of the total costs) (Jin *et al.*, 2019).

One of the main drawbacks for REE bioleaching is the radionuclides (i.e., U and Th) that co-occur with REE, particularly in mine wastes. Due to the similar characteristics shared between REE, U and Th, the processes of leaching, separation, purification and recovery carry the radionuclides and create a solution highly concentrated in both REE and radionuclides. Biomineralization associated with the biosorption process has been proposed to overcome this problem. Reactive biominerals have been used for the removal of toxic metals. For instance, hydroxyapatite (HA) precipitation induced by *Serratia* sp. has been successfully able to uptake both Eu^{3+} and UO_2^{2+}, separately. In this experiment, the immobilized biofilm-HA in a flow-through column was able to recover 95% of Eu^{3+} from 0.5 mM $Eu(NO_3)_{2(aq)}$ at 4 mL/h (0.5 column fluid vol/h) from 2 L (Macaskie *et al.*, 2017). Although, the selective uptake of REE from a cocktail of metals in solution has not been achieved using this method.

Several studies suggest that the efficacy and selectivity of REE may be improved by adjusting the most important factor, pH (García *et al.*, 2020). Likewise, bacterial cell surfaces can also be engineered to enhance selectivity (Park *et al.*, 2016). For example, in a cell surface bioengineering study, metal-binding peptides or proteins were displayed to selectively bind REE from competing metals. The sorption of Tb^{3+} in the biomass of *Caulobacter crescentus* was 50 μM at a cell concentration of 8×10^8 cells/mL which compares favorably to other biosorption systems (Park *et al.*, 2016).

Recently discovered proteins such as lanthanophores, which might contribute to the solubility and uptake of REE, as well as pyrroloquinoline quinone (PQQ)-dependent MDH (XoxF) and lanmodulin (LanM), which have potential applications for the selective REE uptake, might be more sustainable alternatives for REE metallurgy in the near future. The biological processes may offer specificity, low energy consumption and minimal waste generation, although the complexity of waste streams, toxicity and competing side reactions might be challenges for the optimum functionality of these proteins. A better understanding of the complexity of selective metal uptake must be achieved and a feasibility study should be carried out considering that the high value of the product might justify the additional cost of the biorecovery process via proteins.

REFERENCES

Abbas S. H., Ismail I. M., Mostafa T. M. and Sulaymon A. H. (2014). Biosorption of heavy metals: a review. *Journal of Chemical Science and Technology*, **3**, 74–102.

Adewuyi A. (2020). Chemically modified biosorbents and their role in the removal of emerging pharmaceuticalwaste in the water system. *Water*, **12**, 1–31.

Ambaye T. G., Vaccari M., van Hullebusch E. D., Amrane A. and Rtimi S. (2021). Mechanisms and adsorption capacities of biochar for the removal of organic and inorganic pollutants from industrial wastewater. *International Journal of Environmental Science and Technology*, **18**, 3273–3294.

Andres Y., MacCordick H. J. and Hubert J. C. (1993). Adsorption of several actinide (Th, U) and lanthanide (La, Eu, Yb) ions by Mycobacterium smegmatis. *Applied Microbiology and Biotechnology*, **39**(3), 413–417.

Andrès Y., Texier A. C. and Le Cloirec P. (2003). Rare earth elements removal by microbial biosorption: a review. *Environmental Technology*, **24**(11), 1367–1375.

Ayangbenro A. S., Olanrewaju O. S. and Babalola O. O. (2018). Sulfate-reducing bacteria as an effective tool for sustainable acid mine bioremediation. *Frontiers in Microbiology*, **9**, 1–10.

Bader M., Moll H., Steudtner R., Lösch H., Drobot B., Stumpf T. and Cherkouk A. (2019). Association of Eu (III) and Cm (III) onto an extremely halophilic archaeon. *Environmental Science and Pollution Research*, **26**(9), 9352–9364.

Bae W., Chen W., Mulchandani A. and Mehra R. K. (2000). Enhanced bioaccumulation of heavy metals by bacterial cells displaying synthetic phytochelatins. *Biotechnology and Bioengineering*, **70**, 518–524.

Balaram V. (2019). Rare earth elements: a review of applications, occurrence, exploration, analysis, recycling, and environmental impact. *Geoscience Frontiers*, **10**, 1285–1303.

Ballinger B., Schmeda-Lopez D., Kefford B., Parkinson B., Stringer M., Greig C. and Smart S. (2020). The vulnerability of electric-vehicle and wind-turbine supply chains to the supply of rare-earth elements in a 2-degree scenario. *Sustainable Production and Consumption*, **22**, 68–76.

Barber-Zucker S., Shaanan B. and Zarivach R. (2017). Transition metal binding selectivity in proteins and its correlation with the phylogenomic classification of the cation diffusion facilitator protein family. *Scientific Reports*, **7**(1), 1–12.

Barmettler F., Castelberg C., Fabbri C. and Brandl H. (2016). Microbial mobilization of rare earth elements (REE) from mineral solids — a mini review. *AIMS Microbiology*, **2**, 190–204.

Barnett M. J., Palumbo-Roe B. and Gregory S. P. (2018). Comparison of heterotrophic bioleaching and ammonium sulfate ion exchange leaching of rare earth elements from a Madagascan ion-adsorption clay. *Minerals*, **8**, 1–11.

Bau M., Tepe N. and Mohwinkel D. (2013). Siderophore-promoted transfer of rare earth elements and iron from volcanic ash into glacial meltwater, river and ocean water. *Earth and Planetary Science Letters*, **364**, 30–36.

Bayer M. E. and Bayer M. H. (1991). Lanthanide accumulation in the periplasmic space of *Escherichia coli* B. *Journal of Bacteriology*, **173**(1), 141–149.

Beveridge T. J. (1999). Structures of gram-negative cell walls and their derived membrane vesicles. *Journal of Bacteriology*, **181**(16), 4725–4733.

Binnemans K., Jones P. T., Blanpain B., Van Gerven T., Yang Y., Walton A. and Buchert M. (2013). Recycling of rare earths: a critical review. *Journal of Cleaner Production*, **51**, 1–22.

Binnemans K., Jones P. T., Blanpain B., Van Gerven T. and Pontikes Y. (2015). Towards zero-waste valorisation of rare-earthcontaining industrial process residues: a critical review. *Journal of Cleaner Production*, **99**, 17–38.

Bonificio W. D. and Clarke D. R. (2016). Rare-earth separation using bacteria. *Environmental Science & Technology Letters*, **3**(4), 180–184.

Boradia V. M., Malhotra H., Thakkar J. S., Tillu V. A., Vuppala B., Patil P., Sheokand N., Sharma P., Chauhan A. S., Raje M. and Raje C. I. (2014). Mycobacterium tuberculosis acquires iron by cell-surface sequestration and internalization of human holo-transferrin. *Nature Communications*, **5**(1), 1–13.

Boyanov M. I., Kelly S. D., Kemner K. M., Bunker B. A., Fein J. B. and Fowle D. A. (2003). Adsorption of cadmium to *Bacillus subtilis* bacterial cell walls: a pH-dependent X-ray absorption fine structure spectroscopy study. *Geochimica et Cosmochimica Acta*, **67**(18), 3299–3311.

Breuker A., Ritter S. F. and Schippers A. (2020). Biosorption of Rare Earth Elements by Different Microorganisms in Acidic Solutions. *Metals*, **10**(7), 954.

Brisson V. L., Zhuang W.-Q. and Alvarez-Cohen L. (2016). Bioleaching of rare earth elements from monazite sand. *Biotechnology and Bioengineering*, **113**(2), 339–348.

Brisson V. L., Zhuang W. Q. and Alvarez-Cohen L. (2020). Metabolomic analysis reveals contributions of citric and citramalic acids to rare earth bioleaching by a *Paecilomyces* fungus. *Frontiers in Microbiology*, **10**, 3008.

Brochu C., Wang J., Roy G., Messier N., Wang X.-Y., Saravia N. G. and Ouellette M. (2003). Antimony uptake systems in the protozoan parasite *Leishmania* and accumulation differences in antimony-resistant parasites. *Antimicrobial Agents and Chemotherapy*, **47**(10), 3073–3079.

Cason E. D., Piater L. A. and van Heerden E. (2012). Reduction of U(VI) by the deep subsurface bacterium, *Thermus scotoductus* SA-01, and the involvement of the ABC transporter protein. *Chemosphere*, **86**, 572–577.

Castro L., Blázquez M. L., González F. and Muñoz J. A. (2020). Bioleaching of phosphate minerals using earth elements. *Metals*, **10**(7), 978.

Chen A., Shi Q., Feng J., Ouyang Y., Chen Y. and Tan S. (2010). Dissociation of outer membrane for *Escherichia coli* cell caused by cerium nitrate. *Journal of Rare Earths*, **28**(2), 312–315.

Chen A., Shi Q., Ouyang Y. and Chen Y. (2012). Effect of Ce^{3+} on membrane permeability of *Escherichia coli* cell. *Journal of Rare Earths*, **30**, 947–951.

Cheng Y.-Y., Li B.-B., Li D.-B., Chen J.-J., Li W.-W., Tong Z.-H., Wu C. and Yu H.-Q. (2013). Promotion of iron oxide reduction and extracellular electron transfer in *Shewanella oneidensis* by DMSO. *PLoS One*, **8**, e78466.

Chistoserdova L. and Kalyuzhnaya M. G. (2018). Current trends in methylotrophy. *Trends in Microbiology*, **26**(8), 703–714.

Choudhary S. and Sar P. (2015). Interaction of uranium (VI) with bacteria: potential applications in bioremediation of U contaminated oxic environments. *Reviews in Environmental Science and Bio/Technology*, **14**(3), 347–355.

Christenson E. A. and Schijf J. (2011). Stability of YREE complexes with the trihydroxamate siderophore desferrioxamine B at seawater ionic strength. *Geochimica et Cosmochimica Acta*, **75**, 7047–7062.

Chu F., Beck D. A. C. and Lidstrom M. E. (2016). MxaY regulates the lanthanide-mediated methanol dehydrogenase switch in *Methylomicrobium buryatense*. *PeerJ*, **4**, e2435.

Cotruvo J. A. (2019). The chemistry of lanthanides in biology: recent discoveries, emerging principles, and technological applications. *ACS Central Science*, **5**, 1496–1506.

Cotruvo J. A., Featherston E. R., Mattocks J. A., Ho J. V. and Laremore T. N. (2018). Lanmodulin: a highly selective lanthanide-binding protein from a lanthanide-utilizing bacterium. *Journal of the American Chemical Society*, **140**, 15056–15061.

Da Silva J. F. and Williams R. J. P. (2001). The biological chemistry of the elements: the inorganic chemistry of life. Oxford University Press.

Daumann L. J. (2019). Essential and ubiquitous: the emergence of lanthanide Metallobiochemistry. *Angewandte Chemie International Edition*, **58**(37), 12795–12802.

Deblonde G. J., Mattocks J. A., Park D. M., Reed D. W., Cotruvo J. A. and Jiao Y. (2020). Selective and efficient biomacromolecular extraction of rare-earth elements using Lanmodulin. *Inorganic Chemistry*, **59**(17), 11855–11867.

Deplanche K. and Macaskie L. E. (2008). Biorecovery of gold by *Escherichia coli* and *Desulfovibrio desulfuricans*. *Biotechnology and Bioengineering*, **99**, 1055–1064.

Deplanche K., Murray A., Mennan C., Taylor S. and Macaskie L. (2011). Biorecycling of precious metals and rare earth elements. In: Nanomaterials, M. M. Rahman (ed.), Intech Open, Rijeka, Croatia, pp. 279–314.

Desouky O. A., El-Mougith A. A., Hassanien W. A., Awadalla G. S. and Hussien S. S. (2016). Extraction of some strategic elements from thorium-uranium concentrate using bioproducts of *Aspergillus ficuum* and *Pseudomonas aeruginosa*. *Arabian Journal of Chemistry*, **9**, S785–S805.

Dev S., Sachan A., Dehghani F., Ghosh T., Briggs B. and Aggarwal S. (2020). Mechanisms of biological recovery of rare-earth elements from industrial and electronic wastes: a review. *Chemical Engineering Journal*, **397**, 124596.

Dhankhar R. and Hooda A. (2011). Fungal biosorption – an alternative to meet the challenges of heavy metal pollution in aqueous solutions. *Environmental Technology*, **32**(5), 467–491.

Edington S. C., Gonzalez A., Middendorf T. R., Halling D. B., Aldrich R. W. and Baiz C. R. (2018). Coordination to lanthanide ions distorts *binding site conformation in calmodulin. Proceedings of the National Academy of Sciences*, **115**(14), E3126–E3134.

Emmanuel E. C., Vignesh V., Anandkumar B. and Maruthamuthu S. (2011). Bioaccumulation of cerium and neodymium by Bacillus cereus isolated from rare earth environments of Chavara and Manavalakurichi, India. *Indian Journal of Microbiology*, **51**(4), 488–495.

Erasmus M., Cason E. D., van Marwijk J., Botes E., Gericke M. and van Heerden E. (2014). Gold nanoparticle synthesis using the thermophilic bacterium *Thermus scotoductus* SA-01 and the purification and characterization of its unusual gold reducing protein. *Gold Bulletin*, **47**, 245–253.

Featherston E. R., Rose H. R., Mcbride M. J., Taylor E. M., Amie K. and Cotruvo J. A., Jr. (2019). Biochemical and structural characterization of XoxG and XoxJ and their roles in activity of the lanthanide-dependent methanol dehydrogenase, XoxF. *Chembiochem*, **20**, 2360–2372.

Fitriyanto N. A., Nakamura M., Muto S., Kato K., Yabe T., Iwama T., Kawai K. and Pertiwiningrum A. (2011a). Ce^{3+}-induced exopolysaccharide production by *Bradyrhizobium* sp. MAFF211645. *Journal of Bioscience and Bioengineering*, **111**, 146–152.

Fitriyanto N. A., Fushimi M., Matsunaga M., Pertiwiningrum A., Iwama T. and Kawai K. (2011b). Molecular structure and gene analysis of Ce^{3+}-induced methanol dehydrogenase of *Bradyrhizobium* sp. MAFF211645. *Journal of Bioscience and Bioengineering*, **111**, 613–617.

Frausto Da Silva J. F. and Williams R. J. P. (2001). The biological chemistry of the elements: the inorganic chemistry of life. Oxford University Press.

Fredrickson J. K., Kostandarithes H. M., Li S. W., Plymale A. E. and Daly M. J. (2000a). Reduction of Fe(III), Cr(VI), U(VI), and Tc(VII) by *Deinococcus radiodurans* R1. *Applied and Environmental Microbiology*, **66**, 2006–2011.

Fredrickson J. K., Zachara J. M., Kennedy D. W., Duff M. C., Gorby Y. A., Li S.-M. W. and Krupka K. M. (2000b). Reduction of U(VI) in goethite (α-FeOOH) suspensions by a dissimilatory metal-reducing bacterium. *Geochimica et Cosmochimica Acta*, **64**, 3085–3098.

Gadd G. M. (2004). Microbial influence on metal mobility and application for bioremediation. *Geoderma*, **122**, 109–119.

García A. C., Latifi M., Amini A. and Chaouki J. (2020). Separation of radioactive elements from rare earth element-bearing minerals. *Metals*, **10**, 1–22.

Giese E. C., Dekker R. F. and Barbosa-Dekker A. M. (2019). Biosorption of lanthanum and samarium by viable and autoclaved mycelium of Botryosphaeria rhodina MAMB-05. *Biotechnology Progress*, **35**(3), e2783.

Giese E. C. and Jordão C. S. (2019). Biosorption of lanthanum and samarium by chemically modified free Bacillus subtilis cells. *Applied Water Science*, **9**(8), 1–8.

Glombitza F., Iske U., Bullmann M. and Dietrich B. (1987). Bacterial leaching of zircon mineral for obtaining trace and rare earth elements. In: Biohydrometallurgy, P. R. Norris and D. P. Kelly (eds), Proceedings of the International Symposium, Warwick, Kew Surry, UK, pp. 407–420.

Good N. M., Vu H. N., Suriano C. J., Subuyuj G. A., Skovran E. and Martinez-Gomez C. N. (2016). Pyrroloquinoline quinone ethanol dehydrogenase in *Methylobacterium extorquens* AM1 extends lanthanide-dependent metabolism to multicarbon substrates. *Journal of Bacteriology*, **198**, 3109–3118.

Hamada T., Asanuma M., Ueki T., Hayashi F., Kobayashi N., Yokoyama S., Michibata H. and Hirota H. (2005). Solution structure of vanabin2, a vanadium(IV)-binding protein from the vanadium-rich ascidian *Ascidia sydneiensis samea*. *Journal of the American Chemical Society*, **127**, 4216–4222.

Harris W. R. (1996). Binding and transport of aluminum by serum proteins. *Coordination Chemistry Reviews*, **149**, 347–365.

Hibi Y., Asai K., Arafuka H., Hamajima M., Iwama T. and Kawai K. (2011). Molecular structure of La^{3+}-induced methanol dehydrogenase-like protein in *Methylobacterium radiotolerans*. *Journal of Bioscience and Bioengineering*, **111**, 547–549.

Hobot J. A., Carlemalm E., Villiger W. and Kellenberger E. (1984). Periplasmic gel: New concept resulting from the reinvestigation of bacterial cell envelope ultrastructure by new methods. *Journal of Bacteriology*, **160**, 143–152.

Holland J. P., Evans M. J., Rice S. L., Wongvipat J., Sawyers C. L. and Lewis J. S. (2012). Annotating MYC status with ^{89}Zr-transferrin imaging. *Nature Medicine*, **18**, 1586–1591.

Honarmand Ebrahimi K., Bill E., Hagedoorn P. L. and Hagen W. R. (2012). The catalytic center of ferritin regulates iron storage via Fe(II)-Fe(III) displacement. *Nature Chemical Biology*, **8**, 941–948.

Hosomomi Y., Baba Y., Kubota F., Kamiya N. and Goto M. (2013). Biosorption of rare earth elements by *Escherichia coli*. *Journal of Chemical Engineering of Japan*, **46**, 450–454.

Hrynkiewicz K., Złoch M., Kowalkowski T., Baum C., Niedojadło K. and Buszewski B. (2014). Strain-specific bioaccumulation and intracellular distribution of Cd^{2+} in bacteria isolated from the rhizosphere, ectomycorrhizae, and fruitbodies of ectomycorrhizal fungi. *Environmental Science and Pollution Research*, **22**(4), 3055–3067.

Huang J., Yu Z., Groom J., Angela J. C., Yasuo T. and Chistoserdova L. (2019). Rare earth element alcohol dehydrogenases widely occur among globally distributed, numerically abundant and environmentally important microbes. *The ISME Journal*, **13**(8), 2005–2017.

Humsa T. Z. and Srivastava R. K. (2015). Impact of rare earth mining and processing on soil and water environment at Chavara, Kollam, Kerala: a case study. *Procedia Earth and Planetary Science*, **11**, 566–581.

Inaoka T. and Ochi K. (2011). Scandium stimulates the production of amylase and bacilysin in *Bacillus subtilis*. *Applied and Environmental Microbiology*, **77**, 8181–8183.

Irving H. and Williams R. J. P. (1948). Order of stability of metal complexes. *Nature*, **162**, 746–747.

Jeremic S., Beškoski V. P., Djokic L., Vasiljevic B., Vrvić M. M., Avdalović J., Cvijović G. G., Beškoski L. S. and Nikodinovic-Runic J. (2016). Interactions of the metal tolerant heterotrophic microorganisms and iron oxidizing autotrophic bacteria from sulphidic mine environment during bioleaching experiments. *Journal of Environmental Management*, **172**, 151–161.

Jiang M., Ohnuki T. and Utsunomiya S. (2018). Biomineralization of middle rare earth element samarium in yeast and bacteria systems. *Geomicrobiology Journal*, **35**(5), 375–384.

Jin H., Reed D. W., Thompson V. S., Fujita Y., Jiao Y., Crain-Zamora M., Fisher J., Scalzone K., Griffel M., Hartley D. and Sutherland J. W. (2019). Sustainable bioleaching of rare earth elements from industrial waste materials using agricultural wastes. *ACS Sustainable Chemistry & Engineering*, **7**(18), 15311–15319.

Johnson D. B., Hedrich S. and Pakostova E. (2017). Indirect redox transformations of iron, copper, and chromium catalyzed by extremely acidophilic bacteria. *Frontiers in Microbiology*, **8**, 211–225.

Jowitt S., Werner T., Weng Z. and Mudd G. (2018). Recycling and secondary sources of the rare earth elements REE Resource. Goldschmidt-AGU Fall Meeting Abstracts, **Boston**, V31G-0204.

Kang X., Csetenyi L. and Gadd G. M. (2020). Monazite transformation into Ce- and La-containing oxalates by *Aspergillus niger*. *Environmental Microbiology*, **22**, 1635–1648.

Kappler A., Benz M., Schink B. and Brune A. (2004). Electron shuttling via humic acids in microbial iron(III) reduction in a freshwater sediment. *FEMS Microbiology Ecology*, **47**, 85–92.

Kawai K., Wang G., Okamoto S. and Ochi K. (2007). The rare earth, scandium, causes antibiotic overproduction in *Streptomyces* spp. *FEMS Microbiology Letters*, **274**, 311–315.

Kawasaki H., Soma N. and Kretsinger R. H. (2019). Molecular dynamics study of the changes in conformation of calmodulin with calcium binding and/or target recognition. *Scientific Reports*, **9**, 1–10.

Kazak E. S., Kalitina E. G., Kharitonova N. A., Chelnokov G. A., Elovskii E. V. and Bragin I. V. (2018). Biosorption of rare-earth elements and yttrium by heterotrophic bacteria in an aqueous environment. *Moscow University Geology Bulletin*, **73**, 287–294.

Kazy S. K., Das S. K. and Sar P. (2006). Lanthanum biosorption by a Pseudomonas sp.: equilibrium studies and chemical characterization. *Journal of Industrial Microbiology and Biotechnology*, **33**(9), 773–783.

Keekan K. K., Jalondhara J. C. and Abhilash (2017). Extraction of Ce and Th from monazite using REE tolerant *Aspergillus niger*. *Mineral Processing and Extractive Metallurgy Review*, **38**(5), 312–320.

Keilhacker M. L. and Minner S. (2017). Supply chain risk management for critical commodities: a system dynamics model for the case of the rare earth elements. Resources, Conservation & Recycling, **125**, 349–362.

Konishi Y. and Noda Y. (2001). Precipitation stripping of rare-earth carbonate powders from rare-earth-loaded carboxylate solutions using carbon dioxide and water. *Industrial & Engineering Chemistry Research*, **40**(8), 1793–1797.

Koraha J., Tsuneyoshi N., Kimoto M., Gauchat J.-F., Nakatake H. and Fukudome K. (2005). Comparison of lipopolysaccharide-binding functions of CD14 and MD-2. *Clinical and Vaccine Immunology*, **12**(11), 1292–1297.

Kotloski N. J. and Gralnick J. A. (2013). Flavin electron shuttles dominate extracellular electron transfer by *Shewanella oneidensis*. *MBio*, **4**, e00553–e005512.

Kucuker M. A., Wieczorek N., Kuchta K. and Copty N. K. (2017). Biosorption of neodymium on Chlorella vulgaris in aqueous solution obtained from hard disk drive magnets. *PLoS One*, **12**(4), e0175255.

Kumar N., Gupta A., Ramteke P., Sahoo H. and Sengupta A. (2019). Biosorption-a green method for the preconcentration of rare earth elements (REEs) from waste solutions: a review. *Journal of Molecular Liquids*, **274**, 148–164.

Lies D. P., Hernandez M. E., Kappler A., Mielke R. E., Gralnick J. A. and Newman D. K. (2005). Shewanella oneidensis MR-1 uses overlapping pathways for iron reduction at a distance and by direct contact under conditions relevant for biofilms. *Applied and Environmental Microbiology*, **71**, 4414–4426.

Lim S. and Franklin S. J. (2004). Lanthanide-binding peptides and the enzymes that might have been. *Cellular and Molecular Life Sciences*, **61**, 2184–2188.

Lloyd J. R. (2003). Microbial reduction of metals and radionuclides. *FEMS Microbiology Reviews*, **27**(2–3), 411–425.

Lovley D. R., Widman P. K., Woodward J. C. and Phillips E. J. (1993). Reduction of uranium by cytochrome c_3 of *Desulfovibrio vulgaris*. *Applied and Environmental Microbiology*, **59**, 3572–3576.

Lumpe H., Pol A., Op Den Camp H. J. M. and Daumann L. J. (2018). Impact of the lanthanide contraction on the activity of a lanthanide-dependent methanol dehydrogenase-a kinetic and DFT study. *Dalton Transactions*, **47**, 10463–10472.

Lumpe H., Menke A., Haisch C., Mayer P., Kabelitz A., Yusenko K. V., Buzanich A. G., Block T., Pöttgen R., Emmerling F. and Daumann L. J. (2020). The earlier the better: structural analysis and separation of lanthanides with pyrroloquinoline quinone. *Chemistry*, **26**(44), 10133–10139.

Macaskie L. E., Moriyama S., Mikheenko I., Singh S. and Murray A. J. (2017). Biotechnology processes for scalable, selective rare earth element recovery. In: Rare Earth Elements, Alfonso Orjuela, J.E. (ed.), IntechOpen, Rijeka, Croatia, pp. 1–39.

MacHalová L., Pipíška M., Trajteaová Z. and Horník M. (2015). Comparison of Cd^{2+} biosorption and bioaccumulation by bacteria-A radiometric study. *Nova Biotechnologica et Chimica*, **14**, 158–175.

Maldonado R. F., Sá-Correia I. and Valvano M. A. (2016). Lipopolysaccharide modification in gram-negative bacteria during chronic infection. *FEMS Microbiology Reviews*, **40**, 480–493.

Maleke M., Valverde A., Gomez-arias A., Cason E. D., Vermeulen J., Coetsee-hugo L., Swart H., van Heerden E. and Castillo J. (2019a). Anaerobic reduction of europium by a *Clostridium* strain as a strategy for rare earth biorecovery. *Scientific Reports*, **9**(1), 1–11.

Maleke M., Valverde A., Vermeulen J.-G., Cason E., Gomez-Arias A., Moloantoa K., Coetsee-Hugo L., Swart H., van Heerden E. and Castillo J. (2019b). Biomineralization and bioaccumulation of europium by a thermophilic metal resistant bacterium. *Frontiers in Microbiology*, **10**, 81.

Mann S., Heywood B. R., Rajam S. and Birchall J. D. (1989). Interfacial control of nucleation of calcium carbonate under organized stearic acid monolayers. *Proceedings of the Royal Society of London. A. Mathematical and Physical Sciences*, **423**(1865), 457–471.

Merroun M. L., Chekroun K. B., Arias J. M. and González-Muñoz M. T. (2003). Lanthanum fixation by *Myxococcus xanthus*: Cellular location and extracellular polysaccharide observation. *Chemosphere*, **52**, 113–120.

Michalak I., Chojnacka K. and Witek-Krowiak A. (2013). State of the art for the biosorption process – a review. *Applied Biochemistry and Biotechnology*, **170**, 1389–1416.

Moriwaki H. and Yamamoto H. (2013). Interactions of microorganisms with rare earth ions and their utilization for separation and environmental technology. *Applied Biochemistry and Biotechnology*, **97**, 1–8.

Moriwaki H., Masuda R., Yamazaki Y., Horiuchi K., Miyashita M., Kasahara J., Tanaka T. and Yamamoto H. (2016). Application of freeze-dried powders of genetically engineered microbial strains as adsorbents for rare earth metal ions. *ACS Applied Materials & Interfaces*, **8**, 26524–26531.

Morozzi G., Cenci G., Scardazza F. and Pitzurra M. (1986). Cadmium uptake by growing cells of gram-positive and gram-negative bacteria. *Microbios*, **48**(194), 27–35.

Mowafy A. M. (2020). Biological leaching of rare earth elements. *World Journal of Microbiology & Biotechnology*, **36**, 1–7.

Murthy S., Bali G. and Sarangi S. (2012). Lead biosorption by a bacterium isolated from industrial effluents. *International Journal of Microbiology Research*, **4**, 196–200.

Nakagawa T., Mitsui R., Tani A., Sasa K., Tashiro S., Iwama T., Hayakawa T. and Kawai K. (2012). A catalytic role of XoxF1 as La^{3+}-dependent methanol dehydrogenase in *Methylobacterium extorquens* strain AM1. *PLoS One*, **7**, 1–7.

Navarro J. and Zhao F. (2014). Life-cycle assessment of the production of rare-earth elements for energy applications: a review. *Frontiers in Energy Research*, **2**, 1–17.

Newsome L., Morris K., Trivedi D., Atherton N. and Lloyd J. R. (2014). Microbial reduction of uranium(VI) in sediments of different lithologies collected from Sellafield. *Applied Geochemistry*, **51**, 55–64.

Newsome L., Morris K., Trivedi D., Bewsher A. and Lloyd J. R. (2015). Biostimulation by glycerol phosphate to precipitate recalcitrant uranium(IV) phosphate. *Environmental Science & Technology*, **49**, 11070–11078.

Nygaard R., Romaniuk J. A. H., Rice D. M. and Cegelski L. (2015). Spectral snapshots of bacterial cell-wall composition and the influence of antibiotics by whole-cell NMR. *Biophysical Journal*, **108**, 1380–1389.

Ochsner A. M., Hemmerle L., Vonderach T., Nüssli R., Bortfeld-Miller M., Hattendorf B. and Vorholt J. A. (2019). Use of rare-earth elements in the phyllosphere colonizer *Methylobacterium extorquens* PA1. *Molecular Microbiology*, **111**, 1152–1166.

Ohmura N., Sasaki K., Matsumoto N. and Saiki H. (2002). Anaerobic respiration using Fe^{3+}, S^0, and H_2 in the chemolithoautotrophic bacterium *Acidithiobacillus ferrooxidans*. *Journal of Bacteriology*, **184**, 2081–2087.

Ojima Y., Kosako S., Kihara M., Miyoshi N., Igarashi K. and Azuma M. (2019). Recovering metals from aqueous solutions by biosorption onto phosphorylated dry baker's yeast. *Scientific Reports*, **9**(1), 1–9.

Okuda S., Sherman D. J., Silhavy T. J., Ruiz N. and Kahne D. (2016). Lipopolysaccharide transport and assembly at the outer membrane: the PEZ model. *Nature Reviews Microbiology*, **14**, 337–345.

Orris G. J. and Grauch R. I. (2002). Rare earth element mines, deposits, and occurrences. Open-File Rep. 02-189 174.

Osman Y., Gebreil A., Mowafy A. M., Anan T. I. and Hamed S. M. (2019). Characterization of *Aspergillus niger* siderophore that mediates bioleaching of rare earth elements from phosphorites. *World Journal of Microbiology & Biotechnology*, **35**, 1–10.

Oves M., Khan M. S. and Zaidi A. (2013). Biosorption of heavy metals by *Bacillus thuringiensis* strain OSM29 originating from industrial effluent contaminated north Indian soil. *Saudi Journal of Biological Sciences*, **20**, 121–129.

Ozaki T., Kimura T., Ohnuki T. and Francis A. J. (2005). Associations of Eu(III) with gram-negative bacteria, *Alcaligenes faecalis*, *Shewanella putrefaciens*, and *Paracoccus denitrificans*. *Journal of Nuclear and Radiochemical Sciences*, **6**(1), 73–76.

Ozaki T., Suzuki Y., Nankawa T., Yoshida T., Ohnuki T., Kimura T. and Francis A. J. (2006). Interactions of rare earth elements with bacteria and organic ligands. *Journal of Alloys and Compounds*, **408**, 1334–1338.

Pan X., Wu W., Lü J., Chen Z., Li L., Rao W. and Guan X. (2017). Biosorption and extraction of europium by Bacillus thuringiensis strain. *Inorganic Chemistry Communications*, **75**, 21–24.

Park S. and Liang Y. (2019). Bioleaching of trace elements and rare earth elements from coal fly ash. *International Journal of Coal Science & Technology*, **6**(1), 74–83.

Park D. M., Reed D. W., Yung M. C., Eslamimanesh A., Lencka M. M., Anderko A., Fujita Y., Riman R. E., Navrotsky A. and Jiao Y. (2016). Bioadsorption of rare earth elements through cell surface display of lanthanide binding tags. *Environmental Science & Technology*, **50**, 2735–2742.

Peng L., Yi L., Zhexue L., Juncheng Z., Jiaxin D., Daiwen P., Ping S. and Songsheng Q. (2004). Study on biological effect of La^{3+} on *Escherichia coli* by atomic force microscopy. *Journal of Inorganic Biochemistry*, **98**, 68–72.

Pessoa J. C., Garribba E., Santos M. F. A. and Santos-Silva T. (2015). Vanadium and proteins: uptake, transport, structure, activity and function. *Coordination Chemistry Reviews*, **301**, 49–86.

Pfeffer C., Larsen S., Song J., Dong M., Besenbacher F., Meyer R. L., Kjeldsen K. U., Schreiber L., Gorby Y. A., El-Naggar M. Y. and Leung K. M. (2012). Filamentous bacteria transport electrons over centimetre distances. *Nature*, **491**, 218–221.

Philip L., Iyengar L. and Venkobachar C. (2000). Original Papers Biosorption of U, La, Pr, Nd, Eu and Dy by Pseudomonas aeruginosa. *Journal of Industrial Microbiology and Biotechnology*, **25**(1), 1–7.

Pidcock E. and Moore G. R. (2001). Structural characteristics of protein binding sites for calcium and lanthanide ions. *Journal of Biological Inorganic Chemistry*, **6**, 479–489.

Pinto M. R., Barreto-Bergter E. and Taborda C. P. (2008). Glycoconjugates and polysaccharides of fungal cell wall and activation of immune system. *Brazilian Journal of Microbiology*, **39**, 195–208.

Pol A., Barends T. R. M., Dietl A., Khadem A. F., Eygensteyn J., Jetten M. S. M. and Op den Camp H. J. M. (2014). Rare earth metals are essential for methanotrophic life in volcanic mudpots. *Environmental Microbiology*, **16**, 255–264.

Qu Y. and Lian B. (2013). Bioleaching of rare earth and radioactive elements from red mud using *Penicillium tricolor* RM-10. *Bioresource Technology*, **136**, 16–23.

Rajkumar M., Ae N., Prasad M. N. V. and Freitas H. (2010). Potential of siderophore-producing bacteria for improving heavy metal phytoextraction. *Trends in Biotechnology*, **28**, 142–149.

Ramos S. J., Dinali G. S., Oliveira C., Martins G. C., Moreira C. G., Siqueira J. O. and Guilherme L. R. G. (2016). Rare earth elements in the soil environment. *Current Pollution Reports*, **2**, 28–50.

Rasoulnia P., Barthen R. and Lakaniemi A.-M. (2020). A critical review of bioleaching of rare earth elements: the mechanisms and effect of process parameters. *Critical Reviews in Environmental Science and Technology*, **51**, 378–427.

Reed D. W., Fujita Y., Daubaras D. L., Jiao Y. and Thompson V. S. (2016). Bioleaching of rare earth elements from waste phosphors and cracking catalysts. *Hydrometallurgy*, **166**, 34–40.

Roszczenko-Jasińska P., Vu H. N., Subuyuj G. A., Crisostomo R. V., Cai J., Lien N. F., Clippard E. J., Ayala E. M., Ngo R. T., Yarza F., Wingett J. P., Raghuraman C., Hoeber C. A., Martinez-Gomez N. C. and Skovran E. (2020). Gene products and processes contributing to lanthanide homeostasis and methanol metabolism in Methylorubrum extorquens AM1. *Sci Rep*, **163**.

Ruming Z., Liu Y. I., Wenhua L. I., Ping S., Songsheng Q. U. and Ziniu Y. U. (2003). Effect of Sm^{3+} ion on growth of *Bacillus thuringiensis* by microcalorimetry. *Biological Trace Element Research*, **95**, 269–277.

Sarvas M., Harwood C. R., Bron S. and Van Dijl J. M. (2004). Post-translocational folding of secretory proteins in gram-positive bacteria. *Biochimica et Biophysica Acta – Molecular Cell Research*, **1694**, 311–327.

Schmid M. (2019). Mitigating supply risks through involvement in rare earth projects: Japan's strategies and what the US can learn. *Resources Policy*, **63**, 1–10.

Shin D., Kim J., Kim B., Jeong J. and Lee J. (2015). Use of phosphate solubilizing bacteria to leach rare earth elements from monazite-bearing ore. *Minerals*, **5**, 189–202.

Silhavy T. J., Kahne D. and Walker S. (2010). The bacterial cell envelope. *Cold Spring Harbor Perspectives in Biology*, **2**(5), a000414.

Skovran E., Raghuraman C. and Martinez-gomez N. C. (2019). Lanthanides in methylotrophy. *Current Issues in Molecular Biology*, **33**, 101–115.

Somasundaran P., Hussien S., Patra P. and Shall H. El. (2018). Environmentally benign bio-leaching extraction of rare earth elements from non-conventional resources. *Annals of Microbiology and Research*, **2**, 54–60.

Su H., Zhang D., Antwi P., Xiao L., Zhang Z., Deng X., Lai C., Zhao J., Deng Y., Liu Z. and Shi M. (2021). Adaptation, restoration and collapse of anammox process to La(III) stress: performance, microbial community, metabolic function and network analysis. *Bioresource Technology*, **325**, 124731.

Takahashi Y., Châtellier X., Hattori K., Kato K. and Fortin D. (2005). Adsorption of rare earth elements onto bacterial cell walls and its implication for REE sorption onto natural microbial mats. *Chemical Geology*, **219**, 53–67.

Takenaka Y., Ozaki T. and Ohnuki T. (2004). Influence of ionic strength on cerium(III) and europium(III) sorption on *Halomonas elongata*. *Journal of Nuclear Science and Technology*, **41**, 1125–1127.

Texier A. C., Andres Y. and Le Cloirec P. (1999). Selective biosorption of lanthanide (La, Eu, Yb) ions by *Pseudomonas aeruginosa*. *Environmental Science & Technology*, **42**, 489–495.

Thompson V. S., Gupta M., Jin H., Vahidi E., Yim M., Jindra M. A., Nguyen V., Fujita Y., Sutherland J. W., Jiao Y. and Reed D. W. (2018). Techno-economic and life cycle analysis for bioleaching rare-earth elements from waste materials. *ACS Sustainable Chemistry & Engineering*, **6**(2), 1602–1609.

Tomchick D. R., Turner R. J., Switzer R. L. and Smith J. L. (1998). Adaptation of an enzyme to regulatory function: Structure of *Bacillus subtilis* PyrR, a *pyr* RNA-binding attenuation protein and uracil phosphoribosyltransferase. *Structure*, **6**, 337–350.

Tsuruta T. (2007). Accumulation of rare earth elements in various microorganisms. *Journal of Rare Earths*, **25**, 526–532.

Ueki T., Adachi T., Kawano S., Aoshima M., Yamaguchi N., Kanamori K. and Michibata H. (2003). Vanadium-binding proteins (vanabins) from a vanadium-rich ascidian *Ascidia sydneiensis samea*. *Biochimica et Biophysica Acta (BBA)-Gene Structure and Expression*, **1626**(1–3), 43–50.

Vardanyan N., Sevoyan G., Navasardyan T. and Vardanyan A. (2018). Recovery of valuable metals from polymetallic mine tailings by natural microbial consortium. *Environmental Technology*, **40**(26), 3467–3472.

Vijayaraghavan K. and Yun Y. S. (2008). Bacterial biosorbents and biosorption. *Biotechnology Advances*, **26**, 266–291.

Villar L. D. and Garcia O. (2006). Effect of anaerobic digestion and initial pH on metal bioleaching from sewage sludge. Journal of Environmental Science and Health, Part A, **41**, 211–222.

Vu H. N., Subuyuj G. A., Vijayakumar S., Good N. M., Martinez-Gomez N. C. and Skovran E. (2016). Lanthanide-dependent regulation of methanol oxidation systems in *Methylobacterium extorquens* AM1 and their contribution to methanol growth. *Journal of Bacteriology*, **198**, 1250–1259.

Wagner J. K., Setayeshgar S., Sharon L. A., Reilly J. P. and Brun Y. V. (2006). A nutrient uptake role for bacterial cell envelope extensions. *Proceedings of the National Academy of Sciences*, **103**(31), 11772–11777.

Waldron K. J. and Robinson N. J. (2009). How do bacterial cells ensure that metalloproteins get the correct metal? *Nature Reviews Microbiology*, **6**, 25–35.

Wang L., Hibino A., Suganuma S., Ebihara A., Iwamoto S., Mitsui R., Tani A., Shimada M., Hayakawa T. and Nakagawa T. (2020). Preference for particular lanthanide species and thermal stability of XoxFs in *Methylorubrum extorquens* strain AM1. *Enzyme and Microbial Technology*, **136**, 109518.

Wehrmann M., Billard P., Martin-Meriadec A., Zegeye A. and Klebensberger J. (2017). Functional role of lanthanides in enzymatic activity and transcriptional regulation of pyrroloquinoline quinone-dependent alcohol dehydrogenases in *Pseudomonas putida* KT2440. *MBio*, **8**, 1–14.

Wehrmann M., Berthelot C., Billard P. and Klebensberger J. (2018). The PedS2/PedR2 two-component system is crucial for the rare earth element switch in Pseudomonas putida KT2440. *Msphere*, **3**(4), e00376-18.

Wenhua L., Ruming Z., Zhixiong X., Xiangdong C. and Ping S. (2003). Effects of La^{3+} on growth, transformation, and gene expression of *Escherichia coli*. *Biological Trace Element Research*, **94**, 167–177.

Wong J. W. C., Xiang L. and Chan L. C. (2002). pH requirement for the bioleaching of the heavy metals from anaerobically digested wastewater sludge. *Water, Air, & Soil Pollution*, **138**, 25–35.

Yavari M., Ebrahimi S. and Aghazadeh V. (2019). Kinetics of different bioreactor systems with *Acidithiobacillus ferrooxidans* for ferrous iron oxidation. *Reaction Kinetics, Mechanisms and Catalysis*, **128**, 611–627.

Yoshida T., Ozaki T., Ohnuki T. and Francis A. J. (2004). Adsorption of rare earth elements by β-Al_2O_3 and *Pseudomonas fluorescens* cells in the presence of desferrioxamine B: implication of siderophores for the Ce anomaly. *Chemical Geology*, **212**, 239–246.

Zhang B., Liu C., Li C. and Jiang M. (2014). A novel approach for recovery of rare earths and niobium from Bayan Obo tailings. *Minerals Engineering*, **65**, 17–23.

Zhang L., Dong H., Liu Y., Bian L., Wang X., Zhou Z. and Huang Y. (2018). Bioleaching of rare earth elements from bastnaesite-bearing rock by actinobacteria. *Chemical Geology*, **483**, 544–557.

Zhao R., Liu Y., Xie Z., Shen P. and Qu S. (2002). Microcalorimetric study of the action of Ce (III) ions on the growth of *E. coli*. *Biological Trace Element Research*, **86**, 167–175.

Zhou G., Shi Q.-S., Huang X.-M. and Xie X.-B. (2015). The three bacterial lines of defense against antimicrobial agents. *International Journal of Molecular Sciences*, **16**, 21711–21733.

Chapter 9

Bioleaching of rare earth elements from industrial and electronic wastes: mechanism and process efficiency

Pema Lhamo and Biswanath Mahanty

9.1 INTRODUCTION

Rare earth elements (REE) are a class of 17 metals consisting of 15 lanthanides (La), yttrium (Y) and scandium (Sc) (Balaram, 2019; Mikołajczak *et al.*, 2017). There are two types of REE: light rare earth elements consisting of lanthanum (La), praseodymium (Pr), cerium (Ce), neodymium (Nd) and samarium (Sa), and heavy rare earth elements consisting of yttrium (Y), europium (Eu), terbium (Tb), dysprosium (Dy), holmium (Ho), erbium (Er), gadolinium (Gd), lutetium (Lu), ytterbium (Yb) and thulium (Tm) (Ambaye *et al.*, 2020). REE are naturally occurring elements and their unique physicochemical properties make them applicable in different sectors, including the electronics, medical, renewable energy and manufacturing sectors (Balaram, 2019; Du & Graedel, 2011). Rare earth elements are present in abundance and global demand is majorly fulfilled by countries like China (61.97%), the United States (12.2%), Myanmar (10.32%), Australia (9.85%) and India (1.41%) (Zhang *et al.*, 2017).

REE are used in the preparation of Mg-alloys or titanium-alloys due to their high solid-solubility, structural integrity, and corrosion resistance (by scavenging undesirable contaminants and stabilizing the oxide/hydroxide surface) – an ideal property in weight-sensitive and biomedical applications (Biesiekierski *et al.*, 2020; Willbold *et al.*, 2015). The Pt-REE alloy has been adopted as an active and durable electrocatalyst as its negative alloying energy or enthalpy of formation

doi: 10.2166/9781789062236_0207

confers resistance against dissolution (Peera *et al.*, 2019). In contrast to conventional materials, REE catalysts have enhanced anti-fatigue life and heat of formation properties. The crystal structure and chemical properties of REE are exploited in the production of glass and ceramics. Their magnetic properties are higher than those of conventional magnets (Zhang *et al.*, 2016).

Although REE are present in abundance they are not in their pure form. They are mostly found as constituents of ores and minerals. The main sources of REE are monazite, bastnäsite and loparite. Therefore, the recovery of REE from these sources is an important aspect for their use in different industrial sectors. Extraction and refining of REE are mostly carried out by hydrometallurgical and pyrometallurgical methods. Though extensively used, the physicochemical methods are often criticized for their high cost and energy demand, and their environmental impacts (Borra *et al.*, 2016a; Priya & Hait, 2017; Zhang *et al.*, 2020). Therefore, efficient, inexpensive and environmentally benign alternative approaches are warranted to meet the increasing demand for REE.

Microorganisms can adapt to extreme environments and are environmentally friendly and cost-efficient, making them a good choice for the recovery of REE (Fathollahzadeh *et al.*, 2019; Rasoulnia *et al.*, 2021). Microorganisms can interact with REE through different bioleaching processes and mobilize them. Mobilization of REE from the solid to the liquid phase is carried out by three bioleaching methods: redoxolysis, acidolysis and complexolysis (Rasoulnia *et al.*, 2021). Microorganisms recover REE through different processes, such as biosorption, bioaccumulation and bioprecipitation. Microbial metabolic processes are exploited by these mechanisms to recover REE. Microorganisms can produce organic acids, which decrease the pH, help with the recovery of REE and also allow microbes to survive in extreme environments (Dev *et al.*, 2020). A wide variety of microorganisms including bacteria, fungi, algae and cyanobacteria can be used for REE recovery from different environmental matrices (Table 9.1) (Rasoulnia *et al.*, 2021; Tsuruta, 2007).

This chapter summarizes the different processes that are used by microorganisms to recover REE from different sources. The role of algal and fungal species in REE recovery is also discussed. The chapter further summarizes studies carried out on microbial recovery of different types of REE.

9.2 MICROBIAL PROCESSES FOR RECOVERY OF RARE EARTH ELEMENTS (REE)

9.2.1 REE mobilization

Biorecovery of rare earth elements (REE) involves microbial metabolic processes (Dev *et al.*, 2020). REE are present in different sources both in the solid and the aqueous phase, where their interaction with microorganisms dictates the efficiency of REE mobilization or immobilization from, respectively, the solid or

Table 9.1 Recovery of rare earth elements by different types of microorganisms.

REE	Microorganism	REE Source	Recovered Amount[a]	Reference
Bacteria				
Y, La,	*Pseudomonas aeruginosa*	Thorium-uranium concentrate	1.2%, 4.3%,	Desouky *et al.* (2016)
Sm	*B. subtilis*	La and Sm oxide	30 mg l ^{1}	Giese and Jordão (2019)
Nd, Eu	*Serratia sp.*	—	90%, 85%	Deplanche *et al.* (2012)
Eu	*Thermus scotoductus*	Eu solution	78%	Maleke *et al.* (2019)
Sc, Lu, Y	*Acetobacter* sp.	Red mud	52%, 53%, 61%	Qu *et al.* (2019)
Nd	*Acidithiobacillus ferrooxidans*	Magnets	86.4%	Auerbach *et al.* (2019a)
Er, Nd, Ce, La	*Leptospirillum ferrooxidans*	Dry discharged incineration slag	100%, 95%, 55%, 54%	Auerbach *et al.* (2019b)
Nd	*L. ferrooxidans*	Magnets	91.3%	Auerbach *et al.* (2019a)
Fungi				
Ce, La, Th, Sa	*A. niger*	Phosphorites	50.1%, 51%, 55%, 66.7%	Osman *et al.* (2019)
Yb, Sc, Eu, La	*A. niger*	Red mud	59%, 44%, 30%, 30%	Qu *et al.* (2015)
Er, Yb	*Candida bombicola*	Coal fly ash	64.6%, 67.7%	Park and Liang (2019)
Dy	*Penidiella sp.* T9	Acidic wastewater	910 μg mg ^{1}	Horiike and Yamashita (2015)

(Continued)

Table 9.1 Recovery of rare earth elements by different types of microorganisms (*Continued*).

REE	Microorganism	REE Source	Recovered Amount[a]	Reference
Yeast				
La	*Saccharomyces cerevisiae*	—	67 mg g ^{1}	Di Caprio *et al.* (2016)
La	*S. cerevisiae rim20Δ*	—	80 mg g ^{1}	Di Caprio *et al.* (2016)
Algae				
Sm	*Turbinaria conoides*	Samarium nitrate hexahydrate	151.6 mg l ^{1}	Vijayaraghavan *et al.* (2017)
Yb, Eu, Ce, La	*T. conoides*	Solution of REE as nitrate derivatives	121.2 mg g ^{1}, 138.2 mg g ^{1}, 152.8 mg g ^{1}, 154.7 mg g ^{1}	Vijayaraghavan *et al.* (2010)
Nd	*Chlorella vulgaris*	Nd-Fe-B magnets	128.21 mg g ^{1}	Kucuker *et al.* (2014)
Sc	*Posidonia oceanica*	Scandium chloride hexahydrate	66.81 mg g ^{1}	Ramasamy *et al.* (2019)
Nd	*Sargassum sp.*	La(III) and Nd(III) oxide	70–80 mg g ^{1}	Oliveira *et al.* (2012)
Microalgae				
La	*Desmodesmus quadricauda*	Red mud	27.3 mg kg ^{1} d ^{1}	Čížková *et al.* (2019)
La	*Chlamydomonas reinhardtii*	Red mud	24.5 mg kg ^{1} d ^{1}	Čížková *et al.* (2019)
La	*Parachlorella kessleri*	Red mud	12.5 mg kg ^{1} d ^{1}	Čížková *et al.* (2019)

[a]The recovery values reported in % are based on the total REE pool.

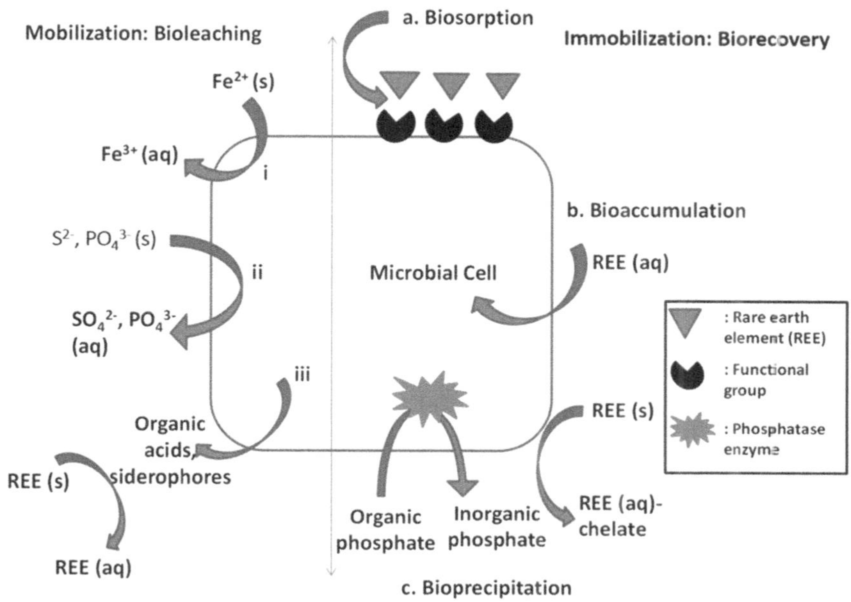

Figure 9.1 Microbial metabolic mechanisms involved in REE recovery. Solid to aqueous phase mobilization occurs through (i) redoxolysis, (ii) acidolysis and (iii) complexolysis methods. REE can either (a) bind with functional groups on microbial surface in biosorption, (b) accumulate inside the cell through channels, or (c) get precipitated with inorganic phosphate liberated through phosphatase enzymes (adapted and modified from Nancharaiah *et al.*, 2016).

the liquid phase (Barmettler *et al.*, 2016). REE in the solid phase are first to be mobilized and this is carried out by processes such as redoxolysis, acidolysis and complexolysis (Barmettler *et al.*, 2016; Brandl and Faramarzi, 2006) as shown in Figure 9.1.

9.2.1.1 Redoxolysis

Mobilization through redoxolysis involves transferring electrons from minerals to microorganisms by oxidation of Fe^{2+} to Fe^{3+}. Microorganisms involved in redoxolysis are *Acidithiobacillus thiooxidans*, *A. ferrooxidans* and *Leptospirillum ferrooxidans* (Auerbach *et al.*, 2019a; Sethurajan *et al.*, 2018).

Microorganisms can interact through either contact mode or non-contact mode. In the contact mode, microorganisms produce extracellular polymeric substances, which serve as a site of the oxidation reaction. Two variants of contact mode, that is one-step and two-step processes, can be relevant depending on the temporal separation between microbial growth and leaching. The REE-containing material and microbial culture are added simultaneously in the same medium in the one-step method. In the two-step process, the REE-containing material is mixed with the pre-grown microbial culture (Qu & Lian, 2013). A one-step process is often preferred for low mineral source concentrations, whereas a two-step process is used for higher REE concentrations (Yang *et al.*, 2008).

The presence of an insoluble matrix containing REE, such as red mud, in a microbial growth environment can exert severe toxicity, downregulating the production of organic acids (Qu *et al.*, 2013). In non-contact mode, the mineral source surface or matrix is not exposed to the microorganisms, rather a growth medium containing leaching agents secreted by the pre-grown culture is added (Zhang *et al.*, 2018).

9.2.1.2 Acidolysis

Acidolysis is carried out by sulfur-oxidizing or phosphate-oxidizing bacteria. Dissolution of REE by sulfur-oxidizing bacteria is carried out by oxidizing sulfide to sulfuric acid. Sulfur-oxidizing bacteria involved in acidolysis are *Alicyclosbacillus disulfidooxidans, A. ferroxidans* and *A. thiooxidans* (Baker & Banfield, 2003). Phosphate-oxidizing bacteria release phosphate from the mineral for dissolution of the REE. Some phosphate-oxidizing bacteria involved in acidolysis include *Enterobacter, Erwinia, Micrococcus, Bacillus, Klebsiella, Pseudomonas, Clavibacter, Thiobacillus, Agrobacterium, Acetobacter, Burkholderia, Serratia, Flavobacterium* and *Streptomyces* (Fathollahzadeh *et al.*, 2019).

9.2.1.3 Complexolysis

In complexolysis, dissolution of REE from the solid phase is carried out by organic acids and siderophores produced by microorganisms (Dev *et al.*, 2020). Microorganisms do not directly interact with the elements, but instead produce organic acids which solubilize the REE and release them from the solid matrix (Hopfe *et al.*, 2017). Effective mobilization of phosphate lattice bound REE from insoluble monazite ore by syntrophic action of native *Firmicutes* and inoculated microbial species (*E. aerogenes*, *P. agglomerans* and *P. putida*) has been observed under laboratory conditions (Corbett *et al.*, 2018). A co-culture of autotrophic *Acidithiobacillus ferrooxidans* and heterotrophic *E. aerogenes* has been very effective in REE bioleaching (up to 40 mg l ^{1}) from monazite, possibly due to the synergic effect of the biogenic organic acids and sulfuric acid (Fathollahzadeh *et al.*, 2018). *Aspergillus niger* can release Ce from bioleaching of monazite in defined synthetic nutrient media (Keekan *et al.*, 2017). Organic acids produced by *Aspergillus niger* (gluconic, citric, oxalic, itaconic and acetic) were tested to solubilize REE, among which the highest solubilization, that is 3 mg l ^{1}, was attained by the use of citric acid (Brisson *et al.*, 2016). Recovery of REE from monazite ore by *A. ficuum* and *P. aeruginosa* amounted to 60.6% and 52.6% in a one-step, and 55% and 47.7% in a two-step process, respectively (Hassanien *et al.*, 2014).

Siderophores are iron carriers produced by microorganisms under low iron availability. Extracellularly secreted microbial siderophores bind with iron present in the environment and traffic it back into the cell. Siderophores form stable complexes with REE which is due to their chelating property, thus resulting in mobilization and extraction of REE (Osman *et al.*, 2019). Siderophores produced by *P. aeruginosa* extracted 1.2% of yttrium (Y), 4.3% of lanthanum (La) and 5.4% of cerium (Ce) (Desouky *et al.*, 2016). Higher removal efficiencies of REE were reported by siderophores produced by *A. niger,* that is 50.1% of Ce, 51% of La, 55% of thorium (Th) and 66.7% of Sa. The authors have suggested that the higher REE recovery can be attributed to the trihydroxymate nature of the siderophore, having better stability with REE complexes (Osman *et al.*, 2019). The difference between the recovery of REE through bacterial and fungal siderophores could be due to the diversity in siderophores produced by bacteria (mixed ligand) and fungi (hydroxamate) and their binding capacity with REE (Ahmed & Holmström, 2014).

9.2.2 REE biorecovery

9.2.2.1 Biosorption

The mechanism of biosorption requires a liquid phase comprising the dissolved species to be sorbed and a solid phase as the sorbent (Gallardo *et al.*, 2020). The biosorption process can be performed by both living and dead microbial cells. Biosorption by living cells is a metabolism dependent process where the elements are adsorbed and transport across the cell membrane. Biosorption by dead cells is a metabolism independent process where the functional groups present on the surface of the microorganisms bind with the REE (Javanbakht *et al.*, 2014) (Figure 9.1). Amino, carboxyl, hydroxyl, phosphate and sulfhydryl groups of glycoproteins, polysaccharides and lipids present on the microbial surface are the binding sites for REE (Giese, 2020). Negatively charged groups such as the carboxyl group can easily bind with metal cations through an ion-exchange mechanism. Amine groups can chelate cations whereas adsorption of anions occurs through either hydrogen bonds or electrostatic interaction.

Common microbial biosorbents used are *Bacillus, Aspergillus, Pseudomonas, Rhizopus* and *Streptomyces*. Biosorbents can be modified to enhance the adsorption capacity in two ways, that is chemical or genetic modification. Chemical modification is used to enhance the binding site using chemical pretreatments, whereas genetic modifications are carried out to increase the accumulating properties and resistance to adverse conditions. The pH of the solution is an important factor affecting the biosorption process where the ideal range is often projected to be between pH 3 and pH 6. A higher pH value can alter the binding sites on the microbial cell surface and often be detrimental to the adsorption process. High temperature facilitates high kinetic energy and surface activity, resulting in higher adsorption (Vijayaraghavan & Yun, 2008).

The biosorption process has many advantages such as simple to operate, no chemical sludge produced and effective in removing contaminants from highly diluted solutions (Das & Das, 2013). Biosorption has been used to recover many REE from various source materials. Pre-treated *Bacillus subtilis* with alkali was used for biosorptive recovery of Sm^{3+} which recovered 30 mg l ^{1} and nearly 100% adsorption at pH 3 (Giese & Jordão, 2019). A biosorption study on Sm recovery using *Turbinaria conoides*, carried out at different pH values 2.0–5.0, suggested a strong influence of pH on the adsorption capacity (Vijayaraghavan *et al.*, 2017). The recovery increased until pH 4 (151.6 mg g ^{1}) but a further increase in pH resulted in a drop in recovery.

The microalgae *Chlorella vulgaris* used for neodymium (Nd) recovery could adsorb 128.21 mg g ^{1} from an initial concentration of 250 mg l ^{1} of Nd at pH 4 and 50°C (Kucuker *et al.*, 2014). Better recovery (157.40 mg g ^{1}) was noted at a lower temperature (30°C) but higher pH (pH 5) with an equivalent Nd concentration (Kucuker *et al.*, 2017). This shows that the recovered amount of Nd is dependent on pH and temperature and a possible interactive effect may be crucial for process performance. Poly-γ-glutamic acid in combination with its sodium salt has been used for the recovery of Nd as a potential biosorbent with a maximum recovery of 215 mg g ^{1} at pH 4 and 305 mg g ^{1} at pH 3 (Hisada & Kawase, 2018). Amino-methyl phosphonic acid has also been reported to recover Nd(IV) but at a significantly lower efficiency, that is 30.32 mg g ^{1} (Elsalamouny *et al.*, 2017).

Two strains of *Saccharomyces cerevisiae,* that is the wild and rim20Δ mutant type, were studied for recovery of lanthanum (La) at two pH (4.0 and 6.0) conditions. Mutant type rim20*Δ* showed better recovery compared to the wild type in both pH conditions, that is 70 mg g ^{1} (pH 4) and 80 mg g ^{1} (pH 6). Improvement in recovery was also observed in the wild type when the solution pH was increased from pH 4 (40 mg g ^{1}) to pH 6 (67 mg g ^{1}), though it was still lower than that of the mutant (Di Caprio *et al.*, 2016). Fast biosorptive enrichment of europium (Eu), samarium (Sm) and neodymium (Nd) from aqueous solution (~0.1 mM) has been reported with live cyanobacterial strains, namely *Anabaena sp.* and *Anabaena cylindrical,* where more than 90% of REE ions could be removed within an hour (Fischer *et al.*, 2019). The result implicated active adsorption of Eu inside the vegetative cyanobacterial cells rather than extracellular polysaccharides (EPS) mediated cell wall precipitation. *Spirulina platensis* biomass has been shown to adsorb rhenium ions from industrial effluents with a maximum biosorption capacity of 142.9 mg g ^{1}. The adsorption process involves either ionic interaction of perrhenate anions (ReO^{4}) with amide/amino-groups or binding to cell surface organic functionalities (Zinicovscaia *et al.*, 2018). Cyanobacteria of the genus *Arthrospira* can adsorb cerium (III) ions from aqueous solutions with a maximum adsorption capacity of 18.1 mg g ^{1} under mild acidic conditions (Sadovsky *et al.*, 2016).

9.2.2.2 Bioaccumulation

Bioaccumulation is another metabolic process that is used to recover REE by its intracellular uptake from the cell surface (Dev *et al.*, 2020). It is a metabolically active mechanism where microorganisms use importer complexes to translocate the REE and accumulate them into their intracellular space. The REE move into the cell through the lipid membrane and once inside the cell, they are sequestered by a storage system, that is peptide ligands and proteins. Channels such as α-helical proteins accumulate REE by facilitating passive diffusion and are energy independent. Porins present on the outer membrane of gram-negative bacteria help to translocate REE across the membrane through β-barrels transport channels to the periplasmic space and then transport them into the cytoplasm for storage.

Another group of single component protein carriers, that is antiporters, uniporters and symporters, is also involved in the bioaccumulation process. These carrier proteins are energy dependent, where uniporters use the charge difference by the proton motive force to translocate cationic metals. Symporters also depend on the proton motive force where the proton creating the charge difference is used as a co-substrate to translocate the target metal. As the bioaccumulation process is carried out by metabolically active living cells, the nutrient and energy requirement of the cells needs to be considered (Diep *et al.*, 2018).

Microorganisms have also been used for the bioaccumulation of REE. *B. cereus* (with 1% soil sample) accumulated 7.05 ($\pm$ 0.01) μmol g^{-1} cell dry weight (CDW) of Ce and 3.17 ($\pm$ 0.01) μmol g^{-1} CDW of Nd (Emmanuel *et al.*, 2011). *Thermus scotoductus* SA-01 recovered europium (Eu) under thermophilic conditions with 78% of it retained on the bacterium's cell surface. The findings from the study are important, not only due to the ability of the microorganism to interact with REE at high temperature (i.e. 65°C), but also its viability at high Eu concentrations, that is up to 1 mM Eu (Maleke *et al.*, 2019). Recovery of REE from 69 species (16 yeasts, 18 fungi, 20 actinomycetes and 22 bacteria) was tested among which *B. lichemiformis* accumulated the highest Sm (316 μmol g^{-1} CDW). The study also reported preferential removal when an equimolar mixture of two REE was used (Tsuruta, 2007).

9.2.2.3 Bioprecipitation

In the bioprecipitation process, the production of inorganic phosphate during microbial metabolism leads to the precipitation of REE as phosphate (Crocket *et al.*, 2018; Feng *et al.*, 2011; Liang *et al.*, 2016). The enzyme phosphatase is translocated to the extracellular polymeric matrix of the microbial cell and when the enzyme is supplied with enough organic phosphate, it secretes out inorganic phosphate. The liberated phosphate binds with the metal ions and causes precipitation (Macaskie *et al.*, 2017). Phosphatase present in the periplasm of

Serratia sp. generates inorganic phosphate allowing the precipitation of REE present in the aqueous solution (Macaskie *et al.*, 2005).

This precipitation process has been used to recover 90% Nd and 85% Eu by *Serratia* sp. (Macaskie *et al.*, 2017). Dysprosium (Dy) has been recovered as $DyPO_4$ by *Penidiella* sp. with a similar mechanism at pH 2.5 (Horiike *et al.*, 2018). The solution pH plays an important role in the precipitative recovery of REE. Recovery of La (as $LaPO_4$) by *Citrobacter* sp. is reduced by 50% due to lower desolubilization of $LaPO_4$ at pH 5 (Tolley *et al.*, 1995). The bioprecipitation process can also be applied for the selective recovery of REE. Selective recovery of REE has been reported during fluorapatite dissolution by *B. megaterium*, where light REE (Nd, La, Sm, Pr and Ce) are precipitated as phosphate salt and heavy REE (Ho, Dy, Tm, Er and Gd) remain dissolved (Feng *et al.*, 2011).

9.3 ROLE OF ALGAL AND FUNGAL SPECIES IN THE RECOVERY OF REE

9.3.1 Algae

Algal species can recover REE through biosorption or bioaccumulation processes where the accumulation process differs with different properties of algal species, medium and the type of element to be recovered (Goecke *et al.*, 2015). Several studies have been reported on biosorption and accumulation of REE by algal species, for example recovery of Yb, Eu, Ce and La by *T. conoides* with a maximum biosorption uptake of 121.2, 138.2, 152.8, 154.7 mg g ^{1}, respectively (Vijayaraghavan *et al.*, 2010). *Posidonia oceanica*, a marine algae grafted with 1-(2-pyridylazo)-2-naphthol adsorbs a maximum of 66.81 mg g ^{1} Sc (III) (Ramasamy *et al.*, 2019). Recovery of Nd(III) and La(III) by *Sargassum* sp. showed a maximum absorption of 70–80 mg g ^{1}, where Nd was better adsorbed by the species than La (Oliveira *et al.*, 2012).

Different components and states of algal and cyanobacterial species have been used in recovering REE, for example tissue homogenate of dehydrated *Ulva lactuca* effectively adsorbs yttrium and other rare earth elements (Zoll & Schijf, 2012). Adsorption of Nd^{3+} can be achieved using the sacran component from *Aphanothece sacrum* and polyvinyl alcohol hydrogel in a ratio of 3:1 (sacran anion/Nd^{3+}) (Okajima *et al.*, 2010). The microalgae *D. quadricauda* accumulates 27.3 mg kg ^{1} of La per day, followed by *C. reinhardtii* and *P. kessleri* accumulating, respectively, 24.5 and 12.5 mg kg ^{1} of La per day (Čížková *et al.*, 2019).

9.3.2 Fungi

The fungal REE recovery mechanism primarily involves bioleaching due to the production of organic acids by the fungal species used. Fungal species, especially

Aspergillus species, recover REE from different sources. For example, *A. niger* and *A. flavus* can recover 86% of the total REE, where *A. niger* shows a higher recovery efficiency and higher oxalic and citric acid content compared to *A. flavus* (Amin *et al.*, 2014). Rare earth elements such as Th, Nd, Pr, La and Ce have been recovered from monazite sand by *A. niger* and *Paecilomyces* sp., which was attributed to the organic acids produced such as citric, oxalic, acetic, gluconic, succinic and itaconic acids (Brisson *et al.*, 2016). Recovery of Dy was carried out by *Penidiella* sp. T9, with the recovered concentration of 910 $\mu g\ mg^{-1}$ (Horiike & Yamashita, 2015). Gluconic acid produced by *Talaromyces* has been effective in the recovery of REE such as Y, Ce, La, Tb and Eu from retorted phosphor powders (Reed *et al.*, 2016a, 2016b). The tea fungus *Kombucha* can effectively recover REE from fluorescent phosphor (Hopfe *et al.*, 2017). The above studies show that fungal species can recover REE through the production of organic acids, which effectively precipitate the REE in solution.

9.4 MICROBIAL RECOVERY OF REE FROM DIFFERENT WASTES

9.4.1 Coal fly ash

Fly ash is treated as a residue arising from thermal coal processing and is considered a concern as solid waste worldwide. Fly ash is a fine-grained solid created during coal combustion from noncombustible coal constituents such as quartz and clay minerals. REE are preserved and enriched in fly ash as coal is burnt, therefore making it a possible source of REE (Dwivedi & Jain, 2014). China with India, Africa, Europe and the Middle East are the main producers of coal fly ash (Sutcu *et al.*, 2019). Extraction of Yb and Er through bioleaching by *Phanerochaete chrysosporium, Candida bombicola* and *Cryptococcus curvatus* has been tested where the highest extraction was achieved by *Candida bombicola,* that is 64.6% of Er and 67.7% of Yb (Park & Liang, 2019). Recovery of REE through *C. bombicola* can also be attributed to its sophorolipid production, which binds with metals through ion exchange leading to mobilization of metals into the solution and easier precipitation (Ayangbenro & Babalola, 2018). Acidophilic chemolithotrophs (*Acidithiobacillus ferrooxidans, A. caldus, A. thiooxidans* and *Sulfobacillus thermosulfidooxidans*) recovered 52.0% of Sc, 52.6% of Y and 59.5% of La at pH 2.0, 10% pulp density and 45°C (Muravyov *et al.*, 2015). Bioleaching by *Leptospirillum ferrooxidans* of fly ash at pH 2.2 and 25°C demonstrated a 100% recovery of Er, 95% of Nd, 55% of Ce and 54% of La (Auerbach *et al.*, 2019b).

9.4.2 Electronic wastes

Electronic wastes (E-waste) include electronic scraps like DVD players, cell phones, cassettes, scanners and electronic equipment, which are generated via an increase in innovation in technology resulting in the replacement of old ones (Abdelbasir *et al.*, 2018). There is a rapid growth in the volume of generated E-waste and the global annual footprint could be about 41.8 million tonnes (based on 2014 estimate) at a compound annual growth rate (CAGR) of 23.5% with a major share from China, the United States and India (Awasthi & Li, 2017). E-wastes contain a higher concentration of REE compared to natural minerals and increasing E-waste quantities make it an efficient source for REE recovery (Tan & Li, 2019).

E-wastes such as batteries and magnets contain REE, which can be extracted and used to make new batteries and magnets. E-waste magnets are important components containing about 20–30% REE (Venkatesan *et al.*, 2018). Dy, Pr and Nd are commonly found REE in magnets with concentrations of 42.1, 3.4 and 259.5 ppm found in NdFeB magnets (Van der Hoogerstraete *et al.*, 2014). The bioleaching process has been used to recover Nd from E-wastes using *A. ferrooxidans* and *L. ferrooxidans* extracting 86.4% and 91.3%, respectively. The authors also recovered 100% of Pr using both microbial strains (Auerbach *et al.*, 2019a).

Phosphors are another rich source of REE predominantly found in products such as LEDs, cathode ray tubes, fluorescent lamps and LCDs (Jha *et al.*, 2016; Tan & Li, 2019). Studies have reported that phosphors contain more than 23% REE (Shin & Kim, 2015; Tan & Li, 2019; Yang *et al.*, 2013). Microorganisms able to recover REE from lighting and electronic displays include *Lactobacillus casei, Komagataeibacter xylins, Gluconobacter oxydans* and *Yarrowia lipolytica* (Hopfe *et al.*, 2018; Reed *et al.*, 2016a, 2016b).

9.4.3 Red mud

The slurry waste produced during the production of aluminum hydroxide from bauxite in the Bayer's process is called red mud. In the bauxite ore, REE ions are present on the surface of the mineral and end up in the slurry waste, that is red mud, during the Bayer's process. Among the REE, Sc holds the highest concentration and value with >95% of the economic value (Borra *et al.*, 2016b). Red mud has an alkaline environment and therefore extraction of REE is carried out by chemoheterotrophic bacteria and fungi as they can grow better in an alkaline environment, whereas chemoautotrophic bacteria cannot extract REE from red mud because of their low tolerance to alkaline environments (Santini *et al.*, 2015).

The fungal strain *Penicillium* tricolor RM-10 isolated from red mud was used to extract REE from red mud at 2% and 10% pulp density. The total extraction concentration of REE was 20–60 mg l^{-1} and the highest yield was obtained from the two-step bioleaching process at 10% pulp density (Qu & Lian, 2013). Another study reported extraction of Yb, Sc, Eu and La at 2% pulp density using *A. niger* from red mud with an extraction efficiency of 59%, 44% and 30% (both Eu and La), respectively (Qu *et al.*, 2015). *Acetobacter* sp. was reported to extract 52% of Sc, 53% of Lu and 61% of Y at 2% pulp density of red mud (Qu *et al.*, 2019). In all three studies mentioned above, it is observed that the microorganisms produce organic acids such as citric, gluconic and oxalic acid which help facilitate precipitation of REE and make the medium acidic. The microalgae species *Desmodesmus quadricauda, Chlamydomonas reinhardtii* and *Parachlorella kessleri* were able to extract La at a 10% pulp density, where the highest extraction yield was achieved by *D. quadricauda,* at 27.3 mg kg^{-1} per day. Red mud contains metals such as magnesium, calcium and iron which support algal growth (Čížková *et al.*, 2019).

9.5 CONCLUSION

An increase in industrial demand for REE has led to their microbial recovery, a better alternative compared to physicochemical methods. Bioleaching is an important method in mobilizing and extracting REE. Bioleaching is performed by many different types of microorganisms to recover REE, and other processes such as biosorption, bioaccumulation and bioprecipitation have been also shown to be effective in recovering REE. Organic acid production by heterotrophic bacteria and fungi is very effective in recovering REE as they facilitate precipitation of REE and also decrease the pH which helps these microorganisms to grow in an extreme environment. These microbial recovery methods are mostly metabolism-dependent, which makes the viability of microorganisms very important in achieving better recovery results. Supply of media, pH and temperature are important factors to be considered as they affect microbial growth, thus affecting the REE recovery efficiency, and optimization of these factors is needed for industrial large-scale production.

REFERENCES

Abdelbasir S. M., Hassan S. S., Kamel A. H. and El-Nasr R. S. (2018). Status of electronic waste recycling techniques: a review. *Environmental Science and Pollution Research*, **25**(17), 16533–16547.

Ahmed E. and Holmström S. J. (2014). Siderophores in environmental research: roles and applications. *Microbial Biotechnology*, **7**(3), 196–208.

Ambaye T. G., Vaccari M., Castro F. D., Prasad S. and Rtimi S. (2020). Emerging technologies for the recovery of rare earth elements (REEs) from the end-of-life electronic wastes: a review on progress, challenges, and perspectives. *Environmental Science and Pollution Research*, **27**, 36052–36074.

Amin M. M., El-Aassy I. E., El-Feky M. G., Sallam A. M., El-Sayed E. M., Nada A. A. and Harpy N. M. (2014). Fungal leaching of rare earth elements from lower carboniferous carbonaceous shales, southwestern Sinai, Egypt. *Romanian Journal of Biophysics*, **24**(1), 25–41.

Auerbach R., Bokelmann K., Stauber R., Gutfleisch O., Schnell S. and Ratering S. (2019a). Critical raw materials–advanced recycling technologies and processes: recycling of rare earth metals out of end of life magnets by bioleaching with various bacteria as an example of an intelligent recycling strategy. *Minerals Engineering*, **134**, 104–117.

Auerbach R., Ratering S., Bokelmann K., Gellermann C., Brämer T., Baumann R. and Schnell S. (2019b). Bioleaching of valuable and hazardous metals from dry discharged incineration slag. An approach for metal recycling and pollutant elimination. *Journal of Environmental Management*, **232**, 428–437.

Awasthi A. K. and Li J. (2017). Management of electrical and electronic waste: A comparative evaluation of China and India. *Renewable and Sustainable Energy Reviews*, **76**, 434–447.

Ayangbenro A. S. and Babalola O. O. (2018). Metal (loid) bioremediation: strategies employed by microbial polymers. *Sustainability*, **10**(9), 3028. doi: 10.3390/su10093028

Baker B. J. and Banfield J. F. (2003). Microbial communities in acid mine drainage. *FEMS Microbiology Ecology*, **44**(2), 139–152.

Balaram V. (2019). Rare earth elements: A review of applications, occurrence, exploration, analysis, recycling, and environmental impact. *Geoscience Frontiers*, **10**(4), 1285–1303.

Barmettler F., Castelberg C., Fabbri C. and Brandl H. (2016). Microbial mobilization of rare earth elements (REE) from mineral solids—A mini review. *AIMS Microbiology*, **2** (2),190–204. doi: 10.3934/microbiol.2016.2.190

Biesiekierski A., Li Y. and Wen C. (2020). The application of the rare earths to magnesium and titanium metallurgy in Australia. *Advanced Materials*, **32**(18), 1901715.

Borra C. R., Mermans J., Blanpain B., Pontikes Y., Binnemans K. and Van Gerven T. (2016a). Selective recovery of rare earths from bauxite residue by combination of sulfation, roasting and leaching. *Minerals Engineering*, **92**, 151–159.

Borra C. R., Blanpain B., Pontikes Y., Binnemans K. and Van Gerven T. (2016b). Recovery of rare earths and other valuable metals from bauxite residue (red mud): a review. *Journal of Sustainable Metallurgy*, **2**(4), 365–386.

Brandl H. and Faramarzi M. A. (2006). Microbe-metal-interactions for the biotechnological treatment of metal-containing solid waste. *China Particuology*, **4**(2), 93–97.

Brisson V. L., Zhuang W. Q. and Alvarez-Cohen L. (2016). Bioleaching of rare earth elements from monazite sand. *Biotechnology and Bioengineering*, **113**(2), 339–348.

Čížková M., Mezricky D., Rucki M., Tóth T. M., Náhlík V., Lanta V., Bišová K., Zachleder V. and Vítová M. (2019). Bio-mining of lanthanides from red mud by green microalgae. *Molecules*, **24**(7), 1356. doi: 10.3390/molecules24071356

Corbett M. K., Eksteen J. J., Niu X. Z. and Watkin E. L. (2018). Syntrophic effect of indigenous and inoculated microorganisms in the leaching of rare earth elements from western Australian monazite. *Research in Microbiology*, **169**(10), 558–568.

Crocket K. C., Hill E., Abell R. E., Johnson C., Gary S. F., Brand T. and Hathorne E. C. (2018). Rare earth element distribution in the NE Atlantic: evidence for benthic sources, longevity of the seawater signal, and biogeochemical cycling. *Frontiers in Marine Science*, **5**, 147. doi: 10.3389/fmars.2018.00147

Das N. and Das D. (2013). Recovery of rare earth metals through biosorption: An overview. *Journal of Rare Earths*, **31**(10), 933–943.

Deplanche K., Merroun M. L., Casadesus M., Tran D. T., Mikheenko I. P., Bennett J. A., Zhu J., Jones I. P., Attard G. A., Wood J. and Selenska-Pobell S. (2012). Microbial synthesis of core/shell gold/palladium nanoparticles for applications in green chemistry. *Journal of the Royal Society Interface*, **9**(72), 1705–1712.

Desouky O. A., El-Mougith A. A., Hassanien W. A., Awadalla G. S. and Hussien S. S. (2016). Extraction of some strategic elements from thorium–uranium concentrate using bioproducts of *Aspergillus ficuum* and *Pseudomonas aeruginosa*. *Arabian Journal of Chemistry*, **9**, S795–S805.

Dev S., Sachan A., Dehghani F., Ghosh T., Briggs B. R. and Aggarwal S. (2020). Mechanisms of biological recovery of rare-earth elements from industrial and electronic wastes: A review. *Chemical Engineering Journal*, **397**, 124596.

Di Caprio F., Altimari P., Zanni E., Uccelletti D., Toro L. and Pagnanelli F. (2016). Lanthanum biosorption by different *Saccharomyces cerevisiae* strains. *Chemical Engineering Transactions*, **49**, 37–42.

Diep P., Mahadevan R. and Yakunin A. F. (2018). Heavy metal removal by bioaccumulation using genetically engineered microorganisms. *Frontiers in Bioengineering and Biotechnology*, **6**, 157. doi: 10.3389/fbioe.2018.00157

Du X. and Graedel T. E. (2011). Global in-use stocks of the rare earth elements: a first estimate. *Environmental Science & Technology*, **45**(9), 4096–4101.

Dwivedi A. and Jain M. K. (2014). Fly ash–waste management and overview: A review. *Recent Research in Science and Technology*, **6**(1), 30–35.

Elsalamouny A. R., Desouky O. A., Mohamed S. A., Galhoum A. A. and Guibal E. (2017). Uranium and neodymium biosorption using novel chelating polysaccharide. *International Journal of Biological Macromolecules*, **104**, 963–968.

Emmanuel E. C., Vignesh V., Anandkumar B. and Maruthamuthu S. (2011). Bioaccumulation of cerium and neodymium by *Bacillus cereus* isolated from rare earth environments of Chavara and Manavalakurichi, India. *Indian Journal of Microbiology*, **51**(4), 488–495.

Fathollahzadeh H., Hackett M. J., Khaleque H. N., Eksteen J. J., Kaksonen A. H. and Watkin E. L. (2018). Better together: potential of co-culture microorganisms to enhance bioleaching of rare earth elements from monazite. *Bioresource Technology Reports*, **3**, 109–118.

Fathollahzadeh H., Eksteen J. J., Kaksonen A. H. and Watkin E. L. (2019). Role of microorganisms in bioleaching of rare earth elements from primary and secondary resources. *Applied Microbiology and Biotechnology*, **103**(3), 1043–1057.

Feng M. H., Ngwenya B. T., Wang L., Li W., Olive V. and Ellam R. M. (2011). Bacterial dissolution of fluorapatite as a possible source of elevated dissolved phosphate in the environment. *Geochimica et Cosmochimica Acta*, **75**(19), 5785–5796.

Fischer C. B., Körsten S., Rösken L. M., Cappel F., Beresko C., Ankerhold G., Schönleber A., Geimer S., Ecker D. and Wehner S. (2019). Cyanobacterial promoted enrichment of rare earth elements europium, samarium and neodymium and intracellular europium particle formation. *RSC Advances*, **9**(56), 32581–32593.

Gallardo K., Castillo R., Mancilla N. and Remonsellez F. (2020). Biosorption of rare-earth elements from aqueous solutions using walnut shell. *Frontiers in Chemical Engineering*, **2**, 4. doi: 10.3389/fceng.2020.00004

Giese E. C. (2020). Biosorption as green technology for the recovery and separation of rare earth elements. *World Journal of Microbiology and Biotechnology*, **36**(4), 1–11.

Giese E. C. and Jordão C. S. (2019). Biosorption of lanthanum and samarium by chemically modified free Bacillus subtilis cells. *Applied Water Science*, **9**(8), 1–8.

Goecke F., Zachleder V. and Vítová M. (2015). Rare earth elements and algae: physiological effects, biorefi nery and recycling. In: Algal Biorefineries: Volume 2: Products and Refinery Design. Springer International Publishing, Berlin, pp. 339–363. doi: 10.1007/978-3-319-20200-6_10

Hassanien W. A. G., Desouky O. A. N. and Hussien S. S. E. (2014). Bioleaching of some rare earth elements from Egyptian monazite using *Aspergillus ficuum* and *Pseudomonas aeruginosa*. *Walailak Journal of Science and Technology (WJST)*, **11**(9), 809–823.

Hisada M. and Kawase Y. (2018). Recovery of rare-earth metal neodymium from aqueous solutions by poly-γ-glutamic acid and its sodium salt as biosorbents: effects of solution pH on neodymium recovery mechanisms. *Journal of Rare Earths*, **36**(5), 528–536.

Hopfe S., Flemming K., Lehmann F., Möckel R., Kutschke S. and Pollmann K. (2017). Leaching of rare earth elements from fluorescent powder using the tea fungus Kombucha. *Waste Management*, **62**, 211–221.

Hopfe S., Konsulke S., Barthen R., Lehmann F., Kutschke S. and Pollmann K. (2018). Screening and selection of technologically applicable microorganisms for recovery of rare earth elements from fluorescent powder. *Waste Management*, **79**, 554–563.

Horiike T. and Yamashita M. (2015). A new fungal isolate, *Penidiella* sp. strain T9, accumulates the rare earth element dysprosium. *Applied and Environmental Microbiology*, **81**(9), 3062–3068.

Horiike T., Kiyono H. and Yamashita M. (2018). Dysprosium biomineralization by *Penidiella* sp. Strain T9. In: Biomineralization. Springer, Singapore, pp. 251–257. doi: 10.1007/978-981-13-1002-7_26

Javanbakht V., Alavi S. A. and Zilouei H. (2014). Mechanisms of heavy metal removal using microorganisms as biosorbent. *Water Science and Technology*, **69**(9), 1775–1787.

Jha M. K., Kumari A., Panda R., Kumar J. R., Yoo K. and Lee J. Y. (2016). Review on hydrometallurgical recovery of rare earth metals. *Hydrometallurgy*, **165**, 2–26.

Keekan K. K., Jalondhara J. C. and Abhilash (2017). Extraction of Ce and Th from monazite using REE tolerant *Aspergillus Niger*. *Mineral Processing and Extractive Metallurgy Review*, **38**(5), 312–320.

Kucuker M. A., Habib H. and Kuchta K. (2014). Biosorption of neodymium (Nd) from Fe-Nd-B magnets. In: *Biotechnological Approach for Recovery Of Rare Earth Elements and Precious Metals From E-Waste-(BIOREEs)*. EurAsia Waste Management Symposium, 28–30 April 2014, YTU 2010 Congress Center, İstanbul, pp. 1–15.

Kucuker M. A., Wieczorek N., Kuchta K. and Copty N. K. (2017). Biosorption of neodymium on *Chlorella vulgaris* in aqueous solution obtained from hard disk drive magnets. *PLoS One*, **12**(4), e0175255.

Liang X., Kierans M., Ceci A., Hillier S. and Gadd G. M. (2016). Phosphatase-mediated bioprecipitation of lead by soil fungi. *Environmental Microbiology*, **18**(1), 219–231.

Macaskie L. E., Yong P., Paterson-Beedle M., Thackray A. C., Marquis P. M., Sammons R. L., Nott K. P. and Hall L. D. (2005). A novel non line-of-sight method for coating hydroxyapatite onto the surfaces of support materials by biomineralization. *Journal of Biotechnology*, **118**(2), 187–200.

Macaskie L. E., Moriyama S., Mikheenko I., Singh S. and Murray A. J. (2017). Biotechnology processes for scalable, selective rare earth element recovery. In: Rare Earth Element. InTech, pp. 3–40. doi: 10.5772/intechopen.68429

Maleke M., Valverde A., Vermeulen J. G., Cason E., Gomez-Arias A., Moloantoa K., Coetsee-Hugo L., Swart H., Van Heerden E. and Castillo J. (2019). Biomineralization and bioaccumulation of europium by a thermophilic metal resistant bacterium. *Frontiers in Microbiology*, **10**, 81. doi: 10.3389/fmicb.2019.00081

Mikołajczak P., Borowiak K. and Niedzielski P. (2017). Phytoextraction of rare earth elements in herbaceous plant species growing close to roads. *Environmental Science and Pollution Research*, **24**(16), 14091–14103.

Muravyov M. I., Bulaev A. G., Melamud V. S. and Kondrat'eva T. F. (2015). Leaching of rare earth elements from coal ashes using acidophilic chemolithotrophic microbial communities. *Microbiology (Reading, England)*, **84**(2), 194–201.

Nancharaiah Y. V., Mohan S. V. and Lens P. N. L. (2016). Biological and bioelectrochemical recovery of critical and scarce metals. *Trends in Biotechnology*, **34**(2), 137–155.

Okajima M. K., Nakamura M., Mitsumata T. and Kaneko T. (2010). Cyanobacterial polysaccharide gels with efficient rare-earth-metal sorption. *Biomacromolecules*, **11**, 1773–1778.

Oliveira R. C., Guibal E. and Garcia O., Jr. (2012). Biosorption and desorption of lanthanum (III) and neodymium (III) in fixed-bed columns with *Sargassum* sp.: perspectives for separation of rare earth metals. *Biotechnology Progress*, **28**(3), 715–722.

Osman Y., Gebreil A., Mowafy A. M., Anan T. I. and Hamed S. M. (2019). Characterization of *Aspergillus Niger* siderophore that mediates bioleaching of rare earth elements from phosphorites. *World Journal of Microbiology and Biotechnology*, **35**(6), 1–10.

Park S. and Liang Y. (2019). Bioleaching of trace elements and rare earth elements from coal fly ash. *International Journal of Coal Science & Technology*, **6**(1), 74–83.

Peera S. G., Lee T. G. and Sahu A. K. (2019). Pt-rare earth metal alloy/metal oxide catalysts for oxygen reduction and alcohol oxidation reactions: an overview. *Sustainable Energy & Fuels*, **3**(8), 1866–1891.

Priya A. and Hait S. (2017). Comparative assessment of metallurgical recovery of metals from electronic waste with special emphasis on bioleaching. *Environmental Science and Pollution Research*, **24**(8), 6989–7008.

Qu Y. and Lian B. (2013). Bioleaching of rare earth and radioactive elements from red mud using *Penicillium tricolor* RM-10. *Bioresource Technology*, **136**, 16–23.

Qu Y., Lian B., Mo B. and Liu C. (2013). Bioleaching of heavy metals from red mud using *Aspergillus niger*. *Hydrometallurgy*, **136**, 71–77.

Qu Y., Li H., Tian W., Wang X., Wang X., Jia X., Shi B., Song G. and Tang Y. (2015). Leaching of valuable metals from red mud via batch and continuous processes by using fungi. *Minerals Engineering*, **81**, 1–4.

Qu Y., Li H., Wang X., Tian W., Shi B., Yao M. and Zhang Y. (2019). Bioleaching of major, rare earth, and radioactive elements from red mud by using indigenous chemoheterotrophic bacterium *Acetobacter* sp. *Minerals*, **9**(2), 67. doi: 10.3390/min9020067

Ramasamy D. L., Porada S. and Sillanpää M. (2019). Marine algae: A promising resource for the selective recovery of scandium and rare earth elements from aqueous systems. *Chemical Engineering Journal*, **371**, 759–768.

Rasoulnia P., Barthen R. and Lakaniemi A. M. (2021). A critical review of bioleaching of rare earth elements: The mechanisms and effect of process parameters. *Critical Reviews in Environmental Science and Technology*, **51**(4), 378–427.

Reed D. W., Fujita Y., Daubaras D. L., Bruhn D. F., Reiss J. H., Thompson V. S. and Jiao Y. (2016a). Microbially Mediated Leaching Of Rare Earth Elements From Recyclable Materials. IMPC 2016 – 28th International Mineral Processing Congress.

Reed D. W., Fujita Y., Daubaras D. L., Jiao Y. and Thompson V. S. (2016b). Bioleaching of rare earth elements from waste phosphors and cracking catalysts. *Hydrometallurgy*, **166**, 34–40.

Sadovsky D., Brenner A., Astrachan B., Asaf B. and Gonen R. (2016). Biosorption potential of cerium ions using *Spirulina* biomass. *Journal of Rare Earths*, **34**(6), 644–652.

Santini T. C., Kerr J. L. and Warren L. A. (2015). Microbially-driven strategies for bioremediation of bauxite residue. *Journal of Hazardous Materials*, **293**, 131–157.

Sethurajan M., van Hullebusch E. D. and Nancharaiah Y. V. (2018). Biotechnology in the management and resource recovery from metal bearing solid wastes: recent advances. *Journal of Environmental Management*, **211**, 138–153.

Shin D. W. and Kim J. G. (2015). Study on the separation and extraction of rare-earth elements from the phosphor recovered from end of life fluorescent lamps. *Archives of Metallurgy and Materials*, **60**, 1257–1260.

Sutcu M., Erdogmus E., Gencel O., Gholampour A., Atan E. and Ozbakkaloglu T. (2019). Recycling of bottom ash and fly ash wastes in eco-friendly clay brick production. *Journal of Cleaner Production*, **233**, 753–764.

Tan Q. and Li J. (2019). Rare earth metal recovery from typical e-waste. In: Waste Electrical and Electronic Equipment (WEEE) Handbook. Woodhead Publishing, Cambridge, pp. 393–421. doi: 10.1016/B978-0-08-102158-3.00015-X

Tolley M. R., Strachan L. F. and Macaskie L. E. (1995). Lanthanum accumalation from acidic solutions using a *Citrobacter* sp. Immobilized in a flow-through bioreactor. *Journal of Industrial Microbiology and Biotechnology*, **14**(3–4), 271–280.

Tsuruta T. (2007). Accumulation of rare earth elements in various microorganisms. *Journal of Rare Earths*, **25**(5), 526–532.

Vander Hoogerstraete T., Blanpain B., Van Gerven T. and Binnemans K. (2014). From NdFeB magnets towards the rare-earth oxides: a recycling process consuming only oxalic acid. *RSC Advances*, **4**(109), 64099–64111.

Venkatesan P., Vander Hoogerstraete T., Binnemans K., Sun Z., Sietsma J. and Yang Y. (2018). Selective extraction of rare-earth elements from NdFeB magnets by a room-temperature electrolysis pretreatment step. *ACS Sustainable Chemistry & Engineering*, **6**(7), 9375–9382.

Vijayaraghavan K. and Yun Y. S. (2008). Bacterial biosorbents and biosorption. *Biotechnology Advances*, **26**(3), 266–291.

Vijayaraghavan K., Sathishkumar M. and Balasubramanian R. (2010). Biosorption of lanthanum, cerium, europium, and ytterbium by a brown marine alga, *Turbinaria conoides*. *Industrial & Engineering Chemistry Research*, **49**(9), 4405–4411.

Vijayaraghavan K., Rangabhashiyam S., Ashokkumar T. and Arockiaraj J. (2017). Assessment of samarium biosorption from aqueous solution by brown macroalga *Turbinaria conoides*. *Journal of the Taiwan Institute of Chemical Engineers*, **74**, 113–120.

Willbold E., Gu X., Albert D., Kalla K., Bobe K., Brauneis M., Janning C., Nellesen J., Czayka W., Tillmann W. and Zheng Y. (2015). Effect of the addition of low rare earth elements (lanthanum, neodymium, cerium) on the biodegradation and biocompatibility of magnesium. *Acta Biomaterialia*, **11**, 554–562.

Yang J., Wang Q., Wang Q. and Wu T. (2008). Comparisons of one-step and two-step bioleaching for heavy metals removed from municipal solid waste incineration fly ash. *Environmental Engineering Science*, **25**(5), 783–789.

Yang F., Kubota F., Baba Y., Kamiya N. and Goto M. (2013). Selective extraction and recovery of rare earth metals from phosphor powders in waste fluorescent lamps using an ionic liquid system. *Journal of Hazardous Materials*, **254**, 79–88.

Zhang J., Zhao B. and Schreiner B. (2016). Rare earth elements and minerals. In: Separation Hydrometallurgy of Rare Earth Elements. Springer, Cham, 1–17, doi: 10.1007/978-3-319-28235-0_1

Zhang K., Kleit A. N. and Nieto A. (2017). An economics strategy for criticality–application to rare earth element yttrium in new lighting technology and its sustainable availability. *Renewable and Sustainable Energy Reviews*, **77**, 899–915.

Zhang L., Dong H., Liu Y., Bian L., Wang X., Zhou Z. and Huang Y. (2018). Bioleaching of rare earth elements from bastnaesite-bearing rock by actinobacteria. *Chemical Geology*, **483**, 544–557.

Zhang Y., Gu F., Su Z., Liu S., Anderson C. and Jiang T. (2020). Hydrometallurgical recovery of rare earth elements from NdFeB permanent magnet scrap: A review. *Metals*, **10**(6), 841. doi: 10.3390/met10060841

Zinicovscaia I., Safonov A., Troshkina I., Demina L. and German K. (2018). Biosorption of Re(VII) from batch solutions and industrial effluents by cyanobacteria *Spirulina platensis*. *CLEAN – Soil, Air, Water*, **46**(7), 1700576. doi: 10.1002/clen.201700576

Zoll A. M. and Schijf J. (2012). A surface complexation model of YREE sorption on *Ulva lactuca* in 0.05–5.0 M NaCl solutions. *Geochimica et Cosmochimica Acta*, **97**, 183–199.

Chapter 10

Biological recovery of rare earth elements from mine drainage using the sulfidogenic process

Elis W. Nogueira, Roseanne B. Holanda, Gunther Brucha and Márcia H. R. Z. Damianovic

10.1 INTRODUCTION

The rare earth elements (REE) include 15 elements of the lanthanide family, scandium (Sc) and yttrium (Y). They are generally classified as light-REE (La, Ce, Pr and Nd), middle-REE (Sm, Eu, Gd, Tb and Dy) and heavy-REE (Ho, Er, Tm, Yb, Lu and Y) (Cao *et al.*, 2019; Lefticariu *et al.*, 2020). Despite what the name suggests, rare earth elements are not rare or earth. They are found in the earth's crust in abundance. However, the term 'rare' is more related to the difficulty in exploiting the deposits, because of the low concentrations of the elements which often make the area economically non-viable for exploration (Edahbi *et al.*, 2018a). These elements have gained importance due to their excellent catalytic, optical, electronic and magnetic properties, in addition to having high economic value.

REE are mainly used in the electronic, catalyst, ceramic and glass, metallurgical, polishing and other industries. In hi-tech industries, these elements are applied for the production of hybrid and electric vehicles, automotive sensors, permanent magnets used in wind turbines, fuel cells, rechargeable batteries, cell phones, fluorescent lamps, computers, catalytic converters, driver discs, plates and screens (Lefticariu *et al.*, 2020; Salo *et al.*, 2020). They are also used in medicinal

doi: 10.2166/9781789062236_0227

science, the pharmaceutical industry, clean energy devices (solar and wind), the zootechnical sector and agriculture (Migaszewski & Gałuszka, 2015).

From 2010, China began to control the REE market, creating a geopolitical problem in which the country holds about 90% of the world's production of REE ore (Mancheri *et al.*, 2019; Schlinkert & van den Boogaart, 2015; Skirrow *et al.*, 2013), followed by countries such as Australia, the United States and India. However, the world's four largest REE reserves are found in China, Brazil, Vietnam and Russia (US Geological Survey, 2020). The European Commission included REE in the list of critical raw materials due to the risk of shortages in the market and the lack of alternatives for their supply. The US Department of Energy also assessed the situation of yttrium and other REE as critical to supply in the medium term (until 2025) and stated the importance of this raw material for the development of clean energy (Amato *et al.*, 2019).

In addition, environmental problems associated with REE mining and its hydrometallurgical processing is of grave concern. For example, for each ton of REE oxides produced in China using the calcination technique at high temperatures with concentrated sulfuric acid, 9600 to 12000 m^3 of gaseous mineral waste including dust, sulfur dioxide, sulfuric acids and hydrofluoric acid, 75 m^3 of acidic water and approximately 1 ton of radioactive material are generated (Edahbi *et al.*, 2019; Hurst, 2010). In Baotou (China), where there are the largest lanthanide mining plants, it is estimated that 10 million tons of all types of wastewater are generated every year and are discharged without any effective treatment, contaminating the water supply, irrigation systems and the surrounding environment (Hurst, 2010). These problems are not only region specific, they are found throughout the world.

Therefore, alternative sources of REE, such as recovery from mining waste, waste/processing waters, electrical and electronic waste and industrial processing by-products, using sustainable technologies are necessary. Acid mine drainage (AMD) and phosphogypsum waste are some examples of secondary sources of REE in which recovery can be achieved by using biological treatments. Among them, sulfate-reducing bacteria (SRB), widely applied in metal recovery processes due to their versatility and metabolic diversity (Costa *et al.*, 2020; Cunha *et al.*, 2020; Hedrich & Johnson, 2014; Johnson & Sánchez-Andrea, 2019; Lens, 2020; Ňancucheo & Johnson, 2012), have also been used in the precipitation of REE and simultaneous sulfate removal from wastewaters (Mäkinen *et al.*, 2017; Nogueira *et al.*, 2019; Salo *et al.*, 2020, 2018). The biological processes encompassing wastewater treatment and REE recovery offer both environmental and economic benefits.

10.2 REACTIVITY OF REE-BEARING MINERALS

Rare earth minerals are integrated in the form of silicate, carbonate and phosphate crystal structures and can be replaced by other elements depending on the size of the ionic radius and charges of the ion. Cations Ca^{2+}, Na^{+}, Th^{2+} and U^{3+} can, for example, be replaced by rare earth cations (Edahbi *et al.*, 2019).

10.2.1 Reactivity of REE-bearing carbonates

The dissolution of carbonate complexes in natural systems occurs according to Reaction 10.1 and when there is the generation of sulfuric acid from the sulfide oxidation, it can promote the dissolution of the carbonate as presented in Reaction 10.2 (Edahbi *et al.*, 2019; Sherlock *et al.*, 1995). Where, M is Ba, Ca, U, Th, F, Sr or rare earth ions.

$$\text{(REE-M)CO}_3 + \text{H}_2\text{O} + \text{CO}_2 \rightarrow \text{REE}^{3+} + \text{M} + 2\,\text{HCO}_3^{-} \qquad (10.1)$$

$$\text{(REE-M)CO}_3 + \text{H}_2\text{SO}_4 \rightarrow \text{REE}^{3+} + \text{SO}_4^{-} + \text{M} + 2\,\text{HCO}_3^{-} \qquad (10.2)$$

The main factors that control the reactivity of carbonates are the following: pH, temperature, the presence of organic matter and the partial pressure of CO_2 in closed systems. The dissolution rate of carbonates is inversely proportional to pH, that is, the lower the pH, the greater the solubility of carbonates.

10.2.2 Reactivity of REE-bearing silicates

The dissolution of silicate minerals depends on the pH, temperature, mineralogical composition, mineral structure and external environmental factors. Silicates tend to solubilize generating secondary silicates as shown in Reaction 10.3 (Edahbi *et al.*, 2019; Sherlock *et al.*, 1995):

$$\text{(Ca, REE)Al}_{12}\text{Si}_2\text{O}_8 + 2\,\text{H}^{+} + \text{H}_2\text{O} \rightarrow \text{Ca}^{2+} + \text{REE}^{3+} + \text{Al}_2\text{Si}_2\text{O}_5\text{(OH)}_4 \qquad (10.3)$$

10.2.3 Reactivity of REE-bearing phosphates

Phosphate minerals commonly rich in REE are as follows: apatite, monazite ((REE, Th) PO_4) and xenotime (YPO_4). The reaction of the dissolution of apatite in water and in response to the generation of acid is shown in Reactions 10.4 and 10.5, respectively (Edahbi *et al.*, 2019):

$$2\,\text{(Ca, REE)}_5\text{(PO}_4\text{)}_3\text{F} + 15\,\text{H}_2\text{O} \rightarrow 10\,\text{Ca}^{2+} + 10\,\text{REE}^{3+} + 6\,\text{PO}_4^{3-} + 2\,\text{F}^{-} + 30\,\text{OH}^{-} \qquad (10.4)$$

$$\text{(Ca, REE)}_5\text{(PO}_4\text{)}_3\text{F} + 5\,\text{H}_2\text{SO}_4 \rightarrow 5\,\text{(Ca, REE)SO}_4 + \text{H}_3\text{PO}_4 + \text{H}^{+} + \text{F}^{-} \qquad (10.5)$$

Once in solution, REE can precipitate as sulfates, phosphates and fluorides due to their low solubility (K_{sp}) under neutral pH conditions (Table 10.1) and also as secondary phosphates in the form of $REEPO_4$ or REE-F_3.

Table 10.1 Solubility product of rare earth compounds (from Edahbi *et al.*, 2019).

REE Compound	Formula	K_{sp} (25°C)
Carbonates	$RE_2(CO_3)_3$	$10^{-28.25}$ to $10^{-35.77}$
Yttrium fluoride	YF_3	8.62×10^{-21}
Yttrium hydroxide	$Y(OH)_3$	10^{-22}
Phosphates	$REEPO_4$	10^{-24}

10.3 CONVENTIONAL METHODS FOR RECOVERY OF REE

The efficiency of REE recovery depends on factors such as the type of technology used to recover the elements, the target elements that are desirable for recovery, types of gangue minerals, their origin, the amount of REE present (ranging from <mg to several g) and the conditions (pH and temperature) of the extraction process (Edahbi *et al.*, 2019; Jowitt *et al.*, 2018). The following are the most commonly used methods for REE extraction from ores or waste resources: chemical extraction with the hydrometallurgical process (Das & Das, 2013), reduction processes (Gupta & Krishnamurthy, 2004), total dissolution, selective dissolution and sequential leaching.

All REE have the same external electronic configuration $[Xe]6s^2\ 5d^1\ 4f^x$, where 'x' ranges from 0 to 14. The elements are trivalent cations (M^{3+}), except for Ce and Eu which have a charge of 2^+ and 4^+ (depending on the redox conditions), (Jaireth *et al.*, 2014). The main difference between each REE is in their ionic size which decreases from 1.2 to 0.7 Å as the atomic number increases from 57 to 71 (Edahbi *et al.*, 2019). The 4f orbitals are protected from the chemical environment by the 5s, 5p and 5d, 6s orbitals. Thus, when a compound is formed they do not directly participate in the bonding with other elements. This leads to the extremely similar physical and chemical properties of the elements, hindering the processes to separate them (Jaireth *et al.*, 2014), and thus making the REE recovery process more challenging.

These characteristics of REE allow their replacement by other metallic elements that have a similar ionic radius, such as Ca, Na and U (Edahbi *et al.*, 2019). When combined with anions, REE can be present in soluble (chlorides and nitrates) or insoluble (sulfides, fluorides, carbides, oxalates and phosphates) form. They can also form stable complexes with organic molecules (organic matter and fulvic acid) (Migaszewski & Gałuszka, 2015). Another difficulty faced in recycling is in the removal of impurities after extraction to yield pure single elements (Ali, 2014; Jowitt *et al.*, 2018), and they are not easy to precipitate and crystallize using a single method (Naidu *et al.*, 2019).

REE behavior in acid mine water depends on several parameters, such as the influence of iron oxides (Edahbi *et al.*, 2018b; Verplanck *et al.*, 2004), presence of aluminum, fluorine, sulfate, phosphate and temperature (Gimeno Serrano *et al.*, 2000), organic matter content (Edahbi *et al.*, 2019) and redox potential. The highly acidic pH of acid mine water also influences the solubility of REE and determines their fate after their release into surface and ground waters (Ayora *et al.*, 2016; Cravotta, 2008). Regarding the composition of rare earths, the reactivity of the compounds follows the order: carbonates > oxy/iron hydroxides > silicates (Linnen *et al.*, 2013).

The redox conditions play a significant role in the dynamics of REE (Linnen *et al.*, 2013): any change in redox potential could directly or indirectly change pH and the Fe (hydr)oxides, Mn (hydr)oxides, Al, S and dissolved organic carbon concentrations (Mihajlovic *et al.*, 2017), which can lead to changes in solubility, precipitation/co-precipitation and adsorption of REE (Edahbi *et al.*, 2019; Linnen *et al.*, 2013). Mihajlovic *et al.* (2017) reported a strong positive correlation between the release of REE and redox potential and a strong negative correlation between REE release and pH.

Table 10.2 summarizes the different recycling methods applied for recovery of targeted REE from various wastes (Jowitt *et al.*, 2018). Among secondary wastes for REE extraction and selective separation, the following technologies have been applied: adsorption, ion exchange, emulsion liquid membrane, supported liquid membrane and solvent extraction (liquid-liquid extraction) processes (Kumar *et al.*, 2020; Parhi *et al.*, 2018). The advantage and disadvantage of each of these extraction processes are shown in Table 10.3. Recycling of discarded electronic materials for REE recovery was reported by Binnemans *et al.* (2013) and Haque *et al.* (2014). Commercial recycling of these elements is still quite limited and it is estimated that less than 1% of all REE consumed is obtained through recycling of REE-containing wastes (Favot & Massarutto, 2019; Ganguli & Cook, 2018). The REE-bearing wastes that are mostly recycled for REE recovery are fluorescent lamps (Yang *et al.*, 2013), rechargeable batteries and permanent magnets (Binnemans *et al.*, 2013; Edahbi *et al.*, 2019; US Geological Survey, 2020).

The recycling processes still face a number of difficulties associated with the collection of material and its processing efficiency (Binnemans *et al.*, 2013). Another major challenge is regarding separation and purification of REE. The recovery of these metals from electronic equipment requires the use of advanced processes, for which many experimental studies are reported in the literature (Binnemans *et al.*, 2013; Favot & Massarutto, 2019), however, they are yet to reach industrial scale.

10.4 REE-RICH WASTEWATER ASSOCIATED WITH ACID MINE DRAINAGE

REE can be found in high concentrations in acidic waters resulting from mining activity, that is acid mine drainage (AMD), making AMD an important source of REE recovery (Ayora *et al.*, 2016). The formation of AMD can occur either inside the mine pit or in tailing deposition areas. AMD is formed when rocks and

Table 10.2 Potential sources of REE and their recycling processes (adapted from Jowitt *et al.* (2018) and Marra *et al.* (2018)).

Source for Recycling	Targeted REE	Primary Recycling Mechanism
Fluid catalytic cracking (FCC) catalysts	La and Ce	Hydrometallurgical processes (leaching, solvent extraction, selective precipitation); Microbial leaching (bioleaching)
Industrial processes and residues	Depending on the source material, the REE recycling process can target different REE	Pyrometallurgical processes (roasting, calcination); Hydrometallurgical processes (leaching, solvent extraction, selective precipitation); Physical separation & microbial leaching (bioleaching)
WEEE & 'End of Life' consumer goods; Fluorescent material (phosphor powder and fluorescent lamps)	La, Ce, Tb and Y	Pyrometallurgy (roasting, calcination); Hydrometallurgy (leaching, solvent extraction, selective precipitation); Gas phase extraction
Magnets	Nd, Dy and the other REE	Hydrometallurgical processes (leaching, solvent REE extraction, selective precipitation)
Batteries	La, Ce, Pr and Nd	Hydrometallurgical or pyrometallurgical recovery routes
Mobile phones	La, Ce, Pr and Dy	No information available
WEEE shredding dust	Ce, Eu, Nd, La and Y	Bio-hydrometallurgical processes (bioleaching)

WEEE: Waste electrical and electronic equipment.

other sulfide minerals are exposed to natural environmental factors such as water and air, resulting in the formation of sulfuric acid and the consequent solubilization of the metals present in rocks (Streten-Joyce *et al.*, 2013). The different processes involved in AMD generation are described in several previous works (Akcil & Koldas, 2006; Johnson & Hallberg, 2005; Rohwerder *et al.*, 2003; Streten-Joyce *et al.*, 2013).

The main factors that influence the AMD generation rate are: pH, temperature, weathering of rocks, chemical activity of Fe^{3+}, surface area of metal sulfides and microbial activity (Akcil & Koldas, 2006). Using pyrite (FeS_2) as an example, Reaction 10.6 expresses the general reaction that occurs in the oxidation process of pyrite (Kefeni *et al.*, 2017) forming $Fe(OH)_3$. When iron is the oxidizing agent (Reaction 10.7), complete oxidation of pyrite occurs more spontaneously

Table 10.3 Advantages and disadvantages of REE extraction from secondary wastes (Kumar *et al.*, 2020; Marra *et al.*, 2018).

Process	Advantage	Disadvantage
Adsorption and ion exchange	An economical process, easy in operation, applicable in a broad pH range and efficient for solutions containing a variety of metal ion concentrations.	Not effective for some metals.
Sorption	Effective and suitable adoption of low-cost materials such as clays, activated carbons, zeolites, agricultural waste, bio sorbents (bacteria, fungi and algae) and metal oxides.	Low selectivity, slow kinetics, complex extraction behavior for various functional groups on the sorbent and the surface – less potential for its application in commercial/field level operations.
Emulsion liquid membrane and supported liquid membrane	Good extractive ability over the solid support membrane phase for the recovery of the targeted metal ion.	High time consumption, low selectivity and low recyclability.
Solvent extraction	Several commercial and green solvents are adopted for clean REE separation from aqueous solutions.	Intermediated compounds could be formed and a secondary process required.
Bioleaching	An economical process, easy in operation and use of microorganisms.	Still limited to bench-scale applications.

compared to Reaction 10.8, in which oxygen is the oxidizing agent.

$$FeS_2 + 15\,O_2 + 7/2\,H_2O \rightarrow Fe(OH)_3 + 2\,SO_4{}^{2-} + 4H^+ \tag{10.6}$$

$$FeS_2 + 14\,Fe^{3+} + 8\,H_2O \rightarrow 15\,Fe^{2+} + 2\,SO_4^{2-} + 16\,H^+ \tag{10.7}$$

$$FeS_2 + 7/2\,O_2 + H_2O \rightarrow Fe^{2+} + 2\,SO_4{}^{2-} + 2H^+ \tag{10.8}$$

The complete oxidation of pyrite also occurs when there is little contact between water and metal sulfide, expressed in Reaction 10.9 (Chen *et al.*, 2015).

$$FeS_2 + 15/4\,O_2 + 1/2\,H_2O \rightarrow Fe^{2+} + 2\,SO_4{}^{2-} + H^+ \tag{10.9}$$

Among the dissolved metals, Fe(II) is the one that is found in high abundance in acid mine drainage. A survey by Kefeni *et al.* (2017) shows variation in the concentration of Fe(II) from 400 to 2135 mg L^{-1}, and also Al (~194 mg L^{-1}), Zn (~460 mg L^{-1}) and sulfate (2853 to 3622 mg L^{-1}) are present in the various types of AMD. The presence of dissolved metals associated with AMD depends on the type of gangue mineral present in the oxidized rock (Akcil & Koldas, 2006).

Different mineral sources contain REE along with various commercially-important elements. For instance, the bastnäsite mineral deposit is estimated at 1.5 million metric tons, where the rare earth oxide content corresponds to 3.3% of the deposit volume (Gupta & Krishnamurthy, 2004). Due to this, in addition to the commonly found metals, REE is also present in AMD at various concentrations. Table 10.4 shows studies in which REE are part of the reported AMD composition.

Among the different sources, mineral coal mines (Xingyu *et al.*, 2013) and deactivated uranium mines (Miekeley *et al.*, 1992; Nogueira *et al.*, 2019) feature prominently. At the Ronneburg and Seelingstädt uranium mines (Germany), the sum of REE concentrations was up to 3 mg L^{-1} of mine drainage (Merten *et al.*, 2005). A large reservoir of AMD is present in the city of Caldas (Minas Gerais state, Brazil), originating from uranium mining, which started operations in 1982 and ceased its activities in 1995. The tailings dam located at the Osamu Utsumi mine, *Industrias Nucleares do Brasil* (Figure 10.1), with a volume of 1.97 million m^3, contains solid and liquid mine effluents, including high concentrations of uranium and thorium. In addition, there are several other lagoons with a high REE content present in the mine unit (Figure 10.2).

The concern with respect to mining activity goes beyond the environmental problems, and includes the loss of significant quantities of material with commercial value that also have mining potential. Consequently, research efforts have been made to recover metals and REE from AMD, converting it into an attractive alternative source for REE. Currently, physico-chemical methods are still the main AMD treatment processes, but they are plagued by problems such as high costs and generation of wastes that are difficult to dispose of AMD

Table 10.4 REE composition present in AMD from different mining sites.

Location	pH	Y (μg L^{-1})	Ce (μg L^{-1})	Nd (μg L^{-1})	La (μg L^{-1})	Sm (μg L^{-1})	Reference
Pennsylvania Coal Mine	2.7–7.3	0.11–530	0.01–370	0.006–260	0.005–140	<0.005–79	Xingyu *et al.* (2013)
Osamu Utsumi (GW47)	3.57	—	6270	6490	13500	764	Miekeley *et al.* (1992)
Waterman, (CMD-L)	4.1	9.47	4.34	3.83	1.13	1.30	Stewart *et al.* (2017)
Strattanville, (CMD-P)	4.4	575	255	86.2	32.9	28.9	Stewart *et al.* (2017)
Osamu Utsumi (BNF)	3.4	4100	24900	9550	40100	—	Nogueira *et al.* (2019)

Figure 10.1 Pit at the Osamu Utsumi Mine, located in Caldas (Minas Gerais, Brazil). On the left side is the location of the pond where the sludge from the treatment plant is deposited. (From the author).

Figure 10.2 Four lagoons with a high REE and uranium content at the Osamu Utsumi Mine unit (Brazil). (From the author).

treatment of this tailing dam consists of active treatment (physico-chemical), which involves addition of alkaline chemicals aiming to raise the pH to the dischargeable limits and also precipitate many of the metallic pollutants. The treated volume can exceed 300 $m^3 h^{-1}$, depending on the volume of rainfall. The material precipitated during the physico-chemical treatment is removed from the treatment unit and deposited inside the pit at the Osamu Utsumi mine (Figure 10.1).

In 2018, the company spent R$ 1.8 million, buying calcium hydroxide with low magnesium content (Nogueira *et al.*, 2019). In addition to the high operational cost, another problem associated with the physico-chemical treatment is the high amount of sludge generated (Sahinkaya *et al.*, 2018), which is still an environmental issue.

Due to these problems, researchers are endeavoring to develop alternatives for efficient and economically viable treatments for the AMD treatment and recovery of metals present in mine tailings. In this regard, anaerobic biological treatment can provide an alternative solution for simultaneous REE recovery and AMD treatment. The sulfate reduction process allows sulfate removal and the generation of alkalinity, important for raising the pH and maintaining it at levels adequate for safe release. Sulfate present in AMD is reduced mainly due to the metabolism of sulfate-reducing bacteria (SRB), which use it as a final electron acceptor, and the sulfide produced is utilized for metal precipitation. The work developed by Nogueira *et al.* (2019) for the biological treatment of the Osamu Utsumi AMD is presented in the following section along with the main challenges (see Section 10.5).

10.5 RECOVERY OF REE THROUGH BIOLOGICAL TREATMENT

10.5.1 SRB treatment of REE-containing mining waste

A limited number of studies have reported on the possibility of recovering REE from waste or wastewater using the sulfidogenic process. Previous studies reported on recovering REE from phosphogypsum (PG) waste leachate (Mäkinen *et al.*, 2017; Salo *et al.*, 2018, 2020) and actual AMD (Nogueira *et al.*, 2019) using sulfate-reducing bacteria.

The application of sulfate-reducing bacteria in the treatment of AMD allows the removal of sulfate, concomitantly with the generation of alkalinity as shown in Reaction 10.10. The product of sulfate reduction, hydrogen sulfide (H_2S), reacts with metals in solution and precipitates them as metal sulfides (Reaction 10.11) (Sahinkaya *et al.*, 2009):

$$SO_4^{2-} + e^- + 2\,H^+ \rightarrow H_2S + 2\,HCO_3^- \qquad (10.10)$$

$$\text{Metal} + H_2S \rightarrow MeS + 2\,H^+ \qquad (10.11)$$

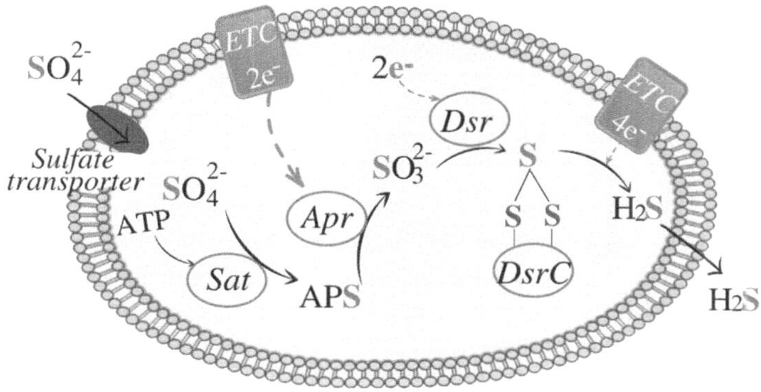

Figure 10.3 Metabolic pathway of dissimilatory sulfate reduction. Adapted from Jørgensen *et al.* (2019). ETC: electron transfer complex; DsrC: dissimilatory (bi) sulfite reductase complex; Dsr: dissimilatory (bi)sulfite reductase; APS: adenosine-5′-phosphosulfate; Apr: adenylyl-sulfate reductase; Sat: ATP sulfurylase.

The cellular mechanism associated with sulfate reduction in SRB is illustrated in Figure 10.3. Sulfate (SO_4^{2-}) gets inside the cell via sulfate transporters and is activated with adenosine triphosphate (ATP) in the cytoplasm by the enzyme ATP sulfurylase (Sat) to form adenosine-5′-phosphosulfate (APS). The enzyme adenylyl-sulfate reductase (Apr) assists in the reduction of APS to sulfite, receiving electrons from a membrane-bound electron transfer complex (ETC). Then, sulfite is reduced to H_2S by the dissimilatory sulfite reductase (Dsr) complex via a DsrC-bound trisulfide and diffuses out of the SRB cell membrane (Jørgensen *et al.*, 2019).

10.5.2 Treatment of phosphogypsum waste leachate

10.5.2.1 Bioreactor performance

Regarding phosphogypsum (PG) waste leachate, it is generated during the wet chemical phosphoric acid treatment process, or 'wet process' from apatite ores using sulfuric acid leaching to produce phosphoric acid and fertilizers (Kovler, 2012). Phosphate fertilizers are mainly obtained from phosphorus ore deposits, where they occur as apatite minerals (Cooper *et al.*, 2011).

It is estimated that approximately 5 tons of PG by-product is generated from each ton of phosphoric acid produced (Mäkinen *et al.*, 2017). This waste is characterized by acidity and elevated metal concentrations, containing radioactive impurities as well as REE. Thus, it requires treatment and is a potential source of valuable elements, such as REE (Salo *et al.*, 2018).

In the study presented by Mäkinen *et al.* (2017), batch experiments and a continuous column bioreactor were used for simultaneous sulfate removal and

Table 10.5 REE-concentrations of prepared PG-saturated water used as the influent (*Source*: Mäkinen *et al.*, 2017).

REE	La	Ce	Y	Nd	Dy	Eu	Gd	Pr	Sm
Concentration (μg L^{-1})	2.87	5.13	0.67	3.32	0.27	0.13	0.33	1.16	0.49

recovery of REE from synthetic PG leachate. The main characteristics of the wastewater solution containing REE used in both experiments are presented in Table 10.5. For the batch experiment, *Desulfovibrio desulfuricans* was used as inoculum and lactate and yeast extract as the carbon and nutrient source. The continuous bioreactor was inoculated with 500 mL anaerobic granular sludge collected from an industrial wastewater treatment plant and filled up with PG-saturated water. Ethanol was used as an electron donor during the column study and a sulfate concentration of 1340–1414 mg L^{-1} SO_4^{2-} was maintained to treat the PG-saturated water. The pH of the influent solution to the reactor was maintained between 5.9 and 6.6.

The sulfate reduction efficiency varied from 40–80%, and at its best performance (80% removal efficiency), a sulfate removal rate of 1080 g SO_4 m^{-3} d^{-1} was achieved and only 280 mg SO_4 L^{-1} in the effluent remained. In both batch and continuous flow bioreactors, the authors confirmed REE precipitation and accumulation in the sludge (Mäkinen *et al.*, 2017) and concluded dissolved REE can be precipitated in biological sulfate reduction systems (Table 10.6).

Initial anaerobic sludge represents the sludge used as inoculum for the continuous experiment; Final anaerobic sludge represents the sludge after the continuous experiment, enriched with REE.

Salo *et al.* (2018) evaluated the treatment potential of PG leachate using biological sulfate reduction and the possibility of REE recovery. In this study, two identical upflow anaerobic sludge blanket (UASB) bioreactors (BR1 and BR2) were operated in a continuous flow mode for 333 days at 20–22°C. The concentration of different REE in the PG leachate is shown in Table 10.7. The leachate also contained an average sulfate concentration of 1608 mg L^{-1}. Sodium

Table 10.6 Measured REE concentrations in precipitates of batch experiment and sludge of a continuous experiment (Mäkinen *et al.*, 2017).

REE (mg kg^{-1})	La	Ce	Y	Nd	Dy	Eu	Gd	Pr	Sm
Batch experiment	30, 400	66, 200	8800	n.a.	2400	1730	6750	10, 000	6310
Initial anaerobic sludge	7.3	13	3.6	7.2	0.8	0.2	1.02	1.69	1.4
Final anaerobic sludge	202	477	48.8	295	12.8	10.7	31.3	67.3	43.2

Table 10.7 REE concentration range (μg L^{-1}) of the PG leachate where the pH solution was 5.5–6.2 (*Source*: Salo *et al.*, 2018).

	Ce	**Dy**	**Er**	**Eu**	**Gd**	**Ho**	**La**	**Lu**
Concentration (μg L^{-1})	BD–11.8	BD–0.15	BD–0.08	BD–0.14	BD–0.46	BD–0.05	BD–2.30	BD–0.03
	Nd	**Pr**	**Sm**	**Tb**	**Tm**	**Y**	**Yb**	
Concentration (μg L^{-1})	0.31–2.42	BD- 0.62	0.02–0.38	BD–0.06	BD–0.03	0.24–0.58	BD–0.06	

BD: below detection level.

lactate was used as the carbon source and electron donor in the study. In both bioreactors, a ~ 100% sulfate reduction efficiency was achieved at the beginning. However, from day 200 until 308, the sulfate reduction efficiency reduced (60–70% sulfate removal) and the sulfate removal rate was ~5000 mg L^{-1} d^{-1}. Alkalinity generated from sulfate reduction increased the effluent pH from 5.5–6.2 to 6.5–7.5, the redox potential was between −270 and −350 mV and the sulfide concentration in the effluent was 120–400 mg L^{-1} (Salo *et al.*, 2018).

This study was the first research to describe how REE precipitate under biological treatment using sulfate reducing conditions. Regarding REE precipitation, the REE content in BR1 precipitates was higher than in BR2 (Salo *et al.*, 2018). The main minerals in the final precipitates were apatite, ardealita, gypsum, gypsum-apatite mix, REE-apatite and gypsum-silicate mix. BR1 produced more apatite and gypsum, while BR2 produced more REE-apatite, sulfur, gypsum-silicate mix and carbonate. The great difference in the REE content between BR1 and BR2 was attributed to the additional inoculum added to BR1 and the slight difference in the effluent pH of the two bioreactors, 6.9–7.3 (BR1) and 6.5–7.1 (BR2), since changes in pH affect the structure of the precipitate (Salo *et al.*, 2018).

10.5.2.2 Mineralogy of the REE precipitates

It should be noted that before this study of Salo *et al.* (2018), the precipitation mechanisms of REE under sulfidogenesis conditions were still unknown and the idea of REE precipitation as metal sulfide was one of the possibilities proposed by Mäkinen *et al.* (2017). However, no REE sulfide was detected and precipitation mechanisms for REE remained unclear and the authors suggested that REE tend to precipitate in the SRB system as phosphates instead of sulfides (Salo *et al.*, 2018).

In 2020, a new study was carried out by the same group, this time focusing on acid leaching of REE from phosphogypsum and subsequent treatment using biological sulfate reduction for REE recovery (Salo *et al.*, 2020). H_2SO_4 was used to leach the REE into solution with two different H_2SO_4 concentrations (0.01 and 0.02 M). The concentrations of REE in the solution were much higher than in the previous research reported by Salo *et al.* (2018), with REE dissolution reaching up to 62%. The REE content in the leachate of 0.02 M H_2SO_4 was higher than in the leachate obtained with the 0.01 M H_2SO_4 solution. The leachate obtained was directly fed to the sulfidogenic reactor. The reactor was operated using sodium lactate as the electron donor and the sulfate concentration varied from 1600 to 3800 mg L^{-1}.

The sulfate removal efficiency was stable at around 80% when 0.01 M H_2SO_4 leachate was used as the feed. Moreover, the performance improved (~95%) when the feed was changed to 0.02 M H_2SO_4 as the influent with only a slight decline in the effluent pH from 7.5 to 6.8, indicating a good stability of the bioreactor. However, with the increase in the feed load, the bioreactor performance in terms of sulfate reduction deteriorated. The authors attributed the decline in performance to the low influent pH and high concentration of sulfate. The sulfide concentration was as high as 600 mg L^{-1} when fed with PG leaching water and 0.01 M H_2SO_4 leachate (Salo *et al.*, 2020).

The average REE removal efficiency was 99%, with an influent concentration of maximum 20 μg L^{-1} and effluent below 1 μg L^{-1} even when the sulfate removal efficiency was as low as 40%, suggesting an efficient system for REE removal and recovery (Salo *et al.*, 2020). Final precipitate analyses showed a very high amount (12.4 g total REE) of REE recovered. Inductively coupled plasma (ICP) analyses showed that the REE concentrations approximately doubled after treatment compared to the initial PG material (Salo *et al.*, 2020). The La and Gd elements were the most enriched REE compared to the initial concentration, showing that this treatment can enrich REE from an acidic PG leachate.

The mineralogy of the original PG sample, before the treatment, consisted mostly of a gypsum phase (Wt. 97.53%). After 208 operational days of biological treatment, the final precipitate diagnosed by energy dispersive (ED) spectrometry showed 14 stand-alone phases in total. The main mineral phases found were apatite-francolite (33.53%), pure Ca (22.20%) and Ca-Fe-Al phosphate-sulfate, with traces of F^- or/and OH^- (18.75%). Eleven other phases together corresponded to 25.52% which are pure Ca with trace sulfur (5.50%), Ca with trace sulfur and silicon (0.57%), Ca with trace sulfur, aluminum and phosphorus (0.57%), pure sulfur (1.88%), pure sulfur with trace calcium (1.07%), Ca-Fe-Al phosphate-sulfate (8.27%), Zn-Fe-Na-Al apatite with trace sulfur (2.5%), gypsum relic (1.61%) and apatite-francolite with trace S (0.66%). REE-containing phases corresponded to 2.54% of the sample: Na-Al-REE apatite with trace sulfur (1.66%) and high Ca phosphate-sulfate of REE with trace aluminum (0.88%) (Salo *et al.*, 2020).

Figure 10.4 shows images from bioreactor precipitates analyzed by a mineral liberation analyzer (MLA) (Salo *et al.*, 2020). Electron probe micro analysis (EPMA) showed that REE are distributed not only in separate phases, but also mixed in apatite-resembling phases containing Ca and P (Table 10.8). The mineralogical analyses indicated possible recoverable REE phases.

According to Edahbi *et al.* (2019), REE could precipitate as carbonates, sulfates, phosphates, hydroxides and fluorides due to the low solubility product (K_{sp}) under neutral pH conditions (Table 10.1). However, there was no evidence of REE-sulfide as MLA analyses were not able to detect any REE in the sulfide phase, suggesting REE are probably precipitated as a result of the pH increase rather than interactions with sulfide (Salo *et al.*, 2020). REE-hydroxide precipitation at a pH higher than 8 was described by Miskufova *et al.* (2018) and as carbonate by Rodriguez-Blanco *et al.* (2014).

The positive results presented by Mäkinen *et al.* (2017) and Salo *et al.* (2018, 2020) were very important in clarifying REE precipitated mineral phases mediated by biological sulfate reduction systems. These studies showed that the

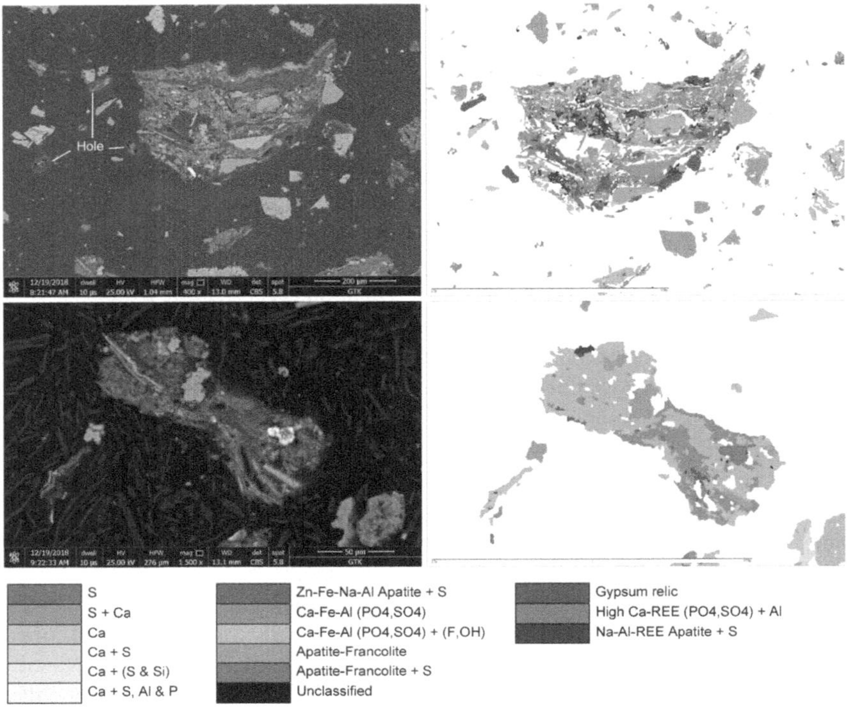

Figure 10.4 Amount of released REE from PG leachate analyzed with a mineral liberation analyzer (MLA) (*Source*: Salo *et al.*, 2020, with permission).

bioreactor treatment accumulated REE, and mineralogical analyses indicated the possibility of REE recovery.

10.5.3 Sulfidic treatment of AMD

10.5.3.1 Bioreactor performance

Another study on REE removal using biological sulfate reduction was reported by Nogueira *et al.* (2019). In contrast to the other studies presented previously, this study treated actual AMD with high concentrations of REE in solution which did not require any leaching process. The AMD used in this study, collected from an acidic lake at the Osamu Utsumi Uranium Mine, has the highest concentration of REE ever described in the literature. Table 10.9 shows the concentration of the main REE in the actual AMD used.

In the Osamu Utsumi mine water, the concentration of dissolved organic carbon is 1–4 mg L^{-1} (Miekeley *et al.*, 1992). Since AMD completely lacks or has a very low concentration of organic matter and other nutrients (Cravotta, 2008; Deng &

Table 10.8 Bioreactor precipitate (BR) REE analysis by ICP spectrometer and EPMA for REE (*Source*: Salo *et al.*, 2020).

	Total ($mg\ kg^{-1}$)	REE content ($mg\ kg^{-1}$)			
REE	**BR**	**All main phases**	**High Ca-REE (PO_4^{3-}, SO_4^{2-}) + Al**	**Na-Al-REE Apatite + S**	**Total**
Ce	1815	61.91	1702.6	3.0	1767.5
Dy	32.52	22.91	16.9	7.3	47.2
Er	7.48	94.73	2.1	5.9	102.7
Eu	33.75	—	74.8	1.7	76.5
Gd	917.8	49.21	43.7	3.7	96.7
Ho	4.22	59.6	0.5	5.1	65.3
La	886.91	31.34	686.5	1.5	719.4
Lu	<1	47.66	0.5	2.6	50.7
Nd	879.1	109.4	664.1	8.0	781.5
Pr	229.54	50.45	183.4	8.2	242.0
Sm	133.42	84.88	91.7	4.3	180.8
Tb	8.70	84.92	1.5	9.4	95.8
Tm	0.52	—	7.6	1.6	9.2
Y	68.17	26.16	58.4	1.9	86.5
Yb	1.99	56.3	0.2	3.1	59.6

Table 10.9 Sulfate and REE concentrations of the actual AMD used for biological treatment (*Source*: Nogueira *et al.*, 2019).

REE	La	Ce	Pr	Nd	U	Y	SO_4^{2-}
Concentration ($mg\ L^{-1}$)	40.1	24.9	3.6	9.55	4.15	4.10	890 (±80)

Lin, 2013; Miekeley *et al.*, 1992), the external addition of electron donor and nutrients is necessary to enable SRB to perform. In this case, sugarcane vinasse was used as the electron donor and source of nutrients (Nogueira *et al.*, 2019).

The main operational parameters of the bioreactor are described by Nogueira *et al.* (2019). It is noteworthy that La and Ce concentrations are in excess of 20 $mg\ L^{-1}$, up to 6 times higher than the Ce concentration and up to 19 times higher than the La concentration reported by Salo *et al.* (2020).

Sulfate reduction was very instable during the treatment and average reduction varied from 10 to 90%. As the volume of treated AMD increased, less sulfate reduction performance was achieved, mostly due to the low influent pH values

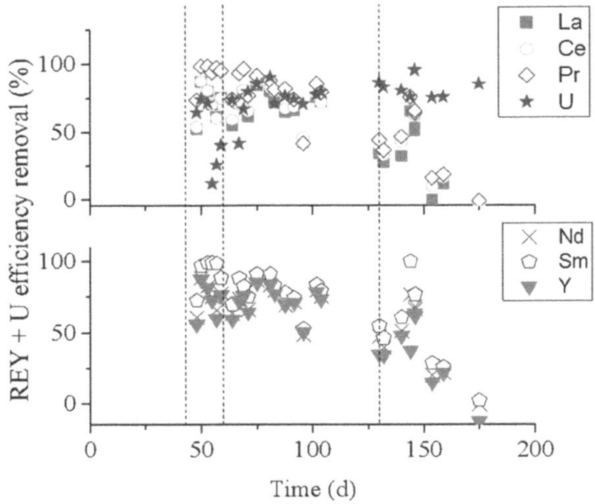

Figure 10.5 Rare earth elements and yttrium (REY) and U removal from the dissolved fraction of a down-flow fixed-structured bed reactor treating actual AMD of the Osamu Utsumi mine (Brazil) (Nogueira *et al.*, 2019).

and AMD characteristics. However, the alkalinity produced due to sulfate reduction increased the average pH of the system from influent pH values of 5.0–7.0 to effluent pH values of 6.4–7.5 (Nogueira *et al.*, 2019).

The REE removal efficiency using the sulfidogenic bioreactor is presented in Figure 10.5. The removal efficiencies for La, Ce, Nd and Y were more than 70%, while Pr and Sm showed more than 80% removal. The results are related to the increase in the pH of the liquid medium, precipitating the REE and other elements in the bioreactor. In addition, the authors suggested that the obtained removal of REE could be due to co-precipitation with aluminum or ferric iron, as a metal sulfide or directly as $REE(OH)_3$ (Nogueira *et al.*, 2019).

10.5.3.2 Mineralogy of REE precipitates

ICP results from four samples of precipitated bioreactor sludge showed significant amounts of several elements had accumulated in the sludges (Table 10.10). The highest concentration of precipitated REE was La which reached around 700 mg L^{-1} in the collected sludge. The concentration of REE precipitated was proportional to the value considered in the wastewater that was fed to the bioreactor (Nogueira *et al.*, 2019), except for U, where the behavior differs from the other elements and a small concentration of U was detected in only one bioreactor precipitate sample.

Table 10.10 Concentration of precipitated REE in the bioreactor (*Source*: Nogueira *et al.*, 2019).

Day	La	Ce	Pr	U	Nd	Sm	Y	Mn	Al	Fe	Si
57	86	67	4	n.d	23	3	9	34	243	236	41
90	245	176	20	n.d	52	6	24	80	702	135	75
137	194	136	17	5	45	5	22	87	620	87	71
110*	696	484	55	n.d	155	15	79	735	2587	406	76

*Sample collected from the conical bottom of the water level equalizer. Values are in mg L^{-1}.

10.5.3.3 Toxicity of REE to bioreactor sludge

This AMD treatment system was efficient for REE recovery, however, when actual AMD fed to the reactor reached 75% of the total volume, the sulfate removal efficiency started to decline and the effluent pH decreased until the bioreactor performance collapsed. Nogueira *et al.* (2019) attributed the failure of the bioreactor system to the low influent pH and the high concentration of toxic metals. The precipitation of the elements in the support material containing biomass can also be a cause of the bioreactor failure, as the precipitated metals could exert inhibitory effects on the SRB and can increase the mass transfer limitation (Nogueira *et al.*, 2019).

Under normal conditions, metals play essential biochemical roles in the cell. However, when metals accumulate above threshold concentrations, they become toxic. In situations where the metal ions are in excess in the microbial cells, they can block the functional groups of enzymes, inhibit transport systems and displace essential metals from their native binding sites, thus disrupting cellular membrane integrity (Dopson *et al.*, 2003). In order to avoid this inhibition, autochthonous microorganisms present in AMD and acidic environments that are adapted (due their heavy metal resistance) could be enriched and used as bioreactor inoculum.

10.6 ECONOMIC FEASIBILITY OF REE RECOVERY FROM SECONDARY SOURCES

The recovery of REE and metals from secondary resources such as AMD and coal mine drainage (CMD) was investigated by Ayora *et al.* (2016), Lefticariu *et al.* (2020) and Nogueira *et al.* (2019) using different processes. Ayora *et al.* (2016) reported that using passive remediation systems, the average ΣREE concentrations treated in the Monte Romero AMD and in the Almagrera AMD were 7.9 and 3.5 mg L^{-1}, respectively. The authors estimated that around 70 to 100 t of REY_2O_3 could be recovered annually from 150 AMD points of the

Iberian Pyrite Belt (IPB, southwest Spain), as the natural processes of AMD generation are expected to continue for centuries or thousands of years (Younger, 1997).

Lefticariu *et al.* (2020) evaluated the potential alternative source of REY and metals from CMD in the Illinois Basin. The average concentration of REE detected in the CMD was 1059 $\mu g\ L^{-1}$ for La, Ce, Gd, Sm, Pr, Ho, Er, Tm, Yb and Lu, while the average concentration of critical REE, *viz*. Nd, Eu, Tb, Dy and Y was 611 $\mu g\ L^{-1}$. Lefticariu *et al.* (2020) concluded that valuable metals could be co-extracted with REE and recovered to enhance the economic values of CMD. Phosphogypsum waste is another source of REE recovery and was reported by Mäkinen *et al.* (2017) and Salo *et al.* (2018, 2020) using biological treatment as discussed in Section 10.5.

Waste electrical and electronic equipment (WEEE) also represents an important secondary REE source (Sahan *et al.*, 2019). Sahan *et al.* (2019) characterized the metal and REE concentration of waste in mobile phones analyzing Blackberry and Nokia devices. Lanthanum, cerium, praseodymium and dysprosium were the REE found in displays. They reported that 0.2 $g\ kg^{-1}$ La and 0.13 $g\ kg^{-1}$ Ce were detected in Blackberry display samples, while 0.48 $g\ kg^{-1}$ La, 0.004 $g\ kg^{-1}$ Ce, 0.03 $g\ kg^{-1}$ Pr and 0.02 $g\ kg^{-1}$ Dy were detected in Nokia display samples. Balaram (2019) reported that 100,000 iPhones have the potential to yield 11 kg of REE. According to Balaram (2019), Apple is committed to using only recycled materials in its supply chain, including REE and metals such as Al, Co, Cu, Au, Ag and W.

Favot and Massarutto (2019) studied the economic viability of recycling and recovering the element yttrium from used lamps, based on the HydroWEEE project by the Italian recycling company, Relight Ltd. The authors investigated the lowest selling price of Y in the market that would make it profitable to recycle the element instead of exploiting the virgin source, which also accounted for the environmental costs of mining.

10.7 FINAL CONSIDERATION

Although there are several studies on metal recovery from wastewater in laboratory-scale systems, recovery of REE is still scarce, and there is a lack of understanding of REE behavior and precipitation in biogenic sulfate-reducing systems. Biological treatment of wastewater containing REE has been shown to be an effective and low-cost treatment system for REE-recovering reactors. However, more research should be conducted in order to improve and optimize the bioreactor system. To achieve proper operation of the REE-recovery reactor, suitability of the organic substrate used, characteristics of the SRB metabolism, their interaction with REE, and optimum REE concentration which can be treated without causing inhibition need to be examined. REE recovery from wastewater could be used to offset the treatment costs, in addition to the environmental benefits.

Recovery and recycling of REE from secondary and waste based sources is thus not only providing an alternative source for these economically-important, critical elements, but also easing the demand for some of the REE and minimizing exploitation of new deposits. However, more efforts need to be put towards improving REE recovery efficiency in actual AMD, besides integrating studies on industrial-scale systems for future commercial applications.

REFERENCES

Akcil A. and Koldas S. (2006). Acid Mine Drainage (AMD): causes, treatment and case studies. *Journal of Cleaner Production*, **14**(12–13), 1139–1145.

Ali S. H. (2014). Social and environmental impact of the rare earth industries. *Resources*, **3**, 123–134.

Amato A., Becci A., Birloaga I., De Michelis I., Ferella F., Innocenzi V., Ippolito N. M., Pillar Jimenez Gomez C., Vegliò F. and Beolchini F. (2019). Sustainability analysis of innovative technologies for the rare earth elements recovery. *Renewable and Sustainable Energy Reviews*, **106**, 41–53.

Ayora C., Macías F., Torres E., Lozano A., Carrero S., Nieto J. M., Pérez-López R., Fernández-Martínez A. and Castillo-Michel H. (2016). Recovery of rare earth elements and yttrium from passive-remediation systems of acid mine drainage. *Environmental Science & Technology*, **50**, 8255–8262.

Balaram V. (2019). Geoscience frontiers rare earth elements: a review of applications, occurrence, exploration, analysis, recycling, and environmental impact. *Geoscience Frontiers*, **10**, 1285–1303.

Binnemans K., Jones P. T., Blanpain B., Van Gerven T., Yang Y., Walton A. and Buchert M. (2013). Recycling of rare earths: a critical review. *Journal of Cleaner Production*, **51**, 1–22.

Cao X., Zhou S., Xie F., Rong R. and Wu P. (2019). The distribution of rare earth elements and sources in Maoshitou reservoir affected by acid mine drainage, Southwest China. *Journal of Geochemical Exploration*, **202**, 92–99.

Chen M., Lu G., Guo C., Yang C., Wu J., Huang W., Yee N. and Dang Z. (2015). Sulfate migration in a river affected by acid mine drainage from the Dabaoshan mining area, South China. *Chemosphere*, **119**, 734–743.

Cooper J., Lombardi R., Boardman D. and Carliell-Marquet C. (2011). The future distribution and production of global phosphate rock reserves. *Resources, Conservation & Recycling*, **57**, 78–86.

Costa R. B., Godoi L. A. G., Braga A. F. M., Delforno T. P. and Bevilaqua D. (2020). Sulfate removal rate and metal recovery as settling precipitates in bioreactors: influence of electron donors. *Journal of Hazardous Materials*, **403**, 123622.

Cravotta C. A. (2008). Dissolved metals and associated constituents in abandoned coal-mine discharges, Pennsylvania, USA. Part 1: constituent quantities and correlations. *Applied Geochemistry*, **23**, 166–202.

Cunha M. P., Fuess L. T., Rodriguez R. P., Lens P. N. L. and Zaiat M. (2020). Sulfidogenesis establishment under increasing metal and nutrient concentrations: An effective approach for biotreating sulfate-rich wastewaters using an innovative structured-bed reactor (AnSTBR). *Bioresource Technology Reports*, **11**, 100458.

Das N. and Das D. (2013). Recovery of rare earth metals through biosorption: an overview. *Journal of Rare Earths*, **31**, 933–943.

Deng D. and Lin L.-S. (2013). Two-Stage combined treatment of acid mine drainage and municipal wastewater. *Water Science and Technology*, **67**(5), 1000–1007.

Dopson M., Baker-Austin C., Koppineedi P. R. and Bond P. L. (2003). Growth in sulfidic mineral environments: metal resistance mechanisms in acidophilic micro-organisms. *Microbiology (Reading, England)*, **149**(8), 1959–1970.

Edahbi M., Plante B. and Benzaazoua M. (2019). Environmental challenges and identification of the knowledge gaps associated with REE mine wastes management. *Journal of Cleaner Production*, **212**, 1232–1241.

Edahbi M., Plante B., Benzaazoua M., Kormos L. and Pelletier M. (2018a). Rare earth elements (La, Ce, Pr, Nd, and Sm) from a carbonatite deposit: mineralogical characterization and geochemical behavior. *Minerals*, **8**(2), 55.

Edahbi M., Plante B., Benzaazoua M., Ward M. and Pelletier M. (2018b). Mobility of rare earth elements in mine drainage: influence of iron oxides, carbonates, and phosphates. *Chemosphere*, **199**, 647–654.

Favot M. and Massarutto A. (2019). Rare-earth elements in the circular economy: The case of yttrium. *Journal of Environmental Management*, **240**, 504–510.

Ganguli R. and Cook D. R. (2018). Rare earths: A review of the landscape. *MRS Energy & Sustainability*, **5**, 1–16.

Gimeno Serrano M. J., Auqué Sanz L. F. and Nordstrom D. K. (2000). REE Speciation in low-temperature acidic waters and the competitive effects of aluminum. *Chemical Geology*, **165**, 167–180.

Gupta C. K. and Krishnamurthy N. (2004). Extractive Metallurgy of Rare Earths. CRC Press, Florida.

Haque N., Hughes A., Lim S. and Vernon C. (2014). Rare earth elements: overview of mining, mineralogy, uses, sustainability and environmental impact. *Resources*, **3**, 614–635.

Hedrich S. and Johnson D. B. (2014). Remediation and selective recovery of metals from acidic mine waters using novel modular bioreactors. *Environmental Science & Technology*, **48**(20), 12206–12212.

Hurst C. (2010). China's Rare Earth Elements Industry: What can the West Learn? Institute for the Analysis of Global Security, Washington DC.

Jaireth S., Hoatson D. M. and Miezitis Y. (2014). Geological setting and resources of the major rare-earth-element deposits in Australia. *Ore Geology Reviews*, **62**, 72–128.

Johnson D. B. and Hallberg K. B. (2005). Acid mine drainage remediation options: A review. *Science of The Total Environment*, **338**(1–2), 3–14.

Johnson D. B. and Sánchez-Andrea I. (2019). Dissimilatory reduction of sulfate and zero-valent sulfur at low pH and its significance for bioremediation and metal recovery. *Advances in Microbial Physiology*, **75**, 205–231.

Jørgensen B. B., Findlay A. J. and Pellerin A. (2019). The biogeochemical sulfur cycle of marine sediments. *Frontiers in Microbiology*, **10**, 849. doi: 10.3389/fmicb.2019.00849

Jowitt S. M., Werner T. T., Weng Z. and Mudd G. M. (2018). Recycling of the rare earth elements. *Current Opinion in Green and Sustainable Chemistry*, **13**, 1–7.

Kefeni K. K., Msagati T. A. M. and Mamba B. B. (2017). Acid mine drainage: prevention, treatment options, and resource recovery: A review. *Journal of Cleaner Production*, **151**, 475–493.

Kovler K. (2012). Radioactive Materials. Federation Proceedings. Woodhead Publishing Series in Civil and Structural Engineering. pp. 196–240. doi: 10.1533/9780857096357.196

Kumar R., Thenepalli T., Whan J., Kumar P., Woo K. and Lee J. (2020). Review of rare earth elements recovery from secondary resources for clean energy technologies: grand opportunities to create wealth from waste. *Journal of Cleaner Production*, **267**, 122048. doi: 10.1016/j.jclepro.2020.122048

Lefticariu L., Klitzing K. L. and Kolker A. (2020). Rare earth elements and yttrium (REY) in coal mine drainage from the illinois basin, USA. *International Journal of Coal Geology*, **217**, 103327. doi: 10.1016/j.coal.2019.103327

Lens P. N. L. (2020). Environmental Technologies to Treat Sulfur Pollution: Principles and Engineering, 2nd edn. IWA publishing. doi: 10.2166/9781789060966

Linnen R. L., Samson I. M., Williams-Jones A. E. and Chakhmouradian A. R. (2013). Geochemistry of the Rare-Earth Element, Nb, Ta, Hf, and Zr Deposits, Treatise on Geochemistry. Second Edition. Elsevier Ltd. pp. 543–568. doi: 10.1016/B978-0-08-095975-7.01124-4

Mäkinen J., Bomberg M., Salo M., Arnold M. and Koukkari P. (2017). Rare earth elements recovery and sulphate removal from phosphogypsum waste waters with sulphate reducing bacteria. *Solid State Phenomena*, **262**, 573–576.

Mancheri N. A., Sprecher B., Bailey G., Ge J. and Tukker A. (2019). Effect of Chinese policies on rare earth supply chain resilience. *Resources, Conservation & Recycling*, **142**, 101–112.

Marra A., Cesaro A., Rene E. R., Belgiorno V. and Lens P. N. L. (2018). Bioleaching of metals from WEEE shredding dust. *Journal of Environmental Management*, **210**, 180–190.

Merten D., Geletneky J., Bergmann H., Haferburg G., Kothe E. and Büchel G. (2005). Rare earth element patterns: A tool for understanding processes in remediation of acid mine drainage. *Geochemistry*, **65**, 97–114.

Miekeley N., Coutinho de Jesus H., Porto da Silveira C. L. and Degueldre C. (1992). Chemical and physical characterization of suspended particles and colloids in waters from the Osamu Utsumi mine and Morro do Ferro analogue study sites, Poços de Caldas, Brazil. *Journal of Geochemical Exploration*, **45**, 409–437.

Migaszewski Z. M. and Gałuszka A. (2015). The characteristics, occurrence, and geochemical behavior of rare earth elements in the environment: A review. *Critical Reviews in Environmental Science and Technology*, **45**, 429–471.

Mihajlovic J., Hans-Joachim S. and Rinklebe J. (2017). Rare earth elements and their release dynamics under pre-de fi nite redox conditions in a floodplain soil. *Chemosphere*, **181**, 313–319.

Miskufova A., Kochmanova A., Havlik T., Horvathova H. and Kuruc P. (2018). Leaching of yttrium, europium and accompanying elements from phosphor coatings. *Hydrometallurgy*, **176**, 216–228.

Naidu G., Ryu S., Thiruvenkatachari R., Choi Y., Jeong S. and Vigneswaran S. (2019). A critical review on remediation, reuse, and resource recovery from acid mine drainage. *Environmental Pollution*, **247**, 1110–1124.

Ňancucheo I. and Johnson D. B. (2012). Selective removal of transition metals from acidic mine waters by novel consortia of acidophilic sulfidogenic bacteria. *Microbial Biotechnology*, **5**(1), 34–44.

Nogueira E. W., Licona F. M., Godoi L. A. G., Brucha G. and Damianovic M. H. R. Z. (2019). Biological treatment removal of rare earth elements and yttrium (REY) and metals from actual acid mine drainage. *Water Science and Technology*, **80**, 1485–1493.

Parhi P. K., Behera S. S. and Mohapatra R. K. (2018). Separation and recovery of Sc(III) from Mg – Sc alloy scrap solution through hollow fiber supported liquid membrane (HFLM) process supported by Bi- functional ionic liquid as carrier. *Separation Science and Technology*, **54**(9), 1478–1488.

Rodriguez-Blanco J. D., Vallina B., Blanco J. A. and Benning L. G. (2014). The role of REE^{3+} in the crystallization of lanthanites. *Mineralogical Magazine*, **78**, 1373–1380.

Rohwerder T., Gehrke T., Kinzler K. and Sand W. (2003). Bioleaching review part A: progress in bioleaching: fundamentals and mechanisms of bacterial metal sulfide oxidation. *Applied Microbiology and Biotechnology*, **63**, 239–248.

Sahan M., Kucuker M. A., Demirel B., Kuchta K. and Hursthouse A. (2019). Determination of metal content of waste mobile phones and estimation of their recovery potential in Turkey. *International Journal of Environmental Research and Public Health*, **16**(5), 887. doi: 10.3390/ijerph16050887

Sahinkaya E., Gungor M., Bayrakdar A., Yucesoy Z. and Uyanik S. (2009). Separate recovery of copper and zinc from acid mine drainage using biogenic sulfide. *Journal of Hazardous Materials*, **171**(1–3), 901–906.

Sahinkaya E., Yurtsever A., Isler E., Coban I. and Aktaş Ö. (2018). Sulfate reduction and filtration performances of an anaerobic membrane bioreactor (AnMBR). *Chemical Engineering Journal*, **349**, 47–55.

Salo M., Mäkinen J., Yang J., Kurhila M. and Koukkari P. (2018). Continuous biological sulfate reduction from phosphogypsum waste leachate. *Hydrometallurgy*, **180**, 1–6.

Salo M., Knauf O., Mäkinen J., Yang X. and Koukkari P. (2020). Integrated acid leaching and biological sulfate reduction of phosphogypsum for REE recovery. *Minerals Engineering*, **155**, 106408. doi: 10.1016/j.mineng.2020.106408

Schlinkert D. and van den Boogaart K. G. (2015). The development of the market for rare earth elements: insights from economic theory. *Resources Policy*, **46**, 272–280.

Sherlock E. J., Lawrence R. W. and Poulin R. (1995). On the neutralization of acid rock drainage by carbonate and silicate minerals. *Environmental Geology*, **25**, 43–54.

Skirrow R. G., Huston D. L., Mernagh T. P., Thorne J. P., Dulfer H. and Senior A. B. (2013). Critical Commodities for a High-Tech World: Australia's Potential to Supply Global Demand. Geoscience Australia, Canberra.

Stewart B. W., Capo R. C., Hedin B. C. and Hedin R. S. (2017). Rare earth element resources in coal mine drainage and treatment precipitates in the Appalachian basin, USA. *International Journal of Coal Geology*, **169**, 28–39.

Streten-Joyce C., Manning J., Gibb K. S., Neilan B. A. and Parry D. L. (2013). The chemical composition and bacteria communities in acid and metalliferous drainage from the wet-dry tropics are dependent on season. *Science of The Total Environment*, **443**, 65–79.

U.S. Geological Survey (2020). Mineral Commodity Summaries, Reston, VA, p. 200. doi: 10.3133/mcs2020

Verplanck P. L., Nordstrom D. K., Taylor H. E. and Kimball B. A. (2004). Rare earth element partitioning between hydrous ferric oxides and acid mine water during iron oxidation. *Applied Geochemistry*, **19**, 1339–1354.

Xingyu L., Zou G., Wang X., Zou L., Wen J., Ruan R. and Wang D. (2013). A novel low pH sulfidogenic bioreactor using activated sludge as carbon source to treat acid mine drainage (AMD) and recovery metal sulfides: pilot scale study. *Minerals Engineering*, **48**, 51–55.

Yang F., Kubota F., Baba Y., Kamiya N. and Goto M. (2013). Selective extraction and recovery of rare earth metals from phosphor powders in waste fluorescent lamps using an ionic liquid system. *Journal of Hazardous Materials*, **254–255**, 79–88.

Younger P. L. (1997). The longevity of minewater pollution : a basis for. *Science of The Total Environment*, **194/195**, 457–466.

Chapter 11

Plant based removal and recovery of rare earth elements

Kirti Avishek and Moushumi Hazra

11.1 INTRODUCTION

The impact of urbanization and industrialization has accelerated the degradation of environmental components. Industrial wastewater and ore mining act as a hotspot for entry of metals, metalloids and rare earth elements (REE) to freshwater ecosystems. They are found in the form of cations, inorganic or organic complexes of humic and fulvic acid as a part of the dissolved organic matter (DOC) and also bound to the suspended solids. This necessitates the use of technologies that assist in the removal of these heavy metals, metalloids and REE. Active (such as precipitation or ion exchange) (Babel & Kurniawan, 2003) or passive (wetlands or by granulate substances) (Kropfelova *et al.*, 2009) water filtration systems have been used successfully.

Considering wetland ecosystems, which act as a sink for metals, metalloids or REE, both natural and constructed wetlands (CWs) can be employed for the above purpose. CWs mimic natural wetlands and are considered as innovative technology that can treat a wide range of pollutants, including nutrients, organic pollutants, bacteria, antibiotic resistant genes, heavy metals, metalloids and REE (Stefanakis, 2019). Wetlands have the potential to remove pollutants by complex reactions involving the plants, microbes and substrates/media along with the process of sedimentation (Davranche *et al.*, 2016).

The main benefits of nature based solutions to be employed on fields are the low construction and operational cost and low maintenance requirement. Moreover, the energy requirement is almost negligible which shows their suitability in both rural

 The chapter is from the book *Environmental Technologies to Treat Rare Earth Elements Pollution: Principles and Engineering* by Arindam Sinharoy and Piet N. L. Lens (Eds.).
doi: 10.2166/9781789062236_0253

and urban areas. CWs find their application in secondary/tertiary treatment or can be used in a decentralized manner. Hence the main objective of this chapter is to provide a holistic approach for (i) evaluating the performance of CWs for the removal of REE and (ii) recovery through the macrophytic species that are used for enhancing the efficiency of CWs.

11.2 SOURCES AND RELEASE OF REE IN THE ENVIRONMENT

11.2.1 Chemical characteristics of REE

REE denote 17 chemically similar metallic elements, including scandium, yttrium and the lanthanoids. Table 11.1 shows the list of REE and their chemical

Table 11.1 List of REE and their chemical characteristics (Aide & Aide, 2012).

Name	Element	Atomic		Ground State Configuration
		Number	Weight	
Lanthanum	La	57	138.9055	[Xe]5d16s2
Cerium	Ce	58	140.12	[Xe]4f15d16s2
Praseodymium	Pr	59	140.9077	[Xe]4f36s2
Neodymium	Nd	60	144.24	[Xe]4f46s2
Promethium	Pm	61	145	[Xe]4f56s2
Samarium	Sm	62	150.36	[Xe]4f66s2
Europium	Eu	63	151.96	[Xe]4f76s2
Gadolinium	Gd	64	157.25	[Xe]4f75d16s2
Terbium	Tb	65	158.9254	[Xe]4f96s2
Dysprosium	Dy	66	162.50	[Xe]4f106s2
Holmium	Ho	67	164.9304	[Xe]4f116s2
Erbium	Er	68	167.26	[Xe]4f126s2
Thulium	Tm	69	168.93	[Xe]4f136s2
Ytterbium	Yb	70	173.04	[Xe]4f146s2
Lutetium	Lu	71	174.967	[Xe]4f145d16s2
Scandium	Sc^{3+}	21	44.9559	[Ar]3d14s2
Yttrium	Y^{3+}	39	88.9059	[Kr]4d15s2
Calcium	Ca^{2+}	20	40.078	[Ar]4s2
Manganese	Mn^{2+}	25	54.938	[Ar]3d54s2
Strontium	Sr^{2+}	38	87.62	[Kr]5s2
Thorium	Th^{4+}	90	232.0381	[Rn]6d27s2
Uranium	U^{4+}	92	238.0289	[Rn]5f36d7s2

characteristics. The electronic configuration comprises an inner shell with electrons in the 4fn orbital with an outer shell consisting of electrons in orbital number 5s2, 5p6, 5d1-10 and 6s2. They possess unique 4f electrons that give unique properties to the REE. They are thermodynamically stable when in trivalent form and their physicochemical properties are similar due to their electronic configuration. The REE are categorized into (i) light REE (LREE) and (ii) heavy REE (HREE) on the basis of physical and chemical properties. LREE consist of elements from lanthanum to europium (Z from 57–63) and HREE from gadolinium to lutetium (Z from 64–71) (Pereao *et al.*, 2018). REE are abundantly found in nature and extracted by various methods for their application in different industries.

11.2.2 Sources of REE

11.2.2.1 Natural sources

Based on their origin, REE can be categorized as (i) natural and (ii) anthropogenic. Naturally occurring REE are present in aquatic ecosystems, where they form complexes with various ligands with the help of ionic bonds and occupy high energy orbitals (Weber & Chaudhuri, 2008). They are abundantly found in rocks where they can be used as a tracer for wetland studies and fine sorption processes taking place in wetland colloid organic matter (Davranche *et al.*, 2011). Other sources include waterlogged soils and sediments that have anoxic conditions. Figure 11.1 indicates the sources and fate of REE in the environment.

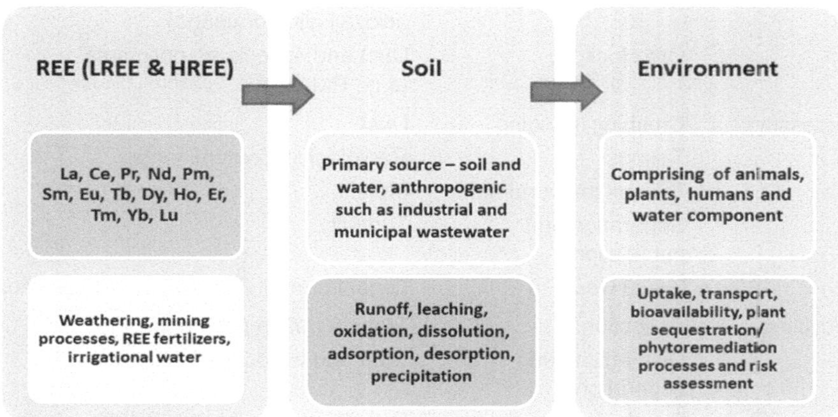

Figure 11.1 Sources and fate of REE in soils. Lanthanum (La), cerium (Ce), praseodymium (Pr), neodymium (Nd), promethium (Pm), samarium (Sm), europium (Eu), gadolinium (Gd), terbium (Tb), dysprosium (Dy), holmium (Ho), erbium (Er), thulium (Tm), ytterbium (Yb) and lutetium (Lu) (*Source*: Aide & Aide, 2012; Gao *et al.*, 2012; Li *et al.*, 2020; Mihajlovic, 2018; Pereao *et al.*, 2018; Prudêncio *et al.*, 2017).

The adsorption of organic matter onto the soil and Fe oxyhydroxide reduction in flooded wetland soil controls the fate and transfer of REE in wetlands (Grybos *et al.*, 2009). It should therefore be mentioned that the source of REE, organic matter, reduction processes and Fe dynamics play a significant role in the REE pattern. The pathway involved for REE migration in the soil involves: (i) erosion, (ii) leaching of REE-inorganic complexes with the percolating water, (iii) organic complexation that results in mobilization/immobilization of REE, (iv) eluviation – illuviation of clay with co-adsorbed REE, (v) plant uptake, (vi) precipitation reactions as a result of removal of REE from percolating water and (vii) adsorption of REE by inorganic colloids such as phyllosilicates and oxyhydroxides. Differences in hydrolysis, inorganic complexation constant and weathering susceptibility of various REE lead to a differential migration potential and fractionation in the wetland environment (Aide & Aide, 2012).

Table 11.2 Activities and emission sources of REE (*Source*: http://www.eurare.org/docs/internalGuidanceReport.pdf).

Type of Activity	Emission Source(s)	Primary Pollutants of Concern
Mining	Overburden	Radiological contaminants
	Waste rock	Metals
	Sub-ore stockpile	Mine influenced waters (e.g., acid/alkaline drainage)
	Ore stockpile	Dust and associated pollutants (e.g., PM 2.5)
Processing	Crushing/grinding	Dust
	Tailings	Radiological contaminants
	Tailings impoundment	Metals
	Separation and purification	Turbidity
	Liquid waste	Organics, dust and associated pollutants
Recycling	Collection	Transportation pollutants
	Dismantling and separation	Dust and associated pollutants
	Scrap waste	Volatile organic compounds
	Landfill	Metals
	Processing	Dust and associated pollutants
		Volatile organic compounds, dioxins, metals Organics

11.2.2.2 Industrial sources

Industries generate large amounts of industrial sludge that enter wastewater treatment plants (WWTPs). Various other activities result in the release of REE as given in Table 11.2. Though the sludge is digested to reduce the amount of organic matter and pathogenic bacteria, they are enriched with inorganic heavy metal and radioactive compounds. Recent reports have highlighted that toxic cadmium (Cd) and nickel (Ni) are also present in the sludge after digestion, which limits its use as a fertilizer (Scancar *et al.*, 2000). Various recovery methods do remove a considerable number of heavy metals from sludge, for example, ultrasound assisted metal recovery though this is costly since it is an energy consuming process. Other methods like bioleaching, electro-reclamation and chemical extraction have also been reported. A combination of two or more techniques may enhance recovery as well as allowing the heavy metals from the sludge to be recycled and reused (Gao *et al.*, 2012). In agriculture, agricultural inputs act as a source of REE to the natural environment and long-term usage can have harmful effects on the soil, groundwater and humans. For example thermophosphate, superphosphates and NPK (consisting of nitrogen, phosphate and potassium) fertilizers may alter the soil macro-fauna community diversity (Jinxia *et al.*, 2010).

11.3 EXTRACTION AND RECOVERY OF REE

11.3.1 Phytoextraction

11.3.1.1 Agromining

The earth's crust has an average concentration of REE up to 189 μg/g as reported by Wei *et al.* (1991). REE tailings contain high concentrations that disperse to the nearby agricultural soil (in the range of 870–1100 μg/g). Lanthanum (La) and cerium (Ce) have been reported in concentrations of 11,100 and 23,600 μg/g (Guo *et al.*, 2013). The REE that are bioavailable can be sequestered using hyperaccumulator plants employing agromining. The availability of REE depends on factors such as pH, organic matter, oxides and the content of adsorbing phases like clay particles (Groenenberg *et al.*, 2010). Therefore, before agromining REE, speciation studies and determination of limiting factors such as lack of organic matter, nutrients, high bulk density and, particularly, presence of clay particles are required to improve plant colonization in tailings.

A recent investigation by Yuan *et al.* (2018) on *Phytolacca americana* (high biomass, fast growing herbaceous plant) growing in an REE mining area, highlighted the fractionation pattern of REE in soil-to-plant absorption and root-to-shoot transport processes. They also investigated the REE accumulation ability of *P. americana* growing naturally in the mining area.

11.3.1.2 REE plant uptake

Wen *et al.* (2015) reported that the concentration of REE accumulated was in the order of roots, stem, leaves and the grains (cereals or legumes). Type of plant species, growth period of the plant, climatic conditions, soil properties, speciation and mobilization process of REE in the soil mainly influence the distribution pattern. Soil pH affects the behavior and availability of metals present in the soil to the plants. Adsorption of metals onto the clay particulates and organic matter increases with increasing soil pH. Hence, the trace elements available for uptake by the plant species are generally greater at low pH than high pH.

11.3.1.3 Sequential extraction of REE from soil

REE present in the soil are fractionated into three constituents comprising (i) (B1) – water soluble, exchangeable and carbonate bound, (ii) (B2) – Fe-Mn oxide bound and (iii) (B3) – organic matter and sulfide bound (Li *et al.*, 1998). The proportion of REE in fraction B1 is large and the REE extracted with CH_3COOH in the first step are water soluble, exchangeable and bound to carbonate forms. It was found that the B1 fraction was the most bioavailable and mobile form of metals in the soil. Metals that were associated with carbonates were easily released to solution when the pH was low. The authors also suggested that the CH_3COOH extractable fraction B1 of REE in mollisol soils played a significant role in the uptake of REE by plants. The grains showed the presence of a large proportion of exogenous REE with low pH in the soil. Whereas soil with a high organic matter and a large B3 fraction (organic matter and sulfide bound forms) demonstrated uptake of REE in a complicated manner.

11.3.2 Other extraction methods

When it comes to separation of REE mixtures into individual elements of concern, the low separation factors between the REE makes it difficult to extract them. This is also due to the similarity of the chemical properties of different REE. The use of organophosphorous extractants and phosphine oxide has been commonly used for extracting REE from different aqueous media (Shaibu *et al.*, 2006; Thakur, 2000). The adsorption process recovers metal ions/REE even from low concentrations of the soil or water matrix. Adsorbents involve a combination of solid phase sorbents. Oxides of iron and manganese as well as certain magnetite nanoparticles are efficient with scalability and are environmentally friendly. Also extraction employing engineered microorganisms (especially mesophilic bacteria) can be used for extraction of REE from aqueous media. Bacteria tend to bind with the metal ions that are assessed for metal binding capacities based on dry biomass which is comparable to the binding capacities of ion exchangers available commercially (see also Chapter 8). Magnetic segregation recovers REE from geothermal fluids with a combination of both physical and magnetic forces, and the separated REE can be collected and recovered from the condensate of the unit (Smith, 2017).

11.4 PHYTOREMEDIATION OF REE

11.4.1 Plant metabolism for REE phytoremediation

11.4.1.1 Plant species selection

Any plant species that is to be selected for phytoextraction should be resistant to high concentrations of REE present in soil or water. The bioavailability of REE can demonstrate the increase or decrease of its toxicity level in particular plant species. Also the concentration and bioavailability of REE and its abundance, along with the amount of water in the soil, has a direct influence on the plant biomass, root growth, and phytoextraction of these elements and minerals (Zhang *et al.*, 2013). Moreover, different plant species that accumulate REE from the environment can be used for both economic and environmental benefits, especially given the increasing price of REE.

Figure 11.2 depicts the general mechanism of REE removal using plant species. Table 11.3 represents REE hyperaccumulator and potential hyperaccumulator plant species. Many previous studies have reported that herbaceous plant species such as *Artemisia vulgaris*, *Achilleamille folium*, *Cichorium intybus*, *Taraxacum officinale*, *Trifolium repens*, *Papaver rhoeas* and *Tripleurospermum inodorum* do help in

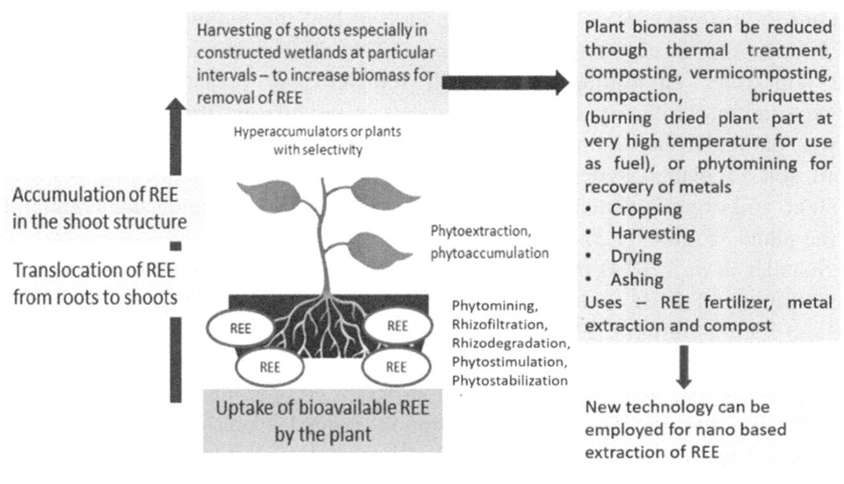

Figure 11.2 Schematic diagram of the rare earth elements uptake mechanism in plants and their subsequent recovery (*Source*: Kadlec & Wallace, 2009; Kurade *et al.*, 2021; Lima & Ottosen, 2021; Stefanakis, 2019).

Table 11.3 Rare earth element (REE) hyperaccumulator plant species and their performance.

Plant Species	Family	Accumulation Characteristics in the Foliage of the Plant Species or BCF Reported	Distribution	Reference
Carya tomentosa	Juglandaceae	ΣREE in foliage: 1350 μg/g (in ash)	America	Thomas (2011)
Pronephrium triphyllum	Thelypteridaceae	ΣREE in foliage: 1027 μg/g	China	Xue (2009)
Pronephrium simplex	Thelypteridaceae	ΣREE in foliage: 1234 μg/g	China	Lai *et al.* (2005)
Blechnum orientale	Blechnaceae	ΣREE in foliage: 1022 μg/g	China	Xiao *et al.* (2003)
Dicranopteris strigose	Gleicheniaceae	BCF of La >1	Japan	Ozaki *et al.* (2000)

Note: ΣREE: represents the sum of 16 rare earth elements; BCF (bio-concentration factor) is calculated as the quotient of the concentration in leaf/shoot of a plant to the concentration in the soil/water.

decontamination of REE and accumulate a significant amount in their root and shoot structure. A study on accumulation of REE contaminated road sites found that among the different plant species *T. inodorum* had the highest concentration of REE in its root structure (Mleczek *et al.*, 2021). The chemical characteristics of the soil/water environment in which the REE is present also affect its transportation, while cation exchange assists in modulating the transport of REE in that particular environment (Dinali Soares *et al.*, 2019). The transport of REE from soil to plants is dependent upon specific factors such as pH, soil clay content, soil organic matter content and redox potential that render its availability to the plants (Figure 11.3).

Schaller *et al.* (2014) indicated the potential use of *Pharagmites australis* for phytomining of REE such as strontium (Sr) and alkaline earth metals (reactive metals) such as lithium (Li) and beryllium (Be). Hence, they employed a CW in combination with a WWTP that treated wastewater from the oil industry loaded with heavy metals and REE. Among the different compounds, Li was not utilized biologically and the residual concentration ranged from 45–896 μg/l in the evaporated pond unit following the CW. Therefore, Li finds its application as a tracer. Similarly, high concentrations of Be and magnesium (Mg) in the evaporation pond indicated that these two elements are harvestable. Be cannot be used as a tracer because it adsorbs in the organic sediments and onto the plants, whereas iron (Fe) and manganese (Mn) were adsorbed by the sediments and

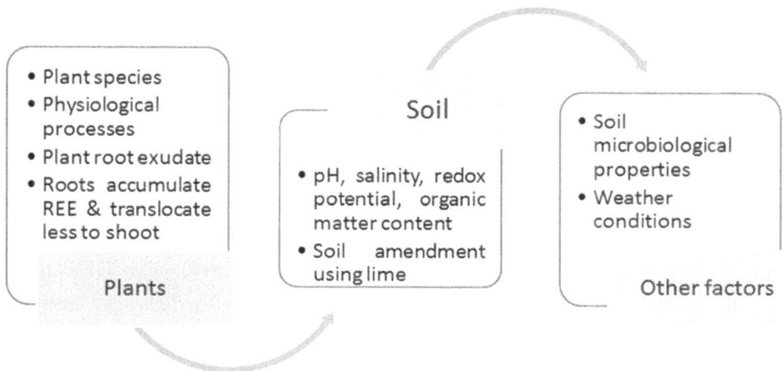

Figure 11.3 Some of the significant factors that influence plant-based uptake of REE (based on general understanding of the present review study).

biologically taken up by *Pharagmites*. It was concluded that the elements having chemical characteristics of inertness and biological inactivity can be recovered and extracted for further utility. The CWs can therefore be employed to harvest REE that can improve remediation of contaminated wastewater loaded with REE.

11.4.1.2 Bioindicator plants

Some plant species can be used as a bioindicator of trace elements and REE, such as *Taraxacum officinale* found in urban areas (Giacomino *et al.*, 2016). These plants possess some peculiar characteristics including that their leaf thickness decreases and their parenchyma palisade tissue becomes less organized when exposed to toxic elements. They have also been found to be efficient in phytoextraction of REE (Maleci *et al.*, 2014). A recent study has also reported *T. officinale* as a good bioindicator of REE (Borowiak *et al.*, 2018). Some of the elements are even found to enhance plant growth, for example the medicinal plant *Salvia miltiorrhiza* showed enhanced plant growth and a higher secondary metabolite content in plant parts when grown in the presence of 100 mM praseodymium (Pr) (Fan *et al.*, 2020). Elongation in root structure and enhanced photosynthetic activity were also observed. The increased secondary metabolite (phenols and flavonoids) production helps in defense responses and elevating plant stress. This is further elicited by enzymatic antioxidant systems such as peroxidase, catalase, ascorbate peroxidase and superoxide dismutase.

11.4.1.3 REE in plant metabolism

Literature does not report REE to be nutritionally essential to plants, but they may compete with calcium for calcium mediated biological processes and this may result in a toxic effect. They have a high charge density (possesses trivalent charges) that allows them to replace divalent Ca that has a lower charge density at the calcium binding sites in any biological molecules as investigated by Brown *et al.* (1990). Where lanthanum replaces calcium and inhibits many plant enzymes and other functional proteins, it is also referred to as 'super calcium'. When REE replace calcium from extracellular binding sites, the efflux of extra and intra cellular calcium gets inhibited which has a negative consequence on plant growth and vigor.

REE do not show toxicity to plants at low concentrations, and they have been used as REE-fertilizer (<0.5 REE mg/kg soil) for treatment of seeds, as sprays or as root fertilizer formulations as reported by Hu *et al.* (2007). The REE treatment increased the yield from 5% to 15% for different crops with different nutrient and soil conditions. REE are generally added as compounds with nitrogen and growth-promoting substances and it is difficult to understand their exact beneficial effects. Long-term usage of these REE fertilizers increases their content in the soil which affects the environment and humans with a significant impact on primary producers and different trophic levels (Turra *et al.*, 2011) and has also demonstrated reduced soil macrofaunal community diversity (Jinxia *et al.*, 2010).

Different plants species, namely *Asclepias syriaca*, *Panicum virgatum*, *Desmodium canadense*, *Raphanus sativus* and *Solanum lycopersicum* were used to study toxicity due to elevated REE (La, Y and Ce) concentrations. The seed germination efficiency was inversely related to the REE concentration, that is, with increased concentration the germination efficiency decreased. La and Ce contamination had a negative impact on germination in *D. canadense* and *S. lycopersium*, while Ce had a harmful effect on *A. syriaca, S. lycopersicum, R. sativus* and *P. virgatum*. A negative effect was observed at the high dose in yttrium soils (102.8 mg/kg of soil) for *D. canadense* and in low pH cerium soil (70.9 mg Ce/kg of soil) for *A. syriaca* (Thomas *et al.*, 2014). Ce can be adsorbed in plants, get into their tissues and may thereby show signs of toxicity. It has been previously reported by Joacir De França *et al.* (2011) that roots accumulate high concentrations of REE providing benefits to the shoots that are the photosynthetic structure of the plants. No noticeable effects have been recorded at 10 mg La/kg dry soil for *Phaseolus vulgaris* or 50 mg La/kg dry soil for *Spinaceaoleracea*, with the highest doses investigated for both plant species (von Tucher & Schmidhalter, 2005).

11.4.2 Plants biomass for biosorption

In addition to the live plants, plant biomass can also be applied as adsorbent for REE removal and recovery. Biosorption is a cost effective method that has been found to separate REE and the plant derived biomass used is able to concentrate REE (Suli *et al.*, 2017). Table 11.4 shows the plant based adsorbents previously reported for REE adsorption. Factors that play a significant role during the biosorption process are pH (4–7), temperature (25–60°C), dose of applied bio-adsorbent (15–200 mg/l), REE concentration (15–300 mg/l) and contact time (300–480

Table 11.4 Various types of plant based adsorbents for biosorption processes (Suli *et al.*, 2017).

Rare Earth Elements	Plant Based Products Used as Adsorbents	Adsorption Capacity (in mg/g)
La	*Sargassum* biomass	40.28
La	*Sargassum fluitans*	101.401
La	*Platamus orientalis* leaf powder	28.65
La	Activated carbon from rice husks	175.4
La	Bamboo charcoal	120
Nd	*Monoraphidium sp.*	1511
Nd	*Kluyveromyces marxiamus, Candida*	12
Nd	*Colliculaosa, Debaromyces hansenii*	10
Ce	*Platamus orientalis* leaf powder	32.05
Er	Activated carbon from rice husks	250
Eu	*Sargassum* biomass	62.30
Sm	*Sargassum* biomass	105
Pr	*Sargassum* biomass	98
Yb	*Sargassum* biomass	48.45

minutes). It should be mentioned that the REE can be extracted from adsorbents by desorption and depending on their purity can be used in suitable applications.

11.5 WETLANDS FOR REE RETENTION AND RECOVERY

11.5.1 Natural wetlands

Wetlands are natural systems that play a significant role in retaining considerable amounts of elements (including REE). However, their performance can vary widely depending upon many operating factors. The main sink in wetlands is the sediment that assists in adsorption/accumulation of molecules. The elements get adsorbed onto the rocks, on the organic/inorganic sediment particles or on the biofilms. Huang *et al.* (2000) described the interaction between biofilms and metal accumulation in wetlands. Generally, the organic/inorganic particles with high elemental load accumulate mainly in the lentic area of the wetland. The heterotrophic microbes that make up the biofilm consume oxygen, which results in a redox gradient within the biofilm. REE strongly bind to the biotic ligands based on the high complex stability and ligand sensitivity. Therefore, the elemental recovery from natural wetlands involves fixation of chemical elements by plant biomass, growing algae or dead/decaying plants that are attached to the litter or are pelagic in form (Axtell *et al.*, 2003). Plant exudates through plant roots and sequestration of REE in their vacuoles are also significant mechanisms involved in REE removal. The elemental recovery from wetlands may be favored depending on the ambient temperature and humidity. Further studies are therefore required to understand the REE recovery potential of the natural wetlands.

11.5.2 Constructed wetlands

Few studies have reported the use and application of constructed wetland for removal and recovery of REE. CWs employ phytoremediation as the extraction technique, where the REE are extracted and transferred to the plants. In China, reports have confirmed that more than 96% of the REE have been removed in the vicinity of a mining area (Chour *et al.*, 2018; Wu *et al.*, 2013). Qin *et al.* (2019) reported that the plants are to be incinerated after recovering REE from mining areas to avoid further contamination of the soil or water compartment. There is a lack of literature confirming the individual uptake of REE and their mechanism in the plants, or the fate of REE or its recovery from waste ashes. A recent study has reported *P. americana* to be efficient enough for accumulation of light REE (Liu *et al.*, 2018).

Altomare *et al.* (2020) reported removal of gadolinium (Gd) that has possible toxic effects on the environment using a CW. The fate of Gd in surface water and sediments of a CW receiving effluents from a wastewater treatment plant was determined. Approximately 25 g of Gd per day entered the wetland, which decreased significantly at the outlet zone of the CW. A higher concentration was found in the CW sediments which possibly acted as a sink for the Gd. It was also evaluated that sediments that had a higher total organic carbon value accumulated higher concentrations of Gd. The mechanism of Gd sequestration in the wetland could be through complexation with the organic biomass, followed by plant uptake or incorporation into the plant biomass to a lower extent.

Braun *et al.* (2018) investigated the impact of the Gd concentration in four aquatic plant species namely *Lemna gibba, Ceratophyllum demersum, Elonea nuttallii* and *E. Canadensis*. These aquatic plant species are potential biological filters used in CWs. The bioconcentration factor (BCF) was <1 inferring that the Gd concentration in the plant tissues was lower than that of the external water. It was concluded that no Gd complexes were accumulated in the plant tissues and whatever the concentration in plants, it reflects the same concentration as that of the water. In comparison, the BCF of the above species for heavy metals such as Cr, Cd, Pb, Ni and Mn are in the range of 100–100,000 (Landolt & Kandeler, 1987), indicating uptake and bioaccumulation of heavy metals and not REE by these species.

Limited attempts have been made to employ CWs for the remediation of REE. However, the potential of CWs for REE removal is significantly high. CWs can be employed for treating REE-containing mine impacted water discharges with appropriate modifications (pretreatments). The respective plant species can be utilized for high purity metals, followed by extractive metallurgy routes. Also the extraction of REE can be carried out from the sediments of CWs where the REE concentrate during their treatment. Furthermore, REE can be utilized as phosphate fertilizers when extracted from mineral rocks. They can be used for seed treatment, leaf sprays, liquid or solid root fertilizer formulations or as plant growth stimulator in the agricultural sector. Concern regarding the ecological and environmental aspects related to the presence of REE in the soil and water environment and their possible transfer in the trophic chain has received less attention, and needs further exploration.

11.6 CONCLUSIONS

Urbanization and industrialization have degraded all components of the environment, especially the water compartment that receives pollutants via effluents. Wetlands serve as an option for recovery of resources including various elements and REE. Therefore, CWs, integrated with various technologies and designs can be efficient for remediation of REE from wastewater/effluent from treatment plants. REE released due to natural and anthropogenic processes end up in the environment and can cause a detrimental effect on the environment. Specific plants such as *Phytolacca Americana*, *Pharagmites australis* and *Taraxacum officinale* have been used for remediation of REE. Various extraction methods such as electroprecipitation, leaching and organic complexations can be used for recovery of REE from wetland plants. The extracted REE can be used as fertilizer as well as seed or plant growth stimulator.

REFERENCES

Aide M. T. and Aide C. (2012). Rare earth elements: their importance in understanding. *International Scholarly Research Notices*, **2012**, 783876. doi: 10.5402/2012/783876

Altomare A. J., Young N. A. and Beazley M. J. (2020). A preliminary survey of anthropogenic gadolinium in water and sediment of a constructed wetland. *Journal of Environmental Management*, **255**, 109897. doi: 10.1016/j.jenvman.2019.109897

Axtell N. R., Sternberg S. P. K. and Claussen K. (2003). Lead and nickel removal using *Microspora* and *Lemna minor*. *Bioresource Technology*, **89**, 41–48.

Babel S. and Kurniawan T. A. (2003). Low-cost adsorbents for heavy metals uptake from contaminated water: a review. *Journal of Hazardous Materials*, **97**(1–3), 219–243.

Borowiak K., Lisiak M., Kanclerz J., Budka A., Adamska A. and Janicka E. (2018). Relations between rare earth elements accumulation in *Taraxacum officinale* L. and land use in an urban area – a preliminary study. *Ecological Indicators*, **94**, 22–27.

Braun M., Zavanyi G., Laczovics A., Berenyi E. and Szabo S. (2018). Can aquatic macrophytes be biofilters for gadolinium based contrasting agents? *Water Research*, **135**, 104–111.

Brown P. H., Rathjen A. H., Graham R. D. and Tribe D. E. (1990). Rare earth elements in biological systems. In: Handbook on the Physics and Chemistry of Rare Earths, K. A. Gschneidner, Jr. and L. Eyring (eds), Elsevier Sciences Publisher B. V., New York, Vol. **13**, pp. 423–453.

Chour Z., Laubie B., Louis J., Tang Y. and Qiu R. (2018). Recovery of rare earth elements from *Dicranopteris dichotoma* by an enhanced ion exchange leaching process. *Chemical Engineering and Processing-Process Intensification*, **130**, 208–213.

Davranche M., Grybos M., Gruau G., Pédrot M., Dia A. and Marsac R. (2011). Rare earth element patterns: a tool for identifying trace metal sources during wetland soil reduction. *Chemical Geology*, **284**, 127–137. doi: 10.1016/j.chemgeo.2011.02.014

Davranche M., Grau G., Dia A., Le Coz-Bouhnik M., Marsac R., Pédrot M. and Pourret O. (2016). Rare earth elements in wetlands. In: Trace Elements in Waterlogged Soils and Sediments, J. Rinklebe, A. S. Knox and M. Paller (eds), Taylor and Francis Group/CRC Press, pp. 135–162.

Dinali G. S., Root R. A., Amistadi M. K., Chorover J., Lopes G. and Guilherme L. R. G. (2019). Rare earth elements (REY) sorption on soils of contrasting mineralogy and texture. *Environment International*, **128**, 279–291.

Fan Z., Zhang K. and Farm F. (2020). Effects of rare earth elements on growth and determination of secondary metabolites under in vitro conditions in *Salvia miltiorrhiza*. *HortScience*, **55**(3), 310–316.

Gao L., Kano N., Sato Y., Li C., Zhang S. and Imaizumi H. (2012). Behavior and distribution of heavy metals including rare earth elements, thorium, and uranium in sludge from industry water treatment plant and recovery method of metals by biosurfactants application. *Bioinorganic Chemistry and Applications*, **2012**, 173819. doi: 10.1155/2012/173819

Giacomino A., Malandrino M., Colombo M. L., Miaglia S., Maimone P., Blancato S., Conca E. and Abollino O. (2016). Metal content in dandelion (*Taraxacum officinale*) leaves: influence of vehicular traffic and safety upon consumption as food. *Journal of Chemistry*, **2016**, 9842987. doi: 10.1155/2016/9842987

Groenenberg J., Romkens P., Comans R., Luster J., Pampura T., Shotbolt L., Tipping E. and De Vries W. (2010). Transfer functions for solid-solution partitioning of cadmium, copper, nickel, lead and zinc in soils: derivation of relationships for free metal ion activities. *European Journal of Soil Science*, **61**(1), 58–73.

Grybos M., Davranche M., Gruau G., Petitjean P. and Pédrot M. (2009). Geoderma Increasing pH drives organic matter solubilization from wetland soils under reducing conditions. *Geoderma*, **154**, 13–19.

Guo W., Fu R. Y., Zhao R. X., Zhao W. J., Guo J. Y. and Zhang J. (2013). Distribution characteristic and current situation of soil rare earth contamination in the Bayan Obo mining area and Baotou tailing reservoir in Inner Mongolia. *Environmental Science*, **34**, 1895–1900.

Hu Z., Richter H., Sparovek G., Schnug E., Hu Z., Richter H. and Sparovek G. (2007). Physiological and biochemical effects of rare earth elements on plants and their agricultural significance: a review. *Journal of Plant Nutrition*, **27**(1), 183–220.

Huang Y. B., Wang W. H. and Peng A. (2000). Accumulation of Cu(II) and Pb(II) by biofilms grown on particulate in aquatic systems. *Journal of Environmental Science & Health Part A*, **35**(4), 575–592.

Jinxia L., Mei H., Xiuqin Y. and Jiliang L. (2010). Effects of the accumulation of the rare earth elements on soil macrofauna community. *Journal of Rare Earths*, **28**, 957–964.

Joacir De França E., De NadaiFernandes E., Turra C., Bacchi M. A., Elias C., Tagliaferro F. S. and Moreira C. F. (2011). Survey of lanthanoids in plants from a tropical region. *International Journal of Environment and Health*, **5**, 32–48.

Kadlec R. H. and Wallace S. D. (2009). Treatment wetlands, second edition TOC and references. In: Treatment Wetlands, Robert H. Kadlec and Scott Wallace (eds), 2nd edn. CRC Press, Taylor and Francis, Boca Raton, 1016 p. doi: 10.1201/9781420012514

Kropfelova L., Jan V., Jaroslav S. and Stichova J. (2009). Removal of trace elements in three horizontal sub-surface flow constructed wetlands in the Czech Republic. *Environmental Pollution*, **157**, 1186–1194.

Kurade M. B., Ha Y., Xiong J., Govindwar S. P., Jang M. and Jeon B. (2021). Phytoremediation as a green biotechnology tool for emerging environmental pollution: a step forward towards sustainable rehabilitation of the environment. *Chemical Engineering Journal*, **415**, 129040. doi: 10.1016/j.cej.2021.129040

Lai Y., Wang Q. Q., Yan W. W., Yang L. M. and Huang B. L. (2005). Preliminary study of the enrichment and fractionation of REEs in a newly discovered REE hyperaccumulator *Pronephrium simplex* by SEC-ICP-MS and MALDI-TOF/ESI-MS. *Journal of Analytical Atomic Spectrometry*, **20**(8), 751–753.

Landolt E. and Kandeler R. (1987). The Family of *Lemnaceae* – a Monographic Study. Ver €off.Geobot. Inst. ETH, StiftungRübel, Zürich, **2**, p. 95.

Li F. L., Shan X. Q., Zhang T. H. and Zhang S. Z. (1998). Evaluation of plant availability of Rare Earth Element in soils by chemical fractionation and multiple regression analysis. *Environmental Pollution*, **102**, 269–277.

Li Q., Zhong H. and Cao Y. (2020). Effective extraction and recovery of rare earth elements (REEs) in contaminated soils using a reusable biosurfactant. *Chemosphere*, **256**, 127070. doi: 10.1016/j.chemosphere.2020.127070

Lima A. T. and Ottosen L. (2021). Recovering rare earth elements from contaminated soils: critical overview of current remediation technologies. *Chemosphere*, **265**, 129163. doi: 10.1016/j.chemosphere.2020.129163

Liu C., Yuan M., Liu W., Guo M., Huot H., Tang Y., Laubie B., Simonnot M. and Morel J. L. (2018). Element case studies: rare earth elements. In: Agromining: Farming for Metals. Mineral Resource Reviews, A. Van der Ent, G. Echevarria, A. Baker and J. Morel (eds), Springer, Cham, pp. 297–308. doi: 10.1007/978-3-319-61899-9_19

Maleci L., Buffa G., Wahsha M. and Bini C. (2014). Morphological changes induced by heavy metals in dandelion (*Taraxacum officinale* Web.) growing on mine soils. *Journal of Soils and Sediments*, **14**(4), 731–743.

Mihajlovic J. (2018). Rare earth elements in German soils – a review. Chemosphere, **205**, 514–523. doi: 10.1016/j.chemosphere.2018.04.059

Mleczek P., Borowiak K. and Budka A. (2021). Possible sources of rare earth elements near different classes of road in Poland and their phytoextraction to herbaceous plant species. *Environmental Research*, **193**, 110580. doi: 10.1016/j.envres.2020.110580

Ozaki T., Enomoto S., Minai Y., Ambe S. and Makide Y. (2000). A survey of trace elements in pteridophytes. *Biological Trace Element Research*, **74**(3), 259–273.

Pereao O., Bode-aluko C. A., Fatoba O. and Petrik L. F. (2018). Rare earth elements removal techniques from water/wastewater: a review. *Desalination and Water Treatment*, **130**, 71–86.

Prudêncio M. I., Valente T., Marques R., Braga M. A. S. and Pamplona J. (2017). Rare earth elements, iron and manganese in ochre-precipitates and wetland soils of a passive treatment system for acid mine drainage. *Procedia Earth and Planetary Science*, **17**, 932–935. doi: 10.1016/j.proeps.2017.01.024

Qin B., Liu W., He E., Li Y., Liu C., Ruan J., Qiu R. and Tang Y. (2019). Vacuum pyrolysis method for reclamation of rare earth elements from hyperaccumulator *Dicranopteris dichotoma* grown in contaminated soil. *Journal of Cleaner Production*, **229**, 480–488.

Scancar J., Milacic R., Strazar M. and Burica O. (2000). Total metal concentrations and partitioning of Cd, Cr, Cu, Fe, Ni and Zn in sewage sludge. *Science of the Total Environment*, **250**(1–3), 9–19.

Schaller J., Headley T., Prigent S. and Breuer R. (2014). Science of the Total Environment Potential mining of lithium, beryllium and strontium from oil field wastewater after enrichment in constructed wetlands and ponds. *Science of the Total Environment*, **493**, 910–913.

Shaibu B. S., Reddy M. L. P., Bhattacharyya A. and Manchanda V. K. (2006). Evaluation of Cyanex 923-coated magnetic particles for the extraction and separation of lanthanides and actinides from nuclear waste streams. *Journal of Magnetism and Magnetic Materials*, **301**(2), 312–318.

Smith Y. R. (2017). On the extraction of rare earth elements from geothermal brines. *Resources*, **6**(3), 1–16.

Stefanakis A. I. (2019). The role of constructed wetlands as green infrastructure for sustainable urban water management. *Sustainability*, **11**(24), 6981. doi: 10.3390/su11246981

Suli L. M., Hanisah W., Ibrahim W., Aziz B. A. and Rizauddin M. (2017). A review of rare earth mineral processing technology. *Chemical Engineering Research Bulletin*, **19**, 20–35.

Thakur N. V. (2000). Separation of rare earths by solvent extraction. *Mineral Processing and Extractive Metullargy Review*, **21**(1–5), 277–306.

Thomas W. A. (2011). Accumulation of rare earths and circulation of cerium by mockernut hickory trees. *Canadian Journal of Botany*, **53**(12), 1159–1165.

Thomas P. J., Carpenter D., Boutin C. and Allison J. E. (2014). Chemosphere Rare earth elements (REEs): effects on germination and growth of selected crop and native plant species. *Chemosphere*, **96**, 57–66.

Turra C., Fernandes E. A. N. and Bacchi M. A. (2011). Evaluation on rare earth elements of Brazilian agricultural supplies. *Journal of Environmental Chemistry and Ecotoxicology*, **3**(4), 86–92.

von Tucher S. and Schmidhalter U. (2005). Lanthanum uptake from soil and nutrient solution and its effects on plant growth. *Journal of Plant Nutrition and Soil Science*, **168**(4), 574–580.

Weber R. J. and Chaudhuri S. (2008). An experimental study of fractionation of the rare earth elements in poplar plants (*Populus eugenei*). Doctoral dissertation, Kansas State University.

Wei F. S., Liu T. L., Teng E. J. and Rui K. S. (1991). A Survey on the background contents of 15 rare earth elements in Chinese soil. *Huanjing Kexue*, **12**, 78–82.

Wen B., Yuan D., Shan X., Li F., Zhang S., Wen B., Yuan D., Shan X., Li F. and Zhang S. (2015). The influence of rare earth element fertilizer application on the distribution and bioaccumulation of rare earth elements in plants under field conditions. *Chemical Speciation & Bioavailability*, **13**(2), 39–48.

Wu J., Chen A. and Peng S. (2013). Identification and application of amino acids as chelators in phytoremediation of rare earth elements lanthanum and yttrium. *Plant and Soil*, **373** (1), 329–338.

Xiao H. Q., Zhang Z. Y, Li F. L. and Chai Z. F. (2003). Study on contents and distribution characteristics of REE in fern by NAA. *Nuclear Techniques*, **26**(6), 420–424.

Xue Y. (2009). Studies of the Hyperaccumulation Ability of Pronephrium Simplex and Pronephrium Triphyllum to Rare Earth Elements and Their Binding Peptides. BSc Thesis: Xiamen University, 99. doi: 10.1039/B501766A

Yuan M., Liu C., Liu W., Guo M., Morel J. L., Huot H., Yu H., Tang Y. and Qiu R. (2018). Accumulation and fractionation of rare earth elements (REEs) in the naturally grown *Phytolacca americana* L. in southern China. *International Journal of Phytoremediation*, **20**, 415–423.

Zhang C., Li Q., Zhang M., Zhang N. and Li M. (2013). Effects of rare earth elements on growth and metabolism of medicinal plants. *Acta Pharmaceutica Sinica B*, **3**(1), 20–24.

Part V

Application of Rare Earth Elements as Nanoparticles

Chapter 12

Rare earth doped nanoparticles and their applications

Tatiana A. Lastovina, Ekaterina O. Podlesnaia and Andriy P. Budnyk

12.1 INTRODUCTION

Oxide materials are experiencing renewed interest from the scientific community as the advancements of nanotechnology have opened new perspectives for their applications in photovoltaics (Ruhle *et al.*, 2012), optoelectronics (Yu *et al.*, 2016), catalysis (Védrine, 2019) and environmental protection (Gusain *et al.*, 2019). In the majority of modern research studies, metals and metal oxides are obtained and examined in the form of nanoparticles (NPs) (Fernández-García & Rodriguez, 2011; Khan *et al.*, 2019). For example, in biomedicine most of the works concern submission of NPs into a body for diagnostics and/or therapy (Kannan *et al.*, 2020b), or the use of nanostructured metal oxides for the development of biosensors (Mohankumar *et al.*, 2021).

The quantum confinement effect drastically tunes the known properties of a material, which at the nanoscale resemble those of the intermediate state between the bulk and molecular/atomic structures (Asha & Ravin, 2020). The matter at the nanoscale becomes an active player interacting with the surrounding environment by surface, the body, or irradiation. In particular, NPs have a much larger surface area to volume ratio than bulk materials, which causes even known noble metals like gold to become reactive (catalytically active) at the nanoscale (Haruta, 2003). The developed surface area of NPs is densely populated with

doi: 10.2166/9781789062236_0273

unsaturated bonds of surface atoms, opening an endless field of surface chemistry (Biener *et al.*, 2009) and relevant applications in catalysis (Moshfegh, 2009). Surface plasmons can greatly accumulate the optical field and dramatically enhance various light-matter interactions (Yu *et al.*, 2019).

A nano-sized body can easily migrate, penetrating pores and membranes, making NPs a deserved carrier for drug delivery and treatment in biomedicine (Mudshinge *et al.*, 2011). The ease of intake of NPs by a living organism raises concerns about their toxicity, and it has been established that some engineered NPs are able to cause toxic effects in living creatures (Jeevanandam *et al.*, 2018). In a broader view, the effects of NPs on the environment, particularly on aquatic and terrestrial ecosystems have received increased attention from the scientific community (Bundschuh *et al.*, 2018). All this suggests that the development of new ('smart') nanomaterials and their applications should comply with the 'Safe- and Sustainable-by-Design' concept, which implies a system approach by integrating safety, circularity and functionality of advanced materials, products and processes throughout their life cycle (Gottardo *et al.*, 2021).

As the name suggests, the concentration of rare earth (RE) elements in minerals seldom exceeds 8–10%, reaching in rare cases 14–15%, and their extraction is typically lengthy, costly and polluting (Eliseeva & Bünzli, 2011). This feature largely determines constraints on their use in applications, that is, generating highly performing functionalities with a minimum loading of the precious elements. This requirement is in full agreement with the opportunities afforded by nanotechnology aimed at creating small-sized materials with high functionality. The RE ions can be incorporated directly into NPs forming oxides/hydroxides, alloys or intermetallic compounds (Davies *et al.*, 2017; Tan *et al.*, 2018).

In a broad view, the RE doped NPs (RENPs) have multiple applications, ranging from photonics (Kakavelakis *et al.*, 2017), sensing (Gadkari *et al.*, 2013), quantum memory and signal processing (Thiel *et al.*, 2011) to catalysis (Anjaneyulu *et al.*, 2015; Ryoo *et al.*, 2020) and biomedicine (Hong *et al.*, 2019). They are also employed in electrochemistry, as cathode and anode electrocatalysts in fuel cells (Antolini & Perez, 2011; Cardoso *et al.*, 2016), as electrode materials in supercapacitors demonstrating higher ionic conductivity, stability and electrochemical activity (Arunachalam *et al.*, 2020) or as perovskite-type oxide bifunctional catalysts in batteries (Ram *et al.*, 2016; Zahoor *et al.*, 2019).

This chapter reviews exemplary reports of RENPs applications in the biomedical and energy sectors (Figure 12.1), which are actively advancing nowadays as evidenced by a growing number of publications. The novel nanostructured materials for biomedical applications are studied as functionalized carriers for targeted drug delivery (Arruebo *et al.*, 2007), contrast agents for magnetic resonance imaging (MRI) (Rümenapp *et al.*, 2012), multimodal response labelling agents (Alcantara & Lee, 2012) and heating bodies in hyperthermic treatment (Jose *et al.*, 2020). In the energy sector, modern electrochemical energy conversion systems employ nanostructured heterogeneous materials as electrode

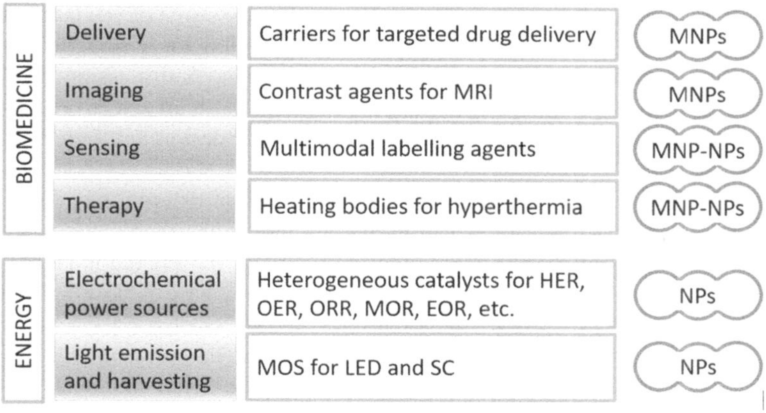

Figure 12.1 Schematic representation of the main fields of applications of RE doped magnetic nanoparticles (MNPs), nanoparticles (NPs) and quantum dots (QDs) composites, reviewed in this Chapter. MRI: Magnetic resonance imaging; HER: hydrogen evolution reaction; OER: oxygen evolution reaction; ORR: oxygen reduction reaction; MOR: methanol oxidation reaction; EOR: ethanol oxidation reaction; MOS: metal oxide semiconductor; LED: light emitting diode; SC: solar cell.

catalysts and additives to electrolytes and polymer membranes (Xu *et al.*, 2019). Sunlight energy harvesting systems are based on a rich variety of semiconductor materials, which employ a rational design at the nano-level to improve their conversion efficiency (Kundu & Patra, 2017). The fluorescent nanostructures have become a basis for the development of bright light-emitting devices (Zhang *et al.*, 2014) as well as for photodynamic therapy applications (Arguinzoniz *et al.*, 2014). The addition of RE elements into the structure of NPs as dopants enhances their magnetic and optical properties (Huang & Zhu, 2019; Qiu *et al.*, 2013; Yu *et al.*, 2015). This chapter provides a comprehensive picture of the demand and prospects for RENPs in biomedical and energy applications.

12.2 BIOMEDICAL APPLICATIONS

12.2.1 Nanoparticles for medical treatment

The purposeful synthesis of NPs for applications in biomedicine can be considered as one of the most challenging scientific directions. Their properties are expected to match strict requirements regarding size distribution, shape, composition, structure and surface functionalization (Khan *et al.*, 2018). Numerous research works have been dedicated to the synthesis of different types of NPs and their laboratory diagnostics. Plenty of inorganic NPs have been examined in the form of stabilized colloidal solutions. Often NPs were combined with a macromolecule into a conjugate (De *et al.*, 2008). A suitable size of NPs is the essential criterion for successful application in biomedicine. NPs in the size range of 10–100 nm with low polydispersity are found residing in the bloodstream for a suitable time for conducting systemic therapies (Yildirimer *et al.*, 2011). A discussion about the optimal size of NPs nevertheless continues, because NPs can selectively interact with biomolecules, cells and tissues depending on their design and function (Kim *et al.*, 2018a).

Because of their ability to respond to the action of an external magnetic field, magnetic NPs (MNPs) are utilized in numerous biomedical applications such as magnetic resonance imaging (MRI) (Zeng *et al.*, 2014), bioseparation (Kim *et al.*, 2018c), targeted delivery (Prijic & Sersa, 2011) and release (Cao *et al.*, 2008), theranostics (Ge *et al.*, 2016), hyperthermia therapy (Kolen'ko *et al.*, 2014), protein purification (Xu *et al.*, 2011) and biosensing (Farzin *et al.*, 2020). The chemical composition, size, morphology and magnetic behaviour determine the efficiency of MNPs. They are typically fabricated from compositions of the Fe, Co and Ni elements, whereas the RE ions are introduced to enhance magnetic properties. Metal oxides are considered more stable *in vivo* and exhibit higher biocompatibility than pure metals. The surface of MNPs is usually covered with a polymer coating (chitosan, dextran or polyethylene glycol (PEG)) for protection and better biocompatibility, or with anchored specific ligands for multiplexed functionality needed in combined hyperthermia-drug delivery and multimodal imaging.

12.2.1.1 Hyperthermal therapy

Hyperthermia as a therapeutic technique against malicious cells is based on their poor ability to dissipate extra heat unlike healthy tissue: the viability of the former significantly decreases in the 41–46°C temperature range (Rajan & Sahu, 2020). In a typical procedure, highly magnetizable MNPs are administered into the cancerous tissue and inductively heated by exposure to an alternating magnetic field (AMF). This leads to irreversible damage of cancer cells loaded with MNPs due to elevated temperature. Biomedically safe frequencies are less than 400 kHz, while the field strength should be at most a few mT (Reyes-Ortega *et al.*, 2021). Specific absorption rate (SAR) is the most commonly used parameter for the estimation of the heat conversion efficacy of MNPs. It is defined as the power produced per sample unit mass. A higher frequency of the applied AMF increases the SAR of MNPs.

The heat from MNPs facilitates the drug penetration into the tumour, enhancing the chemotherapeutic impact. This was illustrated by Kolosnjaj-Tabi *et al.* (2014), who injected PEG-coated 19 nm magnetite nanocubes (700 μg of iron) into epidermoid carcinoma xenografts in mice. In this experiment, AMF treatment was applied to MNPs (magnetic field strength of 23.8 kA m^{-1} and frequency of 111 kHz) and the surface temperature of the injected tumour was increased about 8°C above the temperature of the skin. Looking for maximum heating efficiency in hyperthermia, Kowalik *et al.* (2020) synthesized magnetite MNPs doped with Y^{3+} (up to 10%) by a co-precipitation method. Both size and magnetization of NPs were observed to increase with doping; the maximum magnetization was achieved at 1% Y^{3+} (23 nm NPs). The best value of SAR (194 W/g) was found for the 0.1% Y^{3+} sample (3 mg mL^{-1} in water) exposed to the AMF treatment (16 kA m^{-1} and 413 kHz). Hirosawa *et al.* (2017) prepared Gd-substituted Mg-Zn NPs ($Mg_xZn_{1-x}Gd_yFe_{2-y}O_4$, x is between 0 and 1 and y is between 0 and 0.06) via co-precipitation of metal hydroxides, followed by calcination. The AMF treatment (5 kA m^{-1} and 600 kHz) demonstrated that SAR was dependent on both x and y values; it reached a maximum (~28 W/g) at $x = 0.5$ and $y = 0.02$.

12.2.1.2 Magnetic resonance imaging (MRI)

Contrast-enhanced MRI is an established diagnostic imaging tool, used in about 30 million procedures per year worldwide, with more than 300 million procedures performed over a historical span of 25 years (Lohrke *et al.*, 2016). The determinants of signal intensity and contrast in MRI are spin density, magnetic susceptibility, proton relaxation time and motion (diffusion and perfusion). MNPs generate contrast by shortening the ^1H relaxation times of the locally surrounding water. Relaxation of excited protons occurs via two principal processes: spin-lattice or longitudinal relaxation (characteristic time T_1 [s] and relaxation rate R_1 [s^{-1}]) and spin-spin or transverse relaxation (T_2, R_2). The ability of a contrast agent (CA) to accelerate relaxation is defined by the change in relaxation rate per unit concentration. The proportionality constants are denoted as r_1 and r_2 [mM^{-1} s^{-1}], where $r_2 \geq r_1$ (Strijkers *et al.*, 2007). Thus, the normal contrast in the MRI depends mainly on the proton spin density and relaxation times. The intrinsic differences between (weighted) T_1 and T_2 form a contrast image allowing one to distinguish organs and pathological tissues. Because certain pathologies do not cause significant morphological changes in soft tissues, the pathology may still be detected using a CA that locally changes the relaxation times of the diseased tissue (Strijkers *et al.*, 2007).

The metal ions in a CA are expected to decrease T_1 without causing a substantial line broadening, and Gd^{3+} with a high paramagnetic moment is the best choice (Kim *et al.*, 2018b). Due to the toxicity of a freestanding metal ion, they are normally complexed with protective chelating ligands. Gd-DTPA (gadopentetate dimeglumine) is among the first clinically approved and widely used CAs. Gadolinium retention in multiple organs and the possible clinical consequences are still debated due to incomplete knowledge about the mechanisms of gadolinium tissue deposition and its long-term consequences (Tedeschi *et al.*, 2017). In this regard, paramagnetic transition metal ions Fe^{3+} and Mn^{2+}, which are more common for a human body, have been developed for use as CAs too.

The T_1 CAs in the form of ion complexes have short life spans in the body and work in a nonspecific manner residing within the extracellular space. Since the

MNPs are known to be efficient for molecular and cellular imaging, the development of particulate T_1 CAs has begun. The main studied systems are (i) nanostructured frames of silicas, dendrimers, perfluorocarbons, emulsions and nanotubes with many anchoring sites for Gd-DTPA, (ii) gadolinium-based NPs such as Gd_2O_3, GdF_3 and $GdPO_4$, (iii) MnO NPs and (iv) metal-alloy NPs like FeCo (Na *et al.*, 2009). To achieve the safer application of MRI for early disease diagnosis and image-guided therapy, CAs based on organic nitroxide radicals with improved biocompatibility, ease of functionalization and long blood circulation times have been promoted with the ultimate goal to substitute the metal-containing CAs (Akakuru *et al.*, 2019).

The T_2 CAs are commonly superparamagnetic iron oxide NPs (SPIONs) and variations thereof (Zhao *et al.*, 2020). They are represented by magnetite and maghemite NPs coated with dextran or carboxydextran (Wang, 2011). However, they possess a few disadvantageous features: (i) they give a signal-decreasing effect as a negative imaging agent and (ii) they induce distortion of the magnetic field on neighbouring normal tissues due to the high susceptibility. The contrast efficacy of SPIONs can be enhanced via either NP magnetic properties or coating optimization (Kostevšek, 2020). SPIONs will be discussed in more detail in the following section.

The reduction/oxidation (redox) potential of tissue is disrupted when a disease occurs. Redox-triggered metal ion complexes have great potential for creating large differences in magnetic properties that lead to changes in contrast (Tsitovich *et al.*, 2014). They provide MRI contrast either as T_1 agents or as paramagnetic chemical exchange saturation transfer (paraCEST) CAs. The exogenous action of CEST CAs relies on the magnetization transfer (MT) mechanism, providing an additional contrast mechanism in MRI (Wolff & Balaban, 1989). Although small diamagnetic compounds like sugar or amino acids may serve as CEST CAs, paramagnetic lanthanide (Eu, Ho and Er) chelates are more efficient paraCEST CAs, because they display a large chemical shift (Strijkers *et al.*, 2007).

Further details on the classification, synthesis and application of MRI CAs can be found in dedicated reviews (Avasthi *et al.*, 2020; Caspani *et al.*, 2020; Xiao *et al.*, 2016).

12.2.2 Rare earth doped iron oxide nanoparticles

12.2.2.1 Iron oxide nanoparticles

Iron can form oxides, hydroxides and oxide-hydroxides. The main oxides are Fe_3O_4 (magnetite), α-Fe_2O_3 (hematite), β-Fe_2O_3, γ-Fe_2O_3 (maghemite), ε-Fe_2O_3 and FeO (wüstite) (Arruebo *et al.*, 2007). It is also possible to prepare Fe_4O_5, Fe_5O_6 and Fe_7O_9 at extreme conditions (high pressure–high temperature) (Sinmyo *et al.*, 2016). Among a rich variety of iron oxides, hematite, maghemite and magnetite NPs are mainly used for inhibition of biological activities, which can be specified as antibacterial, antifungal and anticancer activities (Sangaiya & Jayaprakash, 2018).

The crystal structure of magnetite is inverse spinel, where tetrahedral sites are occupied by Fe^{3+} ions, and octahedral sites are occupied by equal numbers of Fe^{3+} and Fe^{2+} ions (the stoichiometric ratio $Fe^{2+/3+}/Fe^{3+} = 2$) (Alcantara & Lee, 2012). Maghemite, which has the same cubic structure but two-thirds of the sites are occupied by Fe^{3+} ions, can be considered as Fe^{2+}-deficient magnetite. The perfect match in crystal structure makes X-ray diffraction (XRD) useless in distinguishing the magnetite and maghemite phases (Kim *et al.*, 2012), and spectroscopic techniques (x-ray based or vibrational) might help to resolve the issue (Soldatov *et al.*, 2018). Hematite crystallizes in the rhombohedral structure, where Fe^{3+} occupies two-thirds of octahedral sites. It is the most stable iron oxide under ambient conditions.

The oxidation of Fe^{2+} in Fe_3O_4 yields γ-Fe_2O_3 (metastable) and α-Fe_2O_3 structures (Nasrazadani & Raman, 1993). This transformation is reversible at high temperature in a reducing atmosphere (Lu & Tsai, 2014) and unlikely to occur naturally. The undesirable Fe^{2+} into Fe^{3+} transformation weakens the magnetic properties and chemical stability of MNPs (Kowalik *et al.*, 2020), enables the occurrence of Fenton and Fenton-like reactions generating reactive oxide species (ROS) (Pereira *et al.*, 2012), and ultimately leading to oxidative cell death (Li *et al.*, 2020). The Fe_3O_4@γ-Fe_2O_3 core/shell NPs with improved stability and a higher magnetic saturation value were proposed as an alternative to a conventional organic or inorganic protection layer against oxidation (Gao *et al.*, 2009).

For *in vivo* administration the hydrodynamic size, which includes the iron oxide particle and coating, has to be considered rather than the size of naked MNPs (Xu *et al.*, 2019). NPs for *in vivo* applications are often covered with more than one layer to ensure biocompatibility, avoid toxicity and prevent obstruction of blood capillaries (Cardoso *et al.*, 2018). Ultra-small (less than 40 nm) SPIONs are used in the form of a multicore system (Lartigue *et al.*, 2020), where the magnetic cores are embedded in a biocompatible organic or inorganic shell: dextran in commercial CAs Feridex®/Endorem® (hydrodynamic size of ~120–180 nm) and Combidex®/Sinerem® (~15–30 nm), carboxydextran in Resovist® (~60 nm) (Jose *et al.*, 2020) or silicon in Lumirem®/Gastromark® (~300 nm) (Kundu & Patra, 2017). The crystallite size of the magnetic core is much smaller: 4.8–5.6 nm for the Feridex®/Endorem® systems and 4.2 nm for Resovist® (Arguinzoniz *et al.*, 2014). Functionalization of MNPs provides control over colloidal stability, the extent of aggregation, surface chemical structures, nontoxicity and biocompatibility.

A living organism reacts to the presence of MNPs in a blood vessel via biofouling and the formation of aggregates, which are swiftly sequestered by cells of the reticular endothelial system such as macrophages (Yu *et al.*, 2015). In this regard, ultra-small SPIONs may escape phagocytosis to some extent with a prolonged circulation time (Qiu *et al.*, 2013). It was also shown that highly crystalline Fe_3O_4 nanocubes are more favourable for sensing applications than polycrystalline Fe_3O_4 nanospheres of the same volume or length/diameter ratio (Kolhatkar *et al.*, 2017).

Variation in the Fe_3O_4 NPs composition by substituting Fe^{2+} with a M^{2+} cation (M = Mg, Zn, Co and Mn) in the tetrahedral or octahedral interstitial sites tunes the properties of $M_xFe_{3-x}O_4$ spinel ferrites (Etemadi & Plieger, 2020). Substitution of Fe^{3+} with an RE ion enhances the magnetic properties to a greater extent because of a large unquenched orbital angular momentum attributed to the *f* electrons, which is characterized by higher spin-orbit coupling (Kershi *et al.*, 2018). Incorporation of the RE ions into the structure of MNPs leads to a stronger temperature dependence of magnetic properties within the meaningful temperature ranges (room temperature, physiological temperature) (Rice *et al.*, 2015). Such modification also increases the oxidation resistance. In particular, lanthanides like the Gd^{3+} ion with seven unpaired electrons or the Eu^{3+} ion with six unpaired electrons can enhance the contrast abilities of SPIONs for MRI (Yang *et al.*, 2015). Zhu *et al.* (2017) modified magnetite NPs with Gd^{3+} chelates via linking peptides to construct NP-substrate (Fe_3O_4-pep-Gd) conjugates for kinetic MMP-2 (matrix metalloproteinase 2) activity assessment *in vitro*, at a cellular level and *in vivo*. MMP-2 is an important cancer marker associated with tumour invasion and metastasis. The authors developed an enzyme responding system based on T_1 relaxivity recovery, schematically presented in Figure 12.2.

12.2.2.2 Rare earth nanoparticles

RENPs are usually prepared by the same methods used for preparing iron oxide NPs, just adding an RE salt in certain quantities. Such a practice is also explained by the need to compare the properties of doped and parent (pristine) MNPs to estimate the effect of the lanthanide. Hence, MNPs can be produced by co-precipitation (Anbarasu *et al.*, 2015), micro-emulsion (Li *et al.*, 2014), microwave-assisted (Lastovina *et al.*, 2017b, 2017c), solvothermal (Lastovina *et al.*, 2017a), ultrasonic-assisted (Jafari Eskandari & Hasanzadeh, 2021), electrochemical (Mazario *et al.* 2014) and polyol (Wagle *et al.*, 2017) synthesis. In addition, high-temperature decomposition (pyrolysis) of Fe-containing compounds (Sun & Zeng, 2002) or eco-friendly (a 'green chemistry' approach) preparation from various plants, usually taken in the form of extracts (Karunakaran *et al.*, 2018), can be used. The main synthesis methods and applications are briefly summarized in Table 12.1. They will be discussed in more detail below.

12.2.2.3 Rare earth nanoparticle synthesis

12.2.2.3.1 Co-precipitation

Co-precipitation in aqueous solutions is the most convenient method for the preparation of iron oxide MNPs, because this method is eco-friendly, uses cheap

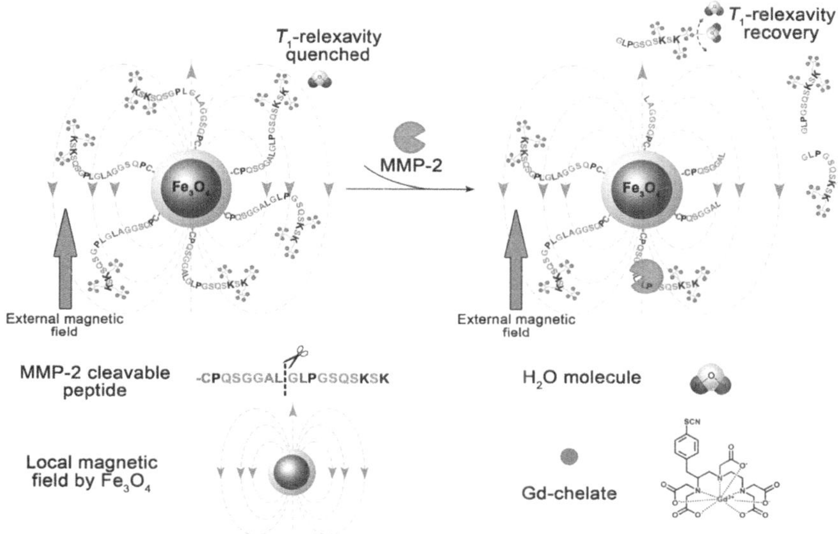

Figure 12.2 Schematic illustration of MMP-2 detection using Fe_3O_4-pepA-Gd conjugates based on T_1 relaxivity recovery. Superparamagnetic Fe_3O_4 nanocrystals induce the local magnetic fields under an external field, and act as a quencher to influence the T_1 spin alignment of attached Gd chelates. MMP-2 cleaves the linking peptide and releases multiple Gd chelates from superparamagnetic Fe_3O_4. Gd chelates regain their T_1 relaxivity as escaping from local magnetic fields induced by superparamagnetic Fe_3O_4. *Reprinted with permission from ACS Applied Materials & Interfaces, Vol. 9 (26), Zhu X., Lin H., Wang L., Tang X., Ma L., Chen Z., Gao J., Activatable T1 Relaxivity Recovery Nanoconjugates for Kinetic and Sensitive Analysis of Matrix Metalloprotease 2, Pages 21688–21696. Copyright (2017) American Chemical Society.*

reagents, proceeds under mild conditions and can be scaled up (Ahn *et al.*, 2012). Precipitation of the Fe^{2+} and Fe^{3+} ions in alkaline water solution as a synthetic approach was presented by Massart (1981). Parameters of the synthesis such as Fe^{2+}/Fe^{3+} ratio, nature of the base, pH value, presence of additional cations like $N(CH_3)_4^+$, $CH_3NH_3^+$, Na^+ and NH_4^+, influence on the MNPs size and the extent of aggregation can be varied (Laurent *et al.*, 2008). Formation of magnetite in a non-oxidizing oxygen environment can occur according to:

$$Fe^{2+} + 2Fe^{3+} + 8\,OH^- \rightarrow Fe_3O_4 + 4H_2O \tag{12.1}$$

Magnetite can transform into maghemite:

$$Fe_3O_4 + 2\,H^+ \rightarrow \gamma\text{-}Fe_2O_3 + Fe^{2+} + H_2O \tag{12.2}$$

Table 12.1 Synthesis methods of the RE doped iron oxide MNPs and their applications.

MNPs Composition	Synthesis Method	NPs Shape and Size	Applications	Reference
Nd-Ce doped Fe_3O_4	Co-precipitation	Spherical ~15 nm (SEM)	Wastewater treatment catalysts	Alimard (2019)
Ho-doped Fe_3O_4	Co-precipitation	Spherical 10–15 nm, Irregular up to 30–40 nm (TEM)	Biomedicine (hyperthermia)	Osial *et al.* (2018)
Er^{3+}-doped Fe_3O_4 Dy^{3+}-doped Fe_3O_4 Gd^{3+}-doped Fe_3O_4	Co-precipitation	Irregular and rod like	Not specified (e.g., photocatalytic)	Jain *et al.* (2018)
Sm^{3+}-doped Fe_3O_4	Polyol method (in tetraethylene glycol)	Spherical 6.0 nm (XRD)/ 5.2 nm (TEM)	Not specified (e.g., MRI contrast agent)	Silva *et al.* (2018)
Sm^{3+}-doped Fe_3O_4	Solvothermal polyol (in ethylene glycol)	Round-like polyhedral shapes, 9.6 and 23.8 nm (TEM)	Biomedicine	Lastovina *et al.* (2017a)
Fe_3O_4:Yb^{3+}:Er^{3+}	Thermal decomposition	Spherical 5.1–10.1 nm (HRTEM)	Simultaneous optical imaging and hyperthermia	de Jesús Ibarra–Sánchez *et al.* (2018)
Sm^{3+}-doped Fe_3O_4	Electrochemical	14.8 nm (XRD) Spherical 20 nm (SEM)	Supercapacitors	Aghazadeh and Ganjali (2018)
Gd^{3+}-doped Fe_3O_4	Surfactant-assisted thermolysis	Octahedral ~12 nm (TEM)	Biomedicine	Douglas *et al.* (2016)
Gd^{3+}-doped Iron oxides (Fe_3O_4/γ-Fe_2O_3, α- Fe_2O_3)	extraction-pyrolytic	21–50 nm (XRD), elongated particles (SEM)	Biomedicine	Serga *et al.* (2020)
Sm-doped Fe_3O_4	Sol-gel	13, 7.3 and 6.7 nm (XRD)	Biomedicine (hyperthermia and MRI)	Guil-López *et al.* (2010)

SEM: scanning electron microscopy; TEM: transmission electron microscopy; HRTEM: high-resolution transmission electron microscopy.

Magnetite NPs can be formed without a protective atmosphere when the hydrazine molecule reacts with the oxygen molecule to form two $[NH_3OH]^+$ cations, which further react with Fe^{2+} ions leading to the formation of magnetite (Hong *et al.*, 2008):

$$N_2H_4 \cdot H_2O + 2H^+ + 0.5\,O_2 \leftrightarrow 2[NH_3OH]^+ \quad (12.3)$$

$$3\,Fe^{2+} + [NH_3OH]^+ + 6\,OH^- \rightarrow Fe_3O_4 + [NH_4]^+ + 3\,H_2O \quad (12.4)$$

This approach was adopted for the preparation of Sm^{3+}, Gd^{3+} and Eu^{3+} doped iron oxide NPs and resulted in the formation of γ-Fe_2O_3 NPs, as had been revealed by Mössbauer spectroscopy (Lastovina *et al.*, 2016). Nucleation of iron oxide NPs in the hydrazine assisted process proceeds through formation of different intermediate compounds including oxyhydroxides α-FeOOH (goethite), β-FeOOH (akaganeite) and γ-FeOOH (lepidocrocite). The presence and concentration of intermediate species depend on such factors as pH, concentration of iron ions and temperature (Yu *et al.*, 2016). The final product can be rod-shaped (Védrine, 2019). For instance, formation of γ-Fe_2O_3 may proceed via the goethite intermediate (Ruhle *et al.*, 2012):

$$N_2H_4 + H_2O \rightarrow [N_2H_5]^+ + OH^- \quad (12.5)$$

$$Fe^{3+} + N_2H_5^+ + O_2 \rightarrow Fe^{2+} + N_2\uparrow + 3\,H_2O \quad (12.6)$$

$$Fe^{3+} + 3\,OH^- \rightarrow Fe(OH)_3\downarrow \quad (12.7)$$

$$2\,Fe^{2+} + O_2 + 2\,OH^- \rightarrow 2(\alpha\text{-}FeOOH) \quad (12.8)$$

$$\alpha\text{-}FeOOH + Fe(OH)_3 \rightarrow \gamma\text{-}Fe_2O_3 + 2\,H_2O \quad (12.9)$$

Co-precipitation can be realized with carbonate ions supplying the hydroxide ions for the occurrence of hydrolysis (Blanco-Andujar *et al.*, 2015). Such an approach allows preparation of spherical MNPs (Kuchma *et al.*, 2017). A small quantity of noble metal introduced during the co-precipitation process can facilitate the formation of NPs providing control over particle size (Lastovina *et al.*, 2018a).

Well-defined MNPs can be prepared using an oleylamine-oleic acid mixture (Sun & Zeng, 2002). Oleylamine can act as either solvent, surfactant or a weak reducing agent depending on the reaction temperature (Mourdikoudis & Liz-Marzán, 2013; Xu *et al.*, 2009). The main drawback of oleylamine-based synthesis is that the prepared MNPs are hydrophobic and the surfactant has to be replaced later for *in vivo* applications.

Polyol synthesis is undoubtedly one of the promising approaches for the preparation of MNPs and RENPs. This elegant technique was developed in the 1980s by Fernand Fievet's group to prepare monodisperse, highly pure, non-agglomerated metal particles from metal oxides, hydroxides or salts in polyols (Fievet *et al.*, 1989). Polyol plays a triple role as a solvent of the solid precursor, a reducing agent and a surfactant. Synthesis in polyols has several advantages: (i) well-crystallized NPs can be obtained during a high-temperature synthesis procedure, (ii) the polyol shell protects NPs from coalescence and oxidation as well as against the action of the reducing medium, (iii) the high viscosity of the polyol medium enables control of the NPs structure and morphology and (iv) the resulting NPs are hydrophilic, which can be critically important for biomedical applications (Fiévet *et al.*, 2018). The use of the polyol method in microwave-assisted or solvothermal syntheses enables preparation of non-toxic RENPs of different sizes (Lastovina *et al.*, 2017a, 2017b).

Gd-doped MNPs are more widespread among RENPs because of their utility as MRI contrast agents. Jain *et al.* (2018) reported on the preparation of $Fe_{3-x}RE_xO_4$ (RE = Er, Dy and Gd) NPs by the co-precipitation method. TEM images demonstrate the formation of spherical Fe_3O_4 NPs of ~8–14 nm and rod-shaped RENPs. The length of the rods increases in the order $Fe_{3-x}Eu_xO_4 < Fe_{3-x}Dy_xO_4 < Fe_{3-x}Gd_xO_4$. Zhang *et al.* (2017) also obtained Gd-doped magnetite NPs by the co-precipitation method under mild conditions. In another study, well-defined octahedral Fe_3O_4 NPs were obtained in a mixture consisting of oleylamine : oleic acid : benzyl ether at 310°C (Douglas *et al.*, 2016).

Co-precipitation was also used to prepare Ho^{3+}-doped magnetite NPs (Osial *et al.*, 2018). The holmium concentration in the samples had important implications on the morphology, size and magnetic properties of the particles. Based on the obtained results, it was concluded that the optimal holmium concentration is 1–2.5%.

12.2.2.3.2 Electrochemical synthesis

Partial substitution of Fe^{3+} with Gd^{3+} ions in the Fe_3O_4 structure was realized by electrochemical synthesis in a two-electrode cell with a stainless steel cathode and a graphite anode (Jouyandeh *et al.*, 2019). The steel cathode, centred between two parallel graphite anodes, was used for the deposition of the Sm^{3+}-doped Fe_3O_4 NPs from the aqueous solution of metals salts (Aghazadeh & Ganjali, 2018). The reaction proceeds through the electrochemical and chemical steps. The electrochemical step includes the generation of the hydroxyl ions according to the equation:

$$2\,H_2O + 2\,\bar{e} \rightarrow 2\,OH^- + H_2 \tag{12.10}$$

The Fe^{2+}, Fe^{3+} and Sm^{3+} ions form the Sm-doped iron oxide via co-precipitation:

$$Fe^{2+} + (2-x)Fe^{3+} + xSm^{3+} + 8\,OH^- \rightarrow\rightarrow Fe(II)Fe(III)_{2-x}Sm_xO_4 + 4\,H_2O \tag{12.11}$$

12.2.2.3.3 Thermal decomposition synthesis

Spherical Yb^{3+}-Er^{3+} co-doped Fe_3O_4 NPs with low polydispersity were prepared by the thermal decomposition method in an oleylamine-oleic acid system and tested for simultaneous optical imaging and hyperthermia applications (de Jesús Ibarra–Sánchez *et al.*, 2018).

12.2.2.3.4 Solvothermal synthesis

Solvothermal synthesis was used for the preparation of spherical Nd-doped Fe_3O_4 NPs with different Nd concentrations (Alimard, 2019). TEM images indicated the formation of large Fe_3O_4 Nd (10%)-Fe_3O_4, (~150 nm) and small Nd_2O_3 (~12 nm) particles, whereas magnetic measurement has shown that Nd (20%)-Fe_3O_4 is the optimum composition.

12.2.2.3.5 Rare earth ion distribution in magnetic nanoparticles

Several works underline that substitution of Fe^{3+} with a larger Gd^{3+} ion leads to a lattice distortion, therefore the presence of a pure spinel phase can indicate that the Gd dopant is distributed uniformly across the lattice without forming a separate phase (Zahmouli *et al.*, 2020). Although high loading of lanthanide and its uniform distribution may cause expansion of the unit cell, as seen in the case of 10% Y^{3+} doped Fe_3O_4 NPs (Kowalik *et al.*, 2020). The expansion was attributed to reaching the maximum Y^{3+} concentration at the Fe^{3+} octahedral sites, taking into account the ionic radius incompatibility: 1.019 Å for Y^{3+} and 0.78 Å for Fe^{3+} at eightfold coordination. Yang *et al.* (2015) demonstrated that, according to the energy dispersive X-ray (EDX) element mapping, the Eu^{3+} ions are homogeneously distributed in iron oxide NPs, whereas XRD data and selected area electron diffraction (SAED) indicated the formation of the mixed-phase magnetite and hexagonal Eu_2O_3.

Petran *et al.* (2018) prepared multi-core iron oxide MNPs doped with Eu^{3+} and Gd^{3+} ions at 5–15 mol.% levels by a solvothermal polyol synthesis. The crystallite size was about 29 nm for undoped magnetite MNPs and roughly one-third of that size for RENPs, decreasing with an increasing level of doping with RE for both elements. This was attributed to a drastic distortion of the crystal due to a significant mismatch in atomic radii between Fe^{3+} and Ln^{3+} (1.078 Å for Gd and 1.087 Å for Eu) substituents of Fe^{3+} octahedral sites. This substitution has also led to a decrement in saturation magnetization from 82 emu/g for pristine MNPs to 55 and 52 emu/g for Gd and Eu doped MNPs, respectively. However, the protein adsorption tests demonstrated a higher affinity of BSA (bovine serum albumin) for RENPs.

The distribution of the RE ions in MNPs can be observed by various techniques. Uniform distribution of the Gd^{3+} ions was confirmed by energy dispersive X-ray spectroscopy in scanning transmission electron microscopy (STEM-EDS) mapping (Perera *et al.*, 2015). In other work from the same group, extended X-ray absorption fine structure (EXAFS) spectroscopy was employed to demonstrate that the Sm–O distance in the Sm^{3+}-doped iron oxide NPs is larger than that in Sm_2O_3 suggesting higher coordination of Sm in the NPs (Soldatov *et al.*, 2018).

12.2.3 Iron oxide composites

Development of new synthesis routes for the preparation of biocompatible RENPs with unique optical and magnetic properties as well as investigation of their properties constitutes an actual direction of scientific progress (Bouzigues *et al.*, 2011). The complexity and heterogeneity of tumour tissues require multimodal diagnosis and multifunctional therapeutic systems. Combining magnetic NPs with various types of nanomaterials broadens their application potential. In this view, heterostructures composed of more than one kind of NPs in different combinations possess unique physicochemical properties derived from constituents and offer 'all-in-one' synergistic performance.

Gold NPs are widely used as biomedical probes and drug carriers. They are also applied in laser-induced photothermal therapy of tumours, as a contrast agent for computed tomography (CT) and in photoacoustic imaging (Nejati *et al.*, 2021). The Au-Fe_3O_4 bi-component nanocomposite has received considerable attention because of its intrinsically peculiar magnetic and optical properties as well as structural designability (Liu *et al.*, 2018).

A core-shell NP consists of a central core and a covering shell, connected through physical and chemical interactions (Zhang, 2020). Due to the synergistic effect between the different components, it is a uniform and symmetrical structural system with balanced and isotropic chemical properties, immanent to both the comprising materials and those originating from their mutual influence. The Au-Fe_3O_4 heterostructure with magnetic core and noble metal shell can be formed by direct coating or connected with an intermediate layer (silicon or polymer). Synthesis of decorated structures is based on the direct deposition of NPs onto the surface of the core particle preliminarily functionalized with organic ligands or silica (Oh & Lee, 2013; Sun *et al.*, 2010; Wang *et al.*, 2004; Zheng *et al.*, 2014). The encapsulation method allows various NPs to be incorporated into a matrix formed from silicon dioxide, cross-linked polymer, liposomes or micelles (Corr *et al.*, 2008; Yi *et al.*, 2005). The main types of engineered MNPs are presented in Figure 12.3. They can be classified into core-shell (single core, reverse core-shell, multicore, yolk-shell, decorated), multilayer core-shell, dumbbell, Janus and heterotrimer, brush-like and cross-linked (Hodges & Schaak, 2017; Mourdikoudis *et al.*, 2021; Wang *et al.*, 2012; Wu *et al.*, 2016).

For instance, Ángeles-Pascual *et al.* (2018) reported on a simple one-pot solvothermal synthesis of chemically stable gold-coated Fe_3O_4 NPs of 20–50 nm size range with considerable saturation magnetization. Kang and Kim (2020) used a sol-gel method to obtain Au-coated (second shell, thickness: ~40 nm) NPs

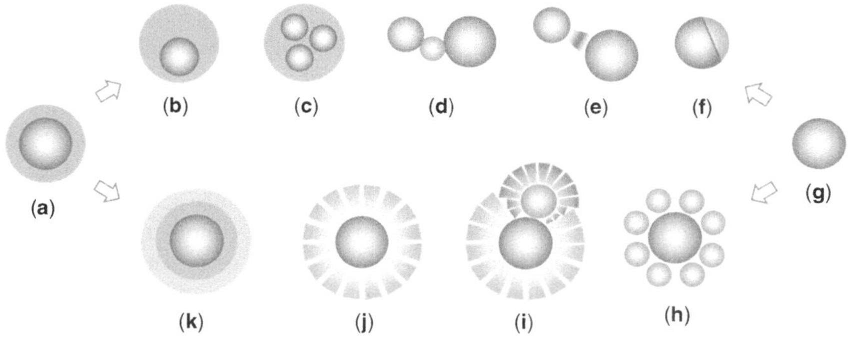

Figure 12.3 Schematic illustration of different types of engineered MNPs: core-shell (a), yolk-shell (b), multi-core (c), heterotrimer (d), cross-linked (e), Janus (f), naked (g), decorated (h), dumbbell (i), brush-like (j) and multi-shell (k).

with core-shell structures, which comprised magnetite cores (diameter ~100 nm) and silica shells (first shell: ~20 nm). Direct growth of composite NPs requires certain compatibility between the involved materials, whereas encapsulation by a coating usually generates large particles, which might be unsuitable for *in vivo* applications. Examples of the RE doping of core-shell nanostructures for biomedical applications can be found in a recent review by Labrador-Páez *et al.* (2018).

In 1989, Casagrande *et al.* (1989) presented spherical glass particles with one hemisphere hydrophilic and the other one hydrophobic. The objects were named 'Janus beads' after the Roman God Janus depicted with twin faces. The surface anisotropy of Janus particles allows spatial-selective bioconjugation with two different antibodies and multiplexed sensing of biological molecules. Inorganic Janus particles have attracted increasing attention owing to their tunable optical, magnetic and catalytic properties within a single particle by variation of materials, domain sizes and morphology (Schick *et al.*, 2014). The introduction of a magnetic component enables them to be driven as drug carriers in the human body utilizing a magnetic field, detecting their location during MRI diagnostics as well as performing many other applications. Among a variety of synthetic strategies the most frequently practised are (i) templating, where a part of the particle is covered by a sacrificial material allowing modification of the other, (ii) direct deposition of the functionalizing material, (iii) phase separation, providing the use of two immiscible precursors kept apart by an interface and (iv) self-assembly, which might be guided by the application of external fields (Campuzano *et al.*, 2019). However, synthetic procedures for Janus particles require complex chemicals and result in low yields of product, hindering scale-up for wide applicability.

Multifunctional nanostructures containing two or more particles of different functionalities can be obtained through cross-linking. Using this approach, Wang *et al.* (2012) cross-linked oleylamine-capped gold and magnetite NPs with a thiol carboxylic polyethylene glycol (HS–PEG–COOH) molecule adopting a phase transfer protocol with sequential ligand exchange. The obtained Au–PEG–Fe_3O_4 conjugates preserved the optical and magnetic properties of both Au and Fe_3O_4 NPs. Brightfield optical microscopy of HeLa cells incubated with the conjugates (1 μM Au) was performed under excitation light with $\lambda =$ 561 nm, chosen to be close to the surface plasmon resonance (SPR) absorption peak at 520 nm for Au NPs. The SPR-enhanced scattering from Au NPs provided high contrast to inspect the cell morphology and location of internalized NPs. In the MRI test, the conjugates showed shortened T_2 relaxation of water molecules and an r_2 relaxivity of 139 $mM^{-1}\ s^{-1}$.

Semiconductor quantum dots (QDs) demonstrate a great utility for fluorescence imaging for both *in vitro* and *in vivo* applications in biomedicine. Combination with magnetic materials in the form of RE ions or as SPIONs enables imaging, magnetic separation and remote manipulation of targeted species (Mahajan *et al.*, 2013). They can be utilized as multimodal contrast agents for *in vivo* fluorescence imaging and MRI (Koole *et al.*, 2009). Thakur *et al.* (2009) reported on the preparation of pH-sensitive iron oxide–CdS conjugates from separately pre-synthesized SPIONs and mercaptopropionic acid-coated CdS QDs then cross-linked with 3-mercaptopropyl trimethoxysilane (Thakur *et al.*, 2009). Conjugates formed clusters of 0.1–1.0 μm diameter and the highest photoluminescent intensity and stability was reached below pH 3.

Magnetic materials frequently appear as part of nanomotors which have drawn great attention in the environmental remediation field (Parmar *et al.*, 2018). Such systems can be driven by light, a bubble or a magnetic field, simultaneously adsorbing target compounds (Su *et al.*, 2019).

12.3 APPLICATION IN ELECTROCHEMICAL DEVICES

RENPs are widely used for different types of electrochemical devices, including fuel cells (FCs), supercapacitors, batteries, as both anode and cathode catalysts or co-catalysts, electrolytes, and additives in polymer electrolyte membranes (Antolini & Perez, 2011). In the following sections, some representative RE doped catalysts, which are currently employed or considered for employment in electrochemical devices, will be described with emphasis on the synthetic procedure, NPs size/shape/composition and characteristic electrochemical properties.

12.3.1 Fuel cells

12.3.1.1 Fuel cell operation

Fuel cells (FCs) are highly efficient electrochemical devices converting the chemical energy of the fuel to electrical energy without intermediate stages and gaining usage as portable power sources (Cassir *et al.*, 2013). FCs are more eco-friendly in comparison with thermal engines and have the significant advantage of unlimited sources of fuel and oxidant (Shao *et al.*, 2016). The common FC consists of an anode, cathode, electrolyte and external circuit. The FC operation is based on the anode and cathode chemical reactions (Sazali *et al.*, 2020) in the presence of catalysts. According to the chosen electrolyte and fuel, the following types of FC can be differentiated: proton exchange membrane FC (PEMFC) including direct methanol FC (DMFC) and direct ethanol FC (DEFC), alkaline FC (AFC) including proton ceramic FC (PCFC) and direct borohydride FC (DBFC), phosphoric acid FC (PAFC), molten carbonate FC (MCFC), solid oxide FC (SOFC) and direct methanol FC (DMFC) (Kirubakaran *et al.*, 2009). The PEMFC is of special interest among them, because such cells operate at low temperature (below 100°C), have a rapid start-up and shut-down, high power density, low weight and volume and no leakage of an electrolyte (Litkohi *et al.*, 2020).

The basic reactions for the main types of FCs are the following ones (Zhong *et al.*, 2008):

PEMFC

$$\text{Anode}: H_2 \rightarrow 2\,H^+ \tag{12.12}$$

$$\text{Cathode}: 1/2\,O_2 + 2\,H^+ + 2\,\bar{e} \rightarrow H_2O \tag{12.13}$$

$$\text{Overall}: H_2 + 1/2\,O_2 \rightarrow H_2O \tag{12.14}$$

DMFC

$$\text{Anode}: CH_3OH + H_2O \rightarrow CO_2 + 6\,H^+ + 6\,\bar{e} \tag{12.15}$$

$$\text{Cathode}: \frac{3}{2}O_2 + 6\,H^+ + 6\,\bar{e} \rightarrow 3\,H_2O \tag{12.16}$$

$$\text{Overall}: CH_3OH + \frac{3}{2}O_2 \rightarrow CO_2 + 2\,H_2O \tag{12.17}$$

DEFC (Burhan *et al.*, 2021)

$$\text{Anode}: CH_3CH_2OH + 3\,H_2O \rightarrow 2\,CO_2 + 12\,H^+ + 12\,\bar{e} \tag{12.18}$$

$$\text{Cathode}: O_2 + 4\,H^+ + 4\,\bar{e} \rightarrow 2\,H_2O \tag{12.19}$$

$$\text{Overall}: CH_3CH_2OH + 3\,O_2 \rightarrow 2\,CO_2 + 3\,H_2O \tag{12.20}$$

12.3.1.2 Electrocatalysts

In the case of the DEFC above, the complete electrooxidation of ethanol to CO_2 and H_2O cannot be achieved with currently employed catalysts. Incomplete electrooxidation leads to the formation of both CH_3CHO and CH_3COOH with a small amount of CO_2 (Burhan *et al.*, 2021; Choudhary & Pramanik, 2020):

$$CH_3CH_2OH + 1/2\,H_2O \rightarrow 1/2\,CH_3COOH + 1/2\,CH_3CHO + 3\,H^+ + 3\,\bar{e} \tag{12.21}$$

Traditional catalysts for FCs typically have a high cost because of noble metals such as platinum and palladium, which are commonly referred to as platinum-group metals (PGM). Unfortunately, rapid degradation and slow reaction kinetics substantially reduce the rate of commercialization of low-temperature FCs (Guterman *et al.*, 2009, 2014). To mitigate these drawbacks, the following approaches have been tried: forming platinum alloys with *d*-metals (Gudko *et al.*, 2009; McKeown & Rhen, 2018), control over the NPs size and shape (Chen *et al.*, 2014, 2020), formation of the core-shell (Alinezhad *et al.*, 2020), Pt-skin or skeleton (Stamenkovic *et al.*, 2006) and yolk (Ao *et al.*, 2018) structures, preparation of PGM-free catalysts like Fe/N/C (Wan *et al.*, 2019), Co/N/C (Haile *et al.*, 2021; Li *et al.*, 2019), Fe/Co/N/C (Lastovina *et al.*, 2018b; Pimonova *et al.*, 2019a, 2019b), metal-free carbons, transition metal carbides (Guil-López *et al.*, 2010) and nitrides (Abdelkareem *et al.*, 2020) and ternary catalysts (Antolini, 2007).

Early research had demonstrated the suitability of RE compounds as electrocatalysts, but the methods for the preparation of stable and active catalysts were missed (Miles, 1975). It was noted that Pt-RE alloys can be active and stable in polycrystalline, single crystal and the NP forms, however, they required a multistep synthesis (Roy *et al.*, 2018). Unlike platinum alloys with *d*-metals (Pt_xCu, Pt_xNi, Pt_xCo and Pt_xAg), the Pt-RE compounds are still imperfectly studied. Therefore, there are no common requirements for particle size, shape, composition and metal content.

12.3.2 Catalysts

RE-based catalysts can be prepared by high-temperature melting, magnetron sputtering (Yoo *et al.*, 2011) or electrodeposition (Han *et al.*, 2016). In the case of non-perovskite structures, RE elements can form alloys, disperse as ions or build heterogeneous structures between catalytic active particles and the RE-based compound (Gao *et al.*, 2019). According to the literature, it is possible to deposit RE metals from RE^{3+} ions in ionic liquids, however, it is often not a single step process (Asen *et al.*, 2018). Pt-RE alloy catalyst prepared by physical metallurgical approaches can exist in the form of Pt_5M (hexagonal $CaCu_5$ or the orthorhombic $AuBe_5$ crystal structure), Pt_3M (Cu_3Au) and Pt_1M (Peera *et al.*, 2019). The Pt_5M and Pt_3Y compositions are better investigated than others.

Carbon-supported Pt-M as well pure Pt NPs can be synthesized by wet chemical synthesis, while the Pt-RE alloys are rarely achieved in this way due to the large difference in the reduction potentials of RE elements (from −1.9 to −2.38 V) and Pt^{2+} (+1.188 V) (Peera *et al.*, 2019). Lanthanides can form strong metal-oxygen bonds (~700–800 kJ mol^{-1}) making reduction more difficult when oxygen-containing precursors are used (Lux & Cairns, 2006). In addition, the RE alloying components are unstable against dissolution in an electrolyte, therefore the formation of the Pt overlayer is favourable (Escudero-Escribano *et al.*, 2016).

Jeon and McGinn (2011) described the synthesis of the Pt–Y electrocatalysts by borohydride reduction. Chloroplatinic acid and yttrium(III) nitrate were dissolved in a mixture of methanol and de-ionized water (1:80, v/v), then carbon support (Vulcan XC72R) was added to get 20 wt% metal loading. The reduction occurred under stirring with 0.2 M of $NaBH_4$. In this way, the Pt_3Y/C and PtY/C samples were obtained from precursors taken in a different ratio. To promote alloying, the catalysts were subsequently annealed at 900°C for 5 min under H_2/Ar flow and labelled as Pt_3Y/C-900 and PtY/C-900. This set of samples was needed to understand the phase formation of Pt–Y alloys on carbon supports. The XRD profiles are presented in Figure 12.4a along with reference phases from a database (Figure 12.4b). The un-annealed samples demonstrated only face-centred cubic (FCC) Pt peaks with a sign of alloying due to a slight shift of the (111) peak toward higher 2θ. After annealing the (111) peak had moved further for Pt_3Y/C-900, suggesting a better alloying. Instead, there was an opposite shift for PtY/C-900, meaning no alloying, and new peaks of the unknown phase had appeared (Figure 12.4c). Thus, the formation of a composite is not a straightforward process.

Formation of Pt–$Y(OH)_3$ NPs by a conventional reduction with $NaBH_4$ was shown by Nishanth *et al.* (2011). The Pt–$Y(OH)_3$ NPs supported on carbon demonstrated an enhanced catalytic activity, which was explained by the

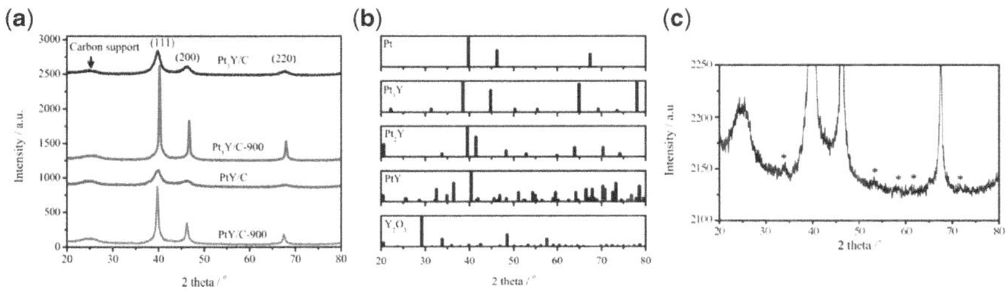

Figure 12.4 (a) XRD spectra of Pt_3Y/C, Pt_3Y/C-900, PtY/C and PtY/C-900 catalysts, (b) XRD peak positions and intensities of pertinent reference phases including Pt (ICDD #03-065-2868), Pt_3Y (ICDD #00-017-0058), Pt_2Y (ICDD #00-012-0333), PtY (ICDD #00-019-0919) and Y_2O_3 (ICDD #01-071-0049) and (c) enlarged image of the XRD result of the PtY/C-900 catalyst. *Reprinted from Journal of Power Sources, Vol. 196 (3), Jeon M. K., McGinn P. J., Carbon supported Pt–Y electrocatalysts for the oxygen reduction reaction, Pages 1127–1131, Copyright (2011), with permission from Elsevier.*

electronic effect and spillover of primary oxides, whereby the oxygen reduction reaction (ORR) is promoted by OH^- ions transferred to the neighbouring metal site (M):

$$Y(OH)_3 + M \rightarrow Y(OH)_2^+ + M - OH + \bar{e} \quad (12.22)$$

$$Y(OH)_2^+ + 2\,H_2O \rightarrow Y(OH)_3 + H_3O^+ \quad (12.23)$$

$$\sum M + 2\,H_2O \rightarrow M - OH + H_3O^+ + \bar{e} \quad (12.24)$$

A possible crossover effect was also suggested for other RE-containing electrocatalysts (Rosalbino *et al.*, 2008). In particular, the contribution of this mechanism was revealed for the Pd-RE electrocatalyst (Sarma *et al.*, 2017). The surface of RE-containing particles bears oxophilic sites, which can interact with OH^- moieties facilitating adsorption of organic molecules like ethanol on the active sites of Pd.

Luo *et al.* (2014) realized a synthesis of the carbon-supported Pt-M_2O_3 (M = Gd, Y) NPs in water-in-oil microemulsions. The as-prepared Pt-Gd_2O_3/C catalyst demonstrated lower ORR activity than Pt/C, however, it was significantly improved after treatment at 100°C under a H_2/N_2 atmosphere.

Neto *et al.* (2009) prepared 20 wt% Pt–RE/C electrocatalysts (RE = La, Ce, Pr, Nd, Sm, Tb, Dy, Ho and Er) by a polyol synthesis in alkaline solution under ambient conditions. The authors underlined that it is impossible to reduce RE(III) ions to the zero-valent state avoiding the formation of RE oxide or hydroxide. XRD demonstrated the presence of RE(III) hydroxides as well as Pt NPs. The average size of the RE-doped crystallite was considerably larger than that of pure Pt. The XRD reflexes of platinum in Pt-RE/C materials were as for pure platinum, indicating a lack of alloying. The values of current density after chronoamperometry experiments at 0.5 V for 30 min were declining in the order PtEr/C > PtTb/C > PtLa/C > PtNd/C > PtDy/C > Pt/C > PtPr/C > PtHo/C > PtSm/C > PtCe/C. Santoro *et al.* (2012) adopted the same synthesis method to produce PtLa/C catalysts with different Pt : La ratios. The XRD patterns evidenced the formation of Pt, La_2O_3 and $La(OH)_3$ phases.

Escudero-Escribano *et al.* (2012, 2016) studied the Pt_5M (M = La, Ce, Sm, Gd, Tb, Dy, Tm and Ca) compositions because of their stability and ability to have several alloys with the same structure. The polycrystalline Pt_5M electrodes were prepared and their electrocatalytic activity in the ORR was evaluated by carrying out cyclic voltammetry measurements in 0.1 M $HClO_4$. Before the catalytic tests, 80–140 cycles in N_2 saturated electrolyte were recorded enabling the formation a thin Pt surface layer. It was shown that the activity of polycrystalline Pt alloys 4wtowardsORRishigherthanpurePtandincreasesintheorder $Pt_5Tb > Pt_5Gd \sim Pt_3Y > Pt_5Sm > Pt_5Ca \sim Pt_5Dy > Pt_5Tm > Pt_5Ce > Pt_5Y \sim Pt_5La >> Pt$. Tafel plots before and after long-term tests (10,000 consecutive cycles from 0.6 to 1.0 V vs. RHE in

O_2-saturated 0.1 M $HClO_4$) confirmed the high stability of the prepared electrodes. The Pt overlayer was preserved in all Pt-lanthanide alloys with no changes in the electrochemically active surface area after the stability test had been done.

Kanady *et al.* (2017) proposed an elegant and painstaking synthesis of carbon-supported Pt_3Y NPs by reducing the Pt and Y salts with trialkylborohydride molten salt ($NaEt_3BH$ and KEt_3BH). The synthesis was completed in two steps: (i) mixing $PtCl_4$, YCl_3 and KEt_3BH, and heating at 200°C for 30–60 min and (ii) annealing at 650°C for 2 h under vacuum. The overall reaction can be described as follows:

$$3PtCl_4 + 3YCl_3 + (21 + x)MEt_3BH \rightarrow Pt_3Y + 21MCl + 21BEt_3 + 10.5H_2 + xMEt_3BH + 2Y^{3+} \quad (12.25)$$

The obtained Pt_3Y NPs of 5–20 nm diameter were loaded on carbon supports as a single particle or particle agglomerates. This synthetic approach requires some additional care: rigorously drying and degassing solvents, precision work in a glovebox, synthesis of KEt_3BH and $NaEt_3BH$ and washing at every stage to remove by-products (borane, Y_2O_3). The obvious merit of such a procedure is the ability to control the particle size by incorporation of an additional salt matrix and stabilization of unsupported colloidal NPs.

Roy *et al.* (2018) reported on the preparation of Pt_xY/C NPs in a stainless steel reactor placed inside a tubular furnace. For the reaction, a dry Pt/C catalyst was mixed with YCl_3 in anhydrous acetonitrile, then left stirring overnight under a cover. The dried powder was distributed evenly across the surface of a graphite foil to avoid contact between the mixture and a metal reactor. It was consistently purged with He, annealed at 800°C during 60 and 360 min in the H_2 flow, and washed with acid. The product of short annealing was composed of Pt_xY and Y_2O_3 NPs, while Pt_3Y NPs were individualized after long-term annealing. The average particle size increased from 7.6–9.9 to 12.1 nm in the course of annealing.

Hu *et al.* (2020) presented a new strategy for the preparation of Pt–RE NPs with tunable alloy composition and particle size. The direct heat treatment of the solid mixture of the platinum and gadolinium salts, CN_2H_2 and carbon support in Ar atmosphere led to a product containing both platinum and gadolinium in oxidized forms. The authors concluded that even platinum cannot be reduced without breaking up the C–N network due to the strong bonding between the metal ions and the N sites of the C–N network. At the same time, heat-treatment of the pre-formed Pt–RE–NC in H_2/Ar atmosphere allowed the C–N network to be broken up, reducing platinum ions to metallic NPs, and transformation of the RE ions into the crystalline phase of RE carbodiimide ($RE_2(CN_2)_3$). Follow-up heat-treatment in 3.3% H_2/Ar at 700°C and acid leaching resulted in the Pt_xGd/C catalysts. Pt_2Gd NPs were predominantly formed with a small portion of Pt_3Gd. Among the number of Pt–RE nanoalloys, Pt_5Ce were characterized by a 5.3 times higher specific activity than the reference Pt/C.

Cui *et al.* (2017) prepared porous Pt-Y NPs by a two-step approach: (i) synthesis of the $Pt_{15}Y_5Al_{80}$ alloy by arc melting of pure metals Pt, Y and Al and (ii) de-alloying of the $Pt_{15}Y_5Al_{80}$ powder in N_2-bubbled 5 M NaOH solution. Prepared Pt-Y NPs exhibited a lower onset potential and higher mass activity (4.1 times higher than that of commercial Pt/C) in ethanol electrooxidation.

Regarding the ternary systems, in an attempt to improve the efficiency of PtSn and PtRu catalysts, which are employed for electrooxidation of alcohols, the PtRuRE and PtSnRE compounds were intensively investigated (Luo *et al.*, 2019; Ponmani *et al.*, 2016).

An *et al.* (2011) investigated the PtRuRE/C catalysts (RE = La, Eu, Gd, Y, Sm and Er) prepared by step-by-step doping of the commercial PtRu/C catalyst. $NaBH_4$ was used as a reducing agent and the treatment was performed at 150°C in Ar atmosphere. It was suggested that the RE elements in catalysts were present in either form of metal or RE(III) oxide. Prepared catalysts were tested toward methanol oxidation in 0.5 mol L^{-1} H_2SO_4 +0.5 mol L^{-1} CH_3OH solutions. Based on the cyclic voltammetry and chronoamperometric experiments, it was established that the addition of RE elements to PtRu/C effectively improved the catalytic activity in all cases, except the PtRuGd/C sample. The catalytic activity in the methanol oxidation reaction decreased in the order PtRuEu/C > PtRuEr/C > PtRuY/C > PtRuSm/C > PtRuLa/C > PtRu/C > PtRuGd/C. The electronic effect of RE significantly affected the catalytic activity, while the average particle size and lattice parameter remained the same. However, the impact from the ratio of RE in the metallic and oxidation state is still to be assessed.

The Pt-Sn-RE/C catalysts can be prepared by a modified polyol synthesis. In particular, Pt-Sn-Eu/C samples were synthesized at 297°C using 1,2-hexadecanediol as a reducing agent, and oleic acid and oleylamine as capping agents (Corradini *et al.*, 2018). After cooling, a carbon carrier was added. The formation of a platinum alloy was evidenced by XRD patterns with no traces of europium oxide. However, the authors suspected the presence of tin and europium as amorphous oxides and/or oxy-hydroxides. Doping with Eu increased the catalyst's activity facilitating adsorption of intermediate products on the Pt sites and favoured the bifunctional EOR (ethanol oxidation reaction) mechanism.

Alternatively, Pt(0), SnO_2 and $Eu(OH)_3$ can be formed by a wet-synthesis method, impregnating the carbon support with metal salts and reducing metals by the HCOOH solution (Wang *et al.*, 2010). During electrochemical tests, the $Pt_3Sn_1Eu_1/C$ catalyst required much lower activation energy than the Pt_3Sn_1/C reference.

A modified treatment with formic acid can be used for the preparation of the Pt–Sn–Ce/C catalyst (Jacob *et al.*, 2015). XPS data indicated the presence of Sn and Ce in the oxidized state. The influence of doping was uneven: stability tests showed that ternary Pt–Sn–Ce/C catalysts were less tolerant to the poisoning by intermediates compared to Pt/C and Pt–Sn/C, but more tolerant than Pt–Ce/C.

Applying the same procedure for the preparation of the Pt–Sn–Pr/C catalysts also leads to the reduction of platinum and oxidizing of Sn and Pr (Corradini *et al.*, 2015). Ternary Pt–Sn–Pr/C catalysts were more tolerant to poisoning by intermediates compared to binary catalysts, wherein the optimal ratio of metals in the Pt–Sn–Pr/C catalysts was (45:45:10).

Synthesis of the PtSmCo NPs by electrodeposition of SmCo NPs in an inert atmosphere and subsequent impregnation and reduction of the K_2PtCl_4 with L-ascorbic acid sodium solution leads to the formation of Pt(0), Co, Co^{2+} and Sm^{3+} with a small amount of metal phase (Gong *et al.*, 2020).

Based on the described synthesis procedure, it can be suggested that the preparation of ternary catalysts in non-oxidized form is possible by high-temperature syntheses in an inert atmosphere (Rhodehouse *et al.*, 2018). However, not every composition can be obtained (Rhodehouse *et al.*, 2020).

12.4 APPLICATIONS IN PHOTOCATALYSIS AND PHOTOVOLTAICS

RE-doped materials demonstrated ability for light emission and harvesting. Photovoltaic (PV) devices represent a fast-developing field due to the availability of solar energy and globally rising demand for alternative energy sources. RE-doped metal oxides are actively commercialized as phosphors for optoelectronic devices (Wang *et al.*, 2015; Yuhua *et al.* 2015). There are plenty of metal oxides, which are used as hosts for RE doping (Yu *et al.*, 2015); the most popular are ZnO, TiO_2, SnO_2 and Y_2O_3 (Table 12.2). All of them are characterized by high catalytic activity and optical stability. The former is due to

Table 12.2 Band gap and unit cell structure of selected metal oxide hosts for RE doping.

Material	Band Gap, eV	Unit Cell Structure	Reference
ZnO	3.2–3.4	Hexagonal wurtzite*, zinc blende, rock salt	Parihar *et al.* (2018); Borysiewicz (2019)
TiO_2	3.0–3.2	Anatase, rutile*, brookite	Gupta and Tripathi (2011); Rahimi *et al.* (2016)
SnO_2	3.6–3.8	Rutile tetragonal	Batzill and Diebold (2005); Bhatnagar *et al.* (2016)
Y_2O_3	5.5–5.6	Cubic	Rajakumar *et al.* (2021); Srinivasan *et al.* (2010)

*The most popular phase due to its stability at room temperature and normal atmospheric pressure.

the available electrons of surface atoms, while the latter is due to the wide bandgap. NPs synthesis methods are commonly grouped into bottom-up and top-down approaches comprising chemical and/or physical techniques (Daksh & Agrawal, 2016; Ramakrishnan & John 2019). A recently emerging trend is the promotion of biosynthesis as an environmentally friendly, cost-effective, biocompatible and safe approach. The biosynthesis of ZnO NPs using plants and microorganisms has been deemed as a promising alternative to traditional approaches (Agarwal *et al.*, 2017; Mirzaei & Darroudi, 2017). The hexagonal wurtzite ZnO NPs are considered phosphors for lighting applications, including white LEDs (Kumar *et al.*, 2014b), colour LEDs (Kumar *et al.*, 2014a) or multicolour emission displays when doped with RE^{3+} ions (Layek *et al.*, 2016).

TiO_2 is the most popular material for photocatalytic, photovoltaic and photoelectrochemical applications due to its low cost, high stability, excellent catalytic activity, non-toxicity and ease of preparation (Xiaobo, 2009). The anatase TiO_2 is preferred over the rutile phase for energy applications, despite having a larger bandgap (3.2 vs. 3.0 eV) (Hanaor & Sorrell, 2011). Stannic oxide, due to its simultaneous transparency and conductivity, is being studied for applications ranging from solar cells (Xiong *et al.*, 2018) to a solid-state chemical sensor (Das & Jayaraman, 2014). Y_2O_3-based nanophosphors with RE^{3+} dopants are employed as fluorophores in bioimaging (Das & Tan, 2008) and nanophosphors for enhancing the photoelectric conversion efficiency of solar cells by exploiting the near infrared (NIR) to a visible up-conversion luminescence process (Yao *et al.*, 2020).

All PV devices exploit the photoelectric effect and three generations of solar cells (SCs) have been developed so far (Zhang *et al.*, 2018). The first generation SCs are comprised of the monocrystalline and polycrystalline silicon PV cells based on the Si wafer technology. The second generation SCs are usually manufactured by thin-film PV technology from either amorphous silicon, CdTe or CIGS ($CuIn_xGa_{(1-x)}Se$, $x = 0..1$). The third generation SCs are solution-processable SCs aiming at developing high-efficiency PV devices by improving the thin-film technology and using a variety of new materials (Table 12.3; Yan & Saunders, 2014). They are a playground for the realization of different concepts, such as multi-junction (sandwich-like architecture), hot carrier collection, multiple exciton generation and up/down conversion. Most of them are built upon metal oxide semiconductors (MOSs) and constitute excitonic SCs covering dye-sensitized SCs (DSSCs), QD-sensitized SCs, perovskite SCs (PSCs) and inverted organic PVs (Tian & Cao, 2016). The conversion efficiency of single-junction SCs is fundamentally limited by the band gap value (E_g) of the semiconductor material from which the PV device is fabricated. Modern advances in material composition and device design are motivated by the aim to approach the Shockley-Queisser (S-Q) limit of 33.7% photovoltaic energy conversion efficiency for a single-junction solar cell (Polman *et al.*, 2016).

Table 12.3 Enhancing PV properties* of MOS SCs by doping with RE ions.

Host	Dopant, Ln^{3+}	V_{OC}, mV	I_{SC}, mA cm^{-2}	Fill Factor	Efficiency, %	Reference
ZnO	0 Ln	494	5.20	0.54	1.37	Lu *et al.* (2011)
	La	499	4.27	0.57	1.22	
	Ce	450	0.66	0.28	0.09	
	Nd	527	4.39	0.58	1.31	
	Sm	530	5.02	0.57	1.52	
	Gd	536	5.64	0.65	1.98	
TiO_2	0% Ln	757	11.2	0.76	6.41	Zhang *et al.* (2012)
	0.1% Ce	741	12.5	0.77	7.12	
	0% Ln	800	8.32	0.64	4.23	Hafez *et al.* (2011)
	5% Eu	770	9.61	0.69	5.16	
	5% Sm	810	10.9	0.67	5.81	
SnO_2	0% Ln	1105	23.09	0.75	19.01	Guo *et al.* (2019)
	3% Sc	1116	23.36	0.77	20.03	
	3% Y	1117	23.61	0.78	20.63	
	3% La	1111	23.21	0.76	19.49	
Y_2O_3	0% Ln	637	13.42	0.64	5.49	Zhao *et al.* (2016)
	Eu	641	14.09	0.64	5.81	
	0% Ln	665	13.75	0.64	5.88	
	Er	680	14.93	0.66	6.68	

*PV cells were illuminated under the standard AM1.5 solar spectrum.

The overall light-to-electrical energy conversion efficiency (η, %) of any PV device is calculated from the short-circuit current density (I_{SC}, mA cm^{-2}), the open-circuit photovoltage (V_{OC}, V), the fill factor (FF) and the intensity of the incident light ($I_{Sun} = 100$ mW cm^{-2}) according to the following equation (Grätzel, 2000):

$$\eta = I_{SC} V_{OC} FF / I_{Sun} \tag{12.26}$$

The estimated conversion efficiency is routinely confronted by the S-Q limit, and the best efficiencies which have been reached so far for single-junction terrestrial cells are 21.6% and 11.9% for a PSC and DSSC, respectively (Green *et al.*, 2021). Moderate values of characteristic parameters of an SC provide insights into its weaknesses, that is, low J_{SC} suggests a poor light absorption or incomplete collection of generated carriers, and a small V_{OC} or FF indicates unwished bulk or interfacial carrier recombination, parasitic resistance, or other electrical nonidealities. Thermalization to the edges of bands and transmission photons with energy less than the bandgap (E_g) both constitute about 70% of the total energy losses encountered by PV devices. These two kinds of losses together with the recombination of electron-hole pairs close to or at the surface result from the spectral mismatch between the incident solar spectrum and the bandgap of MOSs. The spectral mismatch can be compensated either by modifying the PV device to better utilize the spectrum or processing the solar photons to better match the bandgap. The former is reached in a multi-junction SC, where semiconductors with different bandgaps are situated in a series-connect stack. The latter is about spectral modification, which means converting or shifting the incident photons to appropriate energy by up/down conversion to better match with the bandgap of MOSs. Recently, many lanthanide-doped spectral modifiers in a variety of MOS hosts have been employed to reduce the spectral losses and to improve the efficiency of SCs (Lian *et al.*, 2013). The RE ions, namely Ce^{3+}, Tb^{3+}, Eu^{3+}/Eu^{2+}, Gd^{3+}, Dy^{3+}, Sm^{3+}, Tm^{3+}, Er^{3+}, Yb^{3+} and Nd^{3+} are commonly used as dopants to enhance the conversion efficiency of SCs, whose diverse light emission lines are due to the $f \rightarrow f$ transitions.

Excellent progress has been made for DSSCs, which are currently the preferred type due to their low cost, ability to be fabricated on various substrates, structural modifications, excellent transparency and photovoltaic output (Kishore Kumar *et al.*, 2020). The DSSC concept was first presented by O'Regan and Grätzel (1991) using mesoporous TiO_2 NPs and Ru dye for the photoanode. Porous MOS film electrodes from processed NPs were then conventionally used for the fabrication of DSSCs. It was found that integration of TiO_2 with ZnO improves the process of electron-hole transfer between the corresponding conduction and valence bands because the resulting TiO_2/ZnO nanocomposite for the DSSC successfully combines the high reactivity potential of TiO_2 and the high electron mobility of ZnO (Boro *et al.*, 2018). A vast variety of RE element modified TiO_2 photocatalysts is surveyed in a dedicated review of Yu *et al.* (2015).

Lu *et al.* (2011) prepared a DSSC using homemade wurtzite ZnO NPs and Ru $(dcbpy)_2(NCS)_2$ dye. ZnO NPs were cast on fluorine doped tin oxide (FTO) glass and annealed at 400°C for 1 h to form a ZnO film of ~5 μm thickness with crystallites of ~40 electrodes. Those ZnO film electrodes were dipped into 0.2 M aqueous solutions of RE nitrates and annealed as previously. Then the electrode was immersed in a 0.3 mM dye ethanol solution overnight. Finally, the dye-sensitized electrode was assembled in a typical sandwich-type cell with a Pt counter electrode. Modification with Nd, Sm and Gd ions resulted in increased V_{OC} and FF values (unlike modification with Ce-ion), whereas the presence of La, Ce, Nd and Sm ions had a negative impact on J_{SC}. The decrement in parameters for the Ce-ion modified film was explained by the presence of Ce^{4+} ion, which can easily trap the dye excited state electron, preventing it from being injected in the conduction band of ZnO. Similar conclusions regarding the role of the Ce^{4+}/Ce^{3+} dopant have been reached by Zhang *et al.* (2012) who prepared Ce-doped TiO_2 photoanode for a DSSC by a sol-gel synthesis. The cerium(III) nitrate was mixed with the titanium isopropoxide in Ce: Ti molar ratios from 0.05 to 0.9%. The resultant colloid was heated at 220°C for 12 h to obtain the Ce-doped anatase TiO_2 NPs. The photoanodes were deepened in a 40 mM $TiCl_4$ aqueous solution at 70°C and then sintered at 500°C. Sensitizing was done with 0.3 mM N719 acetonitrile and tertbutyl alcohol equal volume ratio solution. The thickness of the TiO_2 photoanode was 8 μm and the average crystalline size was in the 12.16–13.07 nm range. It was established that only a small Ce content (0.1%) improves the performance of the DSSC due to the increased electron injection, while higher Ce loading favours the Ce^{4+} states effectively trapping electrons in the photoanode and suppressing the photocurrent.

Doping of MOSs is a way to create complex nanostructures, as shown by Wang *et al.* (2019) describing Au@mag@TiO_2 (mag = Ni, Fe) micromotors, which can efficiently remove microplastics from water and peroxide under UV illumination.

It is also worth mentioning, that metal oxide NPs are widely employed in photocatalysis. Although, a detailed discussion of this lays beyond the scope of this chapter, a few works have been mentioned here. For instance, nano-ZnO was synthesized by a novel large-scale and environmentally friendly electrochemical method using pulse alternating current and successfully applied for photocatalytic degradation of ciprofloxacin in water (Ulyankina *et al.*, 2021). Using the sol-gel method, TiO_2 was dispersed in mesoporous SiO_2 monoliths from 2 to 40 mol.% to obtain catalytically active systems, as has been demonstrated with activation of hydrogen peroxide (Budnyk *et al.*, 2012). MOSs based on Ti and Zn as well as metal-organic frameworks (MOF) and metal-oxide are actively employed for the photocatalytic degradation of wastewater pollutants such as pharmaceuticals and antibiotics (Gautam *et al.*, 2020). Other types of RE-based mixed metal oxide nanocomposites produced by the wet-chemical route with photocatalytic and antibacterial behaviour are considered in a recent review by Kannan *et al.* (2020a).

12.5 CONCLUSIONS

Advances in nanotechnology, including synthesis of nanostructured composites and development of sophisticated technical means for their characterization, pave the way for efficient use of RE elements as dopants to tune (enhance) certain useful properties of nanostructured materials. Significant efficiency gains justify the use of expensive precursors, which are a known obstacle to the widespread employment of RE elements. The presented examples of their applications in the fields of biomedicine, electrochemistry and photovoltaics show that despite some knowledge gaps, mostly concerning the exact mechanism behind the observed phenomena, there is a clear understanding of the role RE elements play in the composite they have become a part of. Such an understanding rationalizes the purposeful synthesis of RE doped NPs and the main challenge here is to design and optimize the procedure until the required product can be obtained. Another issue to consider is the follow-up of successful scientific findings, which implies a scale-up of the synthesis, which is expected to be 'green' as far as possible.

Multiple studies of nano-sized objects have raised awareness about the potential threats NPs may pose to living organisms if admitted in an uncontrolled manner. A continuously growing level of environmental pollution with engineered NPs increases the safety concerns and encourages the promotion of responsible behaviour among those involved. The circular economy concept and sustainability of technological processes demand integration of safety, circularity and functionality of materials throughout their life cycle. Concerning the RE doped composites, it would require that scarce and valuable resources should not be wasted after a single use (unless there are no other options), implying more efforts are expected from professionals (scientists and engineers) when commercializing developments of nanoscience.

ABBREVIATIONS

AMF	Alternating magnetic field
BSA	Bovine serum albumin
CEST	Chemical exchange saturation transfer
DSSC	Dye-sensitized solar cell
EDX	Energy dispersive X-ray (analysis)
EOR	Ethanol oxidation reaction
EXAFS	Extended X-ray absorption fine structure
FC	Fuel cell
MOS	Metal oxide semiconductor

MNP	Magnetic nanoparticle
MRI	Magnetic resonance imaging
MT	Magnetization transfer
NP	Nanoparticle
ORR	Oxygen reduction reaction
PEG	Polyethylene glycol
PGM	Platinum-group metals
PSC	Perovskite solar cell
PV	Photovoltaic
QD	Quantum dots
RE	Rare earth (element)
ROS	Reactive oxide species
SAED	Selected area electron diffraction
SAR	Specific absorption rate
SPR	Surface plasmon resonance
SC	Solar cell
SPIONs	Superparamagnetic iron oxide nanoparticles
S-Q	Shockley-Queisser (limit)
TEM	Transmission electron microscopy
XPS	X-ray photoemission spectroscopy
XRD	X-ray diffraction

REFERENCES

Abdelkareem M. A., Wilberforce T., Elsaid K., Sayed E. T., Abdelghani E. A. and Olabi A. G. (2020). Transition metal carbides and nitrides as oxygen reduction reaction catalyst or catalyst support in proton exchange membrane fuel cells (PEMFCs). *International Journal of Hydrogen Energy*, **46**(45), 23529–23547.

Agarwal H., Kumar S. V. and Rajeshkumar S. (2017). A review on green synthesis of zinc oxide nanoparticles–an eco-friendly approach. *Resource-Efficient Technologies*, **3**(4), 406–413.

Aghazadeh M. and Ganjali M. R. (2018). Samarium-doped Fe_3O_4 nanoparticles with improved magnetic and supercapacitive performance: a novel preparation strategy and characterization. *Journal of Materials Science*, **53**(1), 295–308.

Ahn T., Kim J. H., Yang H. M., Lee J. W. and Kim J. D. (2012). Formation pathways of magnetite nanoparticles by coprecipitation method. *The Journal of Physical Chemistry C*, **116**(10), 6069–6076.

Akakuru O. U., Iqbal M. Z., Saeed M., Liu C., Paunesku T., Woloschak G., Hosmane N. S. and Wu A. (2019). The transition from metal-based to metal-free contrast agents for T_1 magnetic resonance imaging enhancement. *Bioconjugate Chemistry*, **30**(9), 2264–2286.

Alcantara D. and Lee J. (2012). Chapter 11 – magnetic nanoparticles for application in biomedical sensing. In: Frontiers of Nanoscience, J. M. de la Fuente and V. Grazu (eds), Elsevier, Oxford, pp. 269–89.

Alimard P. (2019). Fabrication and kinetic study of Nd-Ce doped Fe_3O_4-chitosan nanocomposite as catalyst in Fenton dye degradation. *Polyhedron*, **171**, 98–107.

Alinezhad A., Benedetti T. M., Gloag L., Cheong S., Watt J., Chen H. S., Gooding J. J. and Tilley R. D. (2020). Controlling Pt crystal defects on the surface of Ni–Pt Core–Shell nanoparticles for active and stable electrocatalysts for oxygen reduction. *ACS Applied Nano Materials*, **3**(6), 5995–6000.

An X. S., Fan Y. J., Chen D. J., Wang Q., Zhou Z. Y. and Sun S. G. (2011). Enhanced activity of rare earth doped PtRu/C catalysts for methanol electro-oxidation. *Electrochimica Acta*, **56**(24), 8912–8918.

Anbarasu M., Anandan M., Chinnasamy E., Gopinath V. and Balamurugan K. (2015). Synthesis and characterization of polyethylene glycol (PEG) coated Fe_3O_4 nanoparticles by chemical co-precipitation method for biomedical applications. *Spectrochimica Acta Part A: Molecular and Biomolecular Spectroscopy*, **135**, 536–539.

Ángeles-Pascual A., Piñón-Hernández J. R., Estevez-González M., Pal U., Velumani S., Pérez R. and Esparza R. (2018). Structure, magnetic and cytotoxic behaviour of solvothermally grown Fe_3O_4@Au core-shell nanoparticles. *Materials Characterization*, **142**, 237–244.

Anjaneyulu C., Naresh G., Kumar V. V., Tardio J., Rao T. V. and Venugopal A. (2015). Influence of rare earth (La, Pr, Nd, Gd, and Sm) metals on the methane decomposition activity of Ni–Al catalysts. *ACS Sustainable Chemistry & Engineering*, **3**(7), 1298–1305.

Antolini E. (2007). Platinum-based ternary catalysts for low temperature fuel cells: Part II. Electrochemical properties. *Applied Catalysis B: Environmental*, **74**(3–4), 337–350.

Antolini E. and Perez J. (2011). The use of rare earth-based materials in low-temperature fuel cells. *International Journal of Hydrogen Energy*, **36**(24), 15752–15765.

Ao K., Dong J., Fan C., Wang D., Cai Y., Li D., Huang F. and Wei Q. (2018). Formation of yolk–shelled nickel–cobalt selenide dodecahedral nanocages from metal–organic frameworks for efficient hydrogen and oxygen evolution. *ACS Sustainable Chemistry & Engineering*, **6**(8), 10952–10959.

Arguinzoniz A. G., Ruggiero E., Habtemariam A., Hernández-Gil J., Salassa L. and Mareque-Rivas J. C. (2014). Light harvesting and photoemission by nanoparticles for photodynamic therapy. *Particle & Particle Systems Characterization*, **31**(1), 46–75.

Arruebo M., Fernández-Pacheco R., Ibarra M. R. and Santamaría J. (2007). Magnetic nanoparticles for drug delivery. *Nano Today*, **2**(3), 22–32.

Arunachalam S., Kirubasankar B., Pan D., Liu H., Yan C., Guo Z. and Angaiah S. (2020). Research progress in rare earths and their composites based electrode materials for supercapacitors. *Green Energy & Environment*, **5**(3), 259–273.

Asen L., Martens S., Heiz U., Knoll A. C. and Schneider O. (2018). Electrodeposition of Pt and Gd from the Same Ionic Liquid. *ECS Transactions*, **86**(13), 475.

Asha A. B. and Ravin N. (2020). Nanomaterials properties. In: Polymer Science and Nanotechnology, R. Narain (ed.), Elsevier, pp. 343–59. https://www.sciencedirect.com/science/article/pii/B9780128168066099957

Avasthi A., Caro C., Pozo-Torres E., Leal M. P. and García-Martín M. L. (2020). Magnetic nanoparticles as MRI contrast agents. In: Surface-modified Nanobiomaterials for Electrochemical and Biomedicine Applications. Topics in Current Chemistry Collections, A. Puente-Santiago and D. Rodríguez-Padrón (eds), Springer, Cham, pp. 49–91. doi: 10.1007/978-3-030-55502-3_3

Batzill M. and Diebold U. (2005). The surface and materials science of tin oxide. *Progress in Surface Science*, **79**(2–4), 47–154.

Bhatnagar M., Kaushik V., Kaushal A., Singh M. and Mehta B. R. (2016). Structural and photoluminescence properties of tin oxide and tin oxide: C core–shell and alloy nanoparticles synthesised using gas phase technique. *AIP Advances*, **6**(9), 095321.

Biener J., Wittstock A., Baumann T. F., Weissmüller J., Bäumer M. and Hamza A. V. (2009). Surface chemistry in nanoscale materials. *Materials*, **2**(4), 2404–2428.

Blanco-Andujar C., Ortega D., Southern P., Pankhurst Q. A. and Thanh N. T. K. (2015). High performance multi-core iron oxide nanoparticles for magnetic hyperthermia: microwave synthesis, and the role of core-to-core interactions. *Nanoscale*, **7**(5), 1768–1775.

Boro B., Gogoi B., Rajbongshi B. M. and Ramchiary A. (2018). Nano-structured TiO_2/ZnO nanocomposite for dye-sensitized solar cells application: a review. *Renewable and Sustainable Energy Reviews*, **81**, 2264–2270.

Borysiewicz M. A. (2019). ZnO as a functional material, a review. *Crystals*, **9**(10), 505. doi: 10.3390/cryst9100505

Bouzigues C., Gacoin T. and Alexandrou A. (2011). Biological applications of rare-earth based nanoparticles. *ACS Nano*, **5**(11), 8488–8505.

Budnyk A., Damin A., Bordiga S. and Zecchina A. (2012). Synthesis and characterization of high-surface-area silica–titania monoliths. *The Journal of Physical Chemistry C*, **116** (18), 10064–10072.

Bundschuh M., Filser J., Lüderwald S., McKee M. S., Metreveli G., Schaumann G. E., Schulz R. and Wagner S. (2018). Nanoparticles in the environment: where do we come from, where do we go to? *Environmental Sciences Europe*, **30**(1), 1–17.

Burhan H., Cellat K., Yılmaz G. and Şen F. (2021). Direct methanol fuel cells (DMFCs). In: Direct Liquid Fuel Cells, Academic Press, pp. 71–94. https://www.sciencedirect.com/science/article/pii/B9780128186244099933

Campuzano S., Gamella M., Serafín V., Pedrero M., Yáñez-Sedeño P. and Pingarrón J. M. (2019). Magnetic Janus particles for static and dynamic (bio) sensing. *Magnetochemistry*, **5**(3), 47. doi: 10.3390/magnetochemistry5030047

Cao S. W., Zhu Y. J., Ma M. Y., Li L. and Zhang L. (2008). Hierarchically nanostructured magnetic hollow spheres of Fe_3O_4 and γ-Fe_2O_3: preparation and potential application in drug delivery. *The Journal of Physical Chemistry C*, **112**(6), 1851–1856.

Cardoso D. S. P., Santos D. M. F., Šljukić B., Sequeira C. A. C., Macciò D. and Saccone A. (2016). Platinum-rare earth cathodes for direct borohydride-peroxide fuel cells. *Journal of Power Sources*, **307**, 251–258.

Cardoso V. F., Francesko A., Ribeiro C., Bañobre-López M., Martins P. and Lanceros-Mendez S. (2018). Advances in magnetic nanoparticles for biomedical applications. *Advanced Healthcare Materials*, **7**(5), 1700845.

Casagrande C., Fabre P., Raphael E. and Veyssié M. (1989). 'Janus beads': realization and behaviour at water/oil interfaces. *Europhysics Letters*, **9**(3), 251.

Caspani S., Magalhães R., Araújo J. P. and Sousa C. T. (2020). Magnetic nanomaterials as contrast agents for MRI. *Materials*, **13**(11), 2586.

Cassir M., Jones D., Ringuedé A. and Lair V. (2013). Electrochemical devices for energy: fuel cells and electrolytic cells. In: Handbook of Membrane Reactors, A. Basile (ed.), Woodhead Publishing, Cambridge,, pp. 553–606. https://www.sciencedirect.com/science/article/pii/B9780857094155500251

Chen C., Kang Y., Huo Z., Zhu Z., Huang W., Xin H. L., Snyder J. D., Li D., Herron J. A., Mavrikakis M. and Chi M. (2014). Highly crystalline multimetallic nanoframes with three-dimensional electrocatalytic surfaces. *Science*, **343**(6177), 1339–1343.

Chen H., Wu R. and Shen P. K. (2020). One-pot fabrication of site-selective hexapod PtPdCu concave rhombic dodecahedrons as highly efficient catalysts for electrocatalysis. *ACS Sustainable Chemistry & Engineering*, **8**(3), 1520–1526.

Choudhary A. K. and Pramanik H. (2020). Enhancement of ethanol electrooxidation in half cell and single direct ethanol fuel cell (DEFC) using post-treated polyol synthesized Pt-Ru nano electrocatalysts supported on HNO_3-functionalized acetylene black carbon. *International Journal of Hydrogen Energy*, **45**(1), 574–594.

Corr S. A., Rakovich Y. P. and Gun'ko Y. K. (2008). Multifunctional magnetic-fluorescent nanocomposites for biomedical applications. *Nanoscale Research Letters*, **3**(3), 87–104.

Corradini P. G., Santos N. A. and Perez J. (2018). Pt-Sn-Eu/C catalysts: application of rare earth metals as anodes in direct ethanol fuel cells. *Fuel Cells*, **18**(1), 73–81.

Corradini P. G., Antolini E. and Perez J. (2015). Electro-oxidation of ethanol on ternary non-alloyed Pt–Sn–Pr/C catalysts. *Journal of Power Sources*, **275**, 377–383.

Cui R., Mei L., Han G., Chen J., Zhang G., Quan Y., Gu N., Zhang L., Fang Y., Qian B. and Jiang X. (2017). Facile synthesis of nanoporous Pt-Y alloy with enhanced electrocatalytic activity and durability. *Scientific Reports*, **7**(1), 1–10.

Daksh D. and Agrawal Y. K. (2016). Rare earth-doped zinc oxide nanostructures: a review. *Reviews in Nanoscience and Nanotechnology*, **5**(1), 1–27.

Das S. and Jayaraman V. (2014). SnO_2: a comprehensive review on structures and gas sensors. *Progress in Materials Science*, **66**, 112–255.

Das G. K. and Tan T. T. Y. (2008). Rare-earth-doped and codoped Y_2O_3 nanomaterials as potential bioimaging probes. *The Journal of Physical Chemistry C*, **112**(30), 11211–11217.

Davies G. L., O'Brien J. and Gun'ko Y. K. (2017). Rare earth doped silica nanoparticles via thermolysis of a single source metallasilsesquioxane precursor. *Scientific Reports*, **7**(1), 1–8.

de Jesús Ibarra–Sánchez J., López–Luke T., Ramírez–García G., Sidhik S., Córdova–Fraga T., de Jesús Bernal–Alvarado J., Cano M. E., Torres–Castro A. and de la Rosa E. (2018). Synthesis and characterization of Fe_3O_4:Yb^{3+}:Er^{3+} nanoparticles with magnetic and optical properties for hyperthermia applications. *Journal of Magnetism and Magnetic Materials*, **465**, 406–411.

De M., Ghosh P. S. and Rotello V. M. (2008). Applications of nanoparticles in biology. *Advanced Materials*, **20**(22), 4225–4241.

Douglas F. J., MacLaren D. A., Maclean N., Andreu I., Kettles F. J., Tuna F., Berry C. C., Castro M. and Murrie M. (2016). Gadolinium-doped magnetite nanoparticles from a single-source precursor. *RSC Advances*, **6**(78), 74500–74505.

Eliseeva S. V. and Bünzli J. C. G. (2011). Rare earths: jewels for functional materials of the future. *New Journal of Chemistry*, **35**(6), 1165–1176.

Escudero-Escribano M., Verdaguer-Casadevall A., Malacrida P., Grønbjerg U., Knudsen B. P., Jepsen A. K., Rossmeisl J., Stephens I. E. and Chorkendorff I. (2012). Pt_5Gd as a highly active and stable catalyst for oxygen electroreduction. *Journal of the American Chemical Society*, **134**(40), 16476–16479.

Escudero-Escribano M., Malacrida P., Hansen M. H., Vej-Hansen U. G., Velázquez-Palenzuela A., Tripkovic V., Schiøtz J., Rossmeisl J., Stephens I. E. and Chorkendorff I. (2016). Tuning the activity of Pt alloy electrocatalysts by means of the lanthanide contraction. *Science*, **352**(6281), 73–76.

Eskandari M. J. and Hasanzadeh I. (2021). Size-controlled synthesis of Fe_3O_4 magnetic nanoparticles via an alternating magnetic field and ultrasonic-assisted chemical co-precipitation. *Materials Science and Engineering: B*, **266**, 115050.

Etemadi H. and Plieger P. G. (2020). Improvements in the organic-phase hydrothermal synthesis of monodisperse $M_xFe_{3-x}O_4$ (M = Fe, Mg, Zn) spinel nanoferrites for magnetic fluid hyperthermia application. *ACS Omega*, **5**(29), 18091–18104.

Farzin A., Etesami S. A., Quint J., Memic A. and Tamayol A. (2020). Magnetic nanoparticles in cancer therapy and diagnosis. *Advanced Healthcare Materials*, **9**(9), 1901058. doi: 10.1002/adhm.201901058

Fernández-García M., Rodriguez J. A. and Quint J. (2011). Metal Oxide Nanoparticles, Encyclopedia of Inorganic and Bioinorganic Chemistry. doi: 10.1002/9781119951438.eibc0331

Fievet F., Lagier J. P., Blin B., Beaudoin B. and Figlarz M. (1989). Homogeneous and heterogeneous nucleations in the polyol process for the preparation of micron and submicron size metal particles. *Solid State Ionics*, **32**, 198–205.

Fiévet F., Ammar-Merah S., Brayner R., Chau F., Giraud M., Mammeri F., Peron J., Piquemal J. Y., Sicard L. and Viau G. (2018). The polyol process: a unique method for easy access to metal nanoparticles with tailored sizes, shapes and compositions. *Chemical Society Reviews*, **47**(14), 5187–5233.

Gadkari A. B., Shinde T. J. and Vasambekar P. N. (2013). Effect of Sm^{3+} ion addition on gas sensing properties of Mg1-xCdxFe_2O_4 system. *Sensors and Actuators B: Chemical*, **178**, 34–39.

Gao Q., Chen F., Zhang J., Hong G., Ni J., Wei X. and Wang D. (2009). The study of novel Fe_3O_4@ γ-Fe_2O_3 core/shell nanomaterials with improved properties. *Journal of Magnetism and Magnetic Materials*, **321**(8), 1052–1057.

Gao W., Wen D., Ho J. C. and Qu Y. (2019). Incorporation of rare earth elements with transition metal–based materials for electrocatalysis: a review for recent progress. *Materials Today Chemistry*, **12**, 266–281.

Gautam S., Agrawal H., Thakur M., Akbari A., Sharda H., Kaur R. and Amini M. (2020). Metal oxides and metal organic frameworks for the photocatalytic degradation: a review. *Journal of Environmental Chemical Engineering*, **8**(3), 103726.

Ge R., Li X., Lin M., Wang D., Li S., Liu S., Tang Q., Liu Y., Jiang J., Liu L. and Sun H. (2016). Fe_3O_4@ polydopamine composite theranostic superparticles employing preassembled Fe_3O_4 nanoparticles as the core. *ACS Applied Materials & Interfaces*, **8** (35), 22942–22952.

Gong Q., Wang L., Yang N., Tian L., Xie G. and Li B. (2020). Ternary PtSmCo NPs electrocatalysts with enhanced oxygen reduction reaction. *Journal of Rare Earths*, **38** (12), 1305–1311.

Gottardo S., Mech A., Drbohlavova J., Malyska A., Bøwadt S., Sintes J. R. and Rauscher H. (2021). Towards safe and sustainable innovation in nanotechnology: state-of-play for smart nanomaterials. *NanoImpact*, **100297**. doi: 10.1016/j.impact.2021.100297

Grätzel M. (2000). Perspectives for dye-sensitized nanocrystalline solar cells. *Progress in Photovoltaics: Research and Applications*, **8**(1), 171–185.

Green M., Dunlop E., Hohl-Ebinger J., Yoshita M., Kopidakis N. and Hao X. (2021). Solar cell efficiency tables (version 57). *Progress in Photovoltaics: Research and Applications*, **29**(1), 3–15.

Gudko O. E., Lastovina T. A., Smirnova N. V. and Guterman V. E. (2009). Binary Pt-Me/C nanocatalysts: structure and catalytic properties toward the oxygen reduction reaction. *Nanotechnologies in Russia*, **4**(5), 309–318.

Guil-López R., Martínez-Huerta M. V., Guillén-Villafuerte O., Pena M. A., Fierro J. L. G. and Pastor E. (2010). Highly dispersed molybdenum carbide as non-noble electrocatalyst for PEM fuel cells: performance for CO electrooxidation. *International Journal of Hydrogen Energy*, **35**(15), 7881–7888.

Guo Q., Wu J., Yang Y., Liu X., Lan Z., Lin J., Huang M., Wei Y., Dong J., Jia J. and Huang Y. (2019). High-performance and hysteresis-free perovskite solar cells based on rare-earth-doped SnO_2 mesoporous scaffold. *Research*, **2019**, 4049793. doi: 10.34133/2019/4049793

Gupta S. M. and Tripathi M. (2011). A review of TiO_2 nanoparticles. *Chinese Science Bulletin*, **56**(16), 1639–1657.

Gusain R., Gupta K., Joshi P. and Khatri O. P. (2019). Adsorptive removal and photocatalytic degradation of organic pollutants using metal oxides and their composites: a comprehensive review. *Advances in Colloid and Interface Science*, **272**, 102009. doi: 10.1016/j.cis.2019.102009

Guterman V. E., Belenov S. V., Dymnikova O. V., Lastovina T. A., Konstantinova Y. B. and Prutsakova N. V. (2009). Influence of water-organic solvent composition on composition and structure of Pt/C and Pt x Ni/C electrocatalysts in borohydride synthesis. *Inorganic Materials*, **45**(5), 498–505.

Guterman V. E., Lastovina T. A., Belenov S. V., Tabachkova N. Y., Vlasenko V. G., Khodos I. I. and Balakshina E. N. (2014). PtM/C (M = Ni, Cu, or Ag) electrocatalysts: effects of alloying components on morphology and electrochemically active surface areas. *Journal of Solid State Electrochemistry*, **18**(5), 1307–1317.

Hafez H., Saif M. and Abdel-Mottaleb M. S. A. (2011). Down-converting lanthanide doped TiO2 photoelectrodes for efficiency enhancement of dye-sensitized solar cells. *Journal of Power Sources*, **196**(13), 5792–5796.

Haile A. S., Hansen H. A., Yohannes W. and Mekonnen Y. S. (2021). Pyridinic-Type N-Doped graphene on cobalt substrate as efficient electrocatalyst for oxygen reduction reaction in acidic solution in fuel cell. *The Journal of Physical Chemistry Letters*, **12**(14), 3552–3559.

Han S. B., Kwak D. H., Lee Y. W., Kim S. J., Lee J. Y., Lee S., Kwon H. J. and Park K. W. (2016). Electrodeposited nanoporous PtY alloy electrodes with enhanced oxygen reduction reaction. *International Journal of Electrochemical Science*, **11**(5), 3803–3814.

Hanaor D. A. and Sorrell C. C. (2011). Review of the anatase to rutile phase transformation. *Journal of Materials Science*, **46**(4), 855–874.

Haruta M. (2003). When gold is not noble: catalysis by nanoparticles. *The Chemical Record*, **3**(2), 75–87.

Hirosawa F., Iwasaki T. and Watano S. (2017). Synthesis and magnetic induction heating properties of Gd-substituted Mg–Zn ferrite nanoparticles. *Applied Nanoscience*, **7**(5), 209–214.

Hodges J. M. and Schaak R. E. (2017). Controlling configurational isomerism in three-component colloidal hybrid nanoparticles. *Accounts of Chemical Research*, **50** (6), 1433–1440.

Hong R. Y., Li J. H., Li H. Z., Ding J., Zheng Y. and Wei D. G. (2008). Synthesis of Fe_3O_4 nanoparticles without inert gas protection used as precursors of magnetic fluids. *Journal of Magnetism and Magnetic Materials*, **320**(9), 1605–1614.

Hong E., Liu L., Bai L., Xia C., Gao L., Zhang L. and Wang B. (2019). Control synthesis, subtle surface modification of rare-earth-doped upconversion nanoparticles and their applications in cancer diagnosis and treatment. *Materials Science and Engineering: C*, **105**, 110097. doi: 10.1016/j.msec.2019.110097

Hu Y., Jensen J. O., Cleemann L. N., Brandes B. A. and Li Q. (2020). Synthesis of Pt–Rare Earth Metal Nanoalloys. *Journal of the American Chemical Society*, **142**(2), 953–961.

Huang H. and Zhu J. J. (2019). The electrochemical applications of rare earth-based nanomaterials. *Analyst*, **144**(23), 6789–6811.

Jacob J. M., Corradini P. G., Antolini E., Santos N. A. and Perez J. (2015). Electro-oxidation of ethanol on ternary Pt–Sn–Ce/C catalysts. *Applied Catalysis B: Environmental*, **165**, 176–184.

Jain R., Luthra V. and Gokhale S. (2018). Probing influence of rare earth ions (Er^{3+}, Dy^{3+} and Gd^{3+}) on structural, magnetic and optical properties of magnetite nanoparticles. *Journal of Magnetism and Magnetic Materials*, **456**, 179–185.

Jeevanandam J., Barhoum A., Chan Y. S., Dufresne A. and Danquah M. K. (2018). Review on nanoparticles and nanostructured materials: history, sources, toxicity and regulations. *Beilstein Journal of Nanotechnology*, **9**(1), 1050–1074.

Jeon M. K. and McGinn P. J. (2011). Carbon supported Pt–Y electrocatalysts for the oxygen reduction reaction. *Journal of Power Sources*, **196**(3), 1127–1131.

Jose J., Kumar R., Harilal S., Mathew G. E., Prabhu A., Uddin M. S., Aleya L., Kim H. and Mathew B. (2020). Magnetic nanoparticles for hyperthermia in cancer treatment: an emerging tool. *Environmental Science and Pollution Research*, **27**(16), 19214–19225.

Jouyandeh M., Zarrintaj P., Ganjali M. R., Ali J. A., Karimzadeh I., Aghazadeh M., Ghaffari M. and Saeb M. R. (2019). Curing epoxy with electrochemically synthesized $Gd_xFe_{3-x}O_4$ magnetic nanoparticles. *Progress in Organic Coatings*, **136**, 105245. doi: 10.1016/j.porgcoat.2019.105245

Kakavelakis G., Petridis K. and Kymakis E. (2017). Recent advances in plasmonic metal and rare-earth-element upconversion nanoparticle doped perovskite solar cells. *Journal of Materials Chemistry A*, **5**(41), 21604–21624.

Kanady J. S., Leidinger P., Haas A., Titlbach S., Schunk S., Schierle-Arndt K., Crumlin E. J., Wu C. H. and Alivisatos A. P. (2017). Synthesis of Pt_3Y and other early–late intermetallic nanoparticles by way of a molten reducing agent. *Journal of the American Chemical Society*, **139**(16), 5672–5675.

Kang M. and Kim Y. (2020). Au-coated Fe_3O_4@SiO_2 core-shell particles with photothermal activity. *Colloids and Surfaces A: Physicochemical and Engineering Aspects*, **600**, 124957. doi: 10.1016/j.colsurfa.2020.124957

Kannan K., Radhika D., Nesaraj A. S., Sadasivuni K. K., Reddy K. R., Kasai D. and Raghu A. V. (2020a). Photocatalytic, antibacterial and electrochemical properties of novel rare earth metal oxides-based nanohybrids. *Materials Science for Energy Technologies*, **3**, 853–861.

Kannan K., Radhika D., Sadasivuni K. K., Reddy K. R. and Raghu A. V. (2020b). Nanostructured metal oxides and its hybrids for photocatalytic and biomedical applications. *Advances in Colloid and Interface Science*, **281**, 102178.

Karunakaran S., Ramanujam S. and Gurunathan B. (2018). Green synthesised iron and iron-based nanoparticle in environmental and biomedical application: a review. *IET Nanobiotechnology*, **12**(8), 1003–1008.

Kershi R. M., Ali F. M. and Sayed M. A. (2018). Influence of rare earth ion substitutions on the structural, optical, transport, dielectric, and magnetic properties of superparamagnetic iron oxide nanoparticles. *Journal of Advanced Ceramics*, **7**(3), 218–228.

Khan H. A., Sakharkar M. K., Nayak A., Kishore U. and Khan A. (2018). Nanoparticles for biomedical applications: an overview. In: Nanobiomaterials, R. Narayan (ed.), Woodhead Publishing, pp. 357–384.

Khan I., Saeed K. and Khan I. (2019). Nanoparticles: Properties, applications and toxicities. *Arabian Journal of Chemistry*, **12**(7), 908–931.

Kim W., Suh C. Y., Cho S. W., Roh K. M., Kwon H., Song K. and Shon I. J. (2012). A new method for the identification and quantification of magnetite–maghemite mixture using conventional X-ray diffraction technique. *Talanta*, **94**, 348–352.

Kim D., Shin K., Kwon S. G. and Hyeon T. (2018a). Synthesis and biomedical applications of multifunctional nanoparticles. *Advanced Materials*, **30**(49), 1802309.

Kim H. K., Lee G. H. and Chang Y. (2018b). Gadolinium as an MRI contrast agent. *Future Medicinal Chemistry*, **10**(6), 639–661.

Kim J., Tran V. T., Oh S., Kim C. S., Hong J. C., Kim S., Joo Y. S., Mun S., Kim M. H., Jung J. W. and Lee J. (2018c). Scalable solvothermal synthesis of superparamagnetic Fe_3O_4 nanoclusters for bioseparation and theragnostic probes. *ACS Applied Materials & Interfaces*, **10**(49), 41935–41946.

Kirubakaran A., Jain S. and Nema R. K. (2009). A review on fuel cell technologies and power electronic interface. *Renewable and Sustainable Energy Reviews*, **13**(9), 2430–2440.

Kolen'ko Y. V., Bañobre-López M., Rodríguez-Abreu C., Carbó-Argibay E., Sailsman A., Piñeiro-Redondo Y., Cerqueira M. F., Petrovykh D. Y., Kovnir K., Lebedev O. I. and Rivas J. (2014). Large-scale synthesis of colloidal Fe_3O_4 nanoparticles exhibiting high heating efficiency in magnetic hyperthermia. *The Journal of Physical Chemistry C*, **118**(16), 8691–8701.

Kolhatkar A. G., Chen Y. T., Chinwangso P., Nekrashevich I., Dannangoda G. C., Singh A., Jamison A. C., Zenasni O., Rusakova I. A., Martirosyan K. S. and Litvinov D. (2017). Magnetic sensing potential of Fe_3O_4 nanocubes exceeds that of Fe_3O_4 nanospheres. *ACS Omega*, **2**(11), 8010–8019.

Kolosnjaj-Tabi J., Di Corato R., Lartigue L., Marangon I., Guardia P., Silva A. K., Luciani N., Clement O., Flaud P., Singh J. V. and Decuzzi P. (2014). Heat-generating iron oxide nanocubes: subtle 'destructurators' of the tumoral microenvironment. *ACS Nano*, **8**(5), 4268–4283.

Koole R., Mulder W. J., Van Schooneveld M. M., Strijkers G. J., Meijerink A. and Nicolay K. (2009). Magnetic quantum dots for multimodal imaging. *Wiley Interdisciplinary Reviews: Nanomedicine and Nanobiotechnology*, **1**(5), 475–491.

Kostevšek N. (2020). A review on the optimal design of magnetic nanoparticle-based T2 MRI contrast agents. *Magnetochemistry*, **6**(1), 11. doi: 10.3390/magnetochemistry6010011

Kowalik P., Mikulski J., Borodziuk A., Duda M., Kamińska I., Zajdel K., Rybusinski J., Szczytko J., Wojciechowski T., Sobczak K. and Minikayev R. (2020). Yttrium-doped iron oxide nanoparticles for magnetic hyperthermia applications. *The Journal of Physical Chemistry C*, **124**(12), 6871–6883.

Kuchma E. A., Zolotukhin P. V., Belanova A. A., Soldatov M. A., Lastovina T. A., Kubrin S. P., Nikolsky A. V., Mirmikova L. I. and Soldatov A. V. (2017). Low toxic maghemite nanoparticles for theranostic applications. *International Journal of Nanomedicine*, **12**, 6365.

Kumar V., Kumar V., Som S., Duvenhage M. M., Ntwaeaborwa O. M. and Swart H. C. (2014a). Effect of Eu doping on the photoluminescence properties of ZnO nanophosphors for red emission applications. *Applied Surface Science*, **308**, 419–430.

Kumar V., Swart H. C., Gohain M., Kumar V., Som S., Bezuindenhoudt B. C. B. and Ntwaeaborwa O. M. (2014b). Influence of ultrasonication times on the tunable colour emission of ZnO nanophosphors for lighting applications. *Ultrasonics Sonochemistry*, **21**(4), 1549–1556.

Kumar D. K., Kříž J., Bennett N., Chen B., Upadhayaya H., Reddy K. R. and Sadhu V. (2020). Functionalized metal oxide nanoparticles for efficient dye-sensitized solar cells (DSSCs): a review. *Materials Science for Energy Technologies*, **3**, 472–481.

Kundu S. and Patra A. (2017). Nanoscale strategies for light harvesting. *Chemical Reviews*, **117**(2), 712–757.

Labrador-Páez L., Ximendes E. C., Rodríguez-Sevilla P., Ortgies D. H., Rocha U., Jacinto C., Rodríguez E. M., Haro-González P. and Jaque D. (2018). Core–shell rare-earth-doped nanostructures in biomedicine. *Nanoscale*, **10**(27), 12935–12956.

Lartigue L., Coupeau M. and Lesault M. (2020). Luminophore and magnetic multicore nanoassemblies for dual-mode MRI and fluorescence imaging. *Nanomaterials*, **10**(1), 28.

Lastovina T. A., Bugaev A. L., Kubrin S. P., Kudryavtsev E. A. and Soldatov A. V. (2016). Structural studies of magnetic nanoparticles doped with rare-earth elements. *Journal of Structural Chemistry*, **57**(7), 1444–1449.

Lastovina T. A., Budnyk A. P., Kudryavtsev E. A., Nikolsky A. V., Kozakov A. T., Chumakov N. K., Emelyanov A. V. and Soldatov A. V. (2017a). Solvothermal synthesis of Sm^{3+}-doped Fe_3O_4 nanoparticles. *Materials Science and Engineering: C*, **80**, 110–116.

Lastovina T. A., Efimova S. A., Kudryavtsev E. A. and Soldatov A. V. (2017b). Preparation of the Sm^{3+}-doped magnetic nanoparticles via microwave-assisted polyol synthesis. *BioNanoScience*, **7**, 4–10.

Lastovina T. A., Budnyk A. P., Soldatov M. A., Rusalev Y. V., Guda A. A., Bogdan A. S. and Soldatov A. V. (2017c). Microwave-assisted synthesis of magnetic iron oxide nanoparticles in oleylamine–oleic acid solutions. *Mendeleev Communications*, **27**(5), 487–489.

Lastovina T. A., Budnyk A. P., Kubrin S. P. and Soldatov A. V. (2018a). Microwave-assisted synthesis of ultra-small iron oxide nanoparticles for biomedicine. *Mendeleev Communications*, **28**(2), 167–169.

Lastovina T. A., Budnyk A. P., Pimonova Y. A., Bugaev A. L., Fedorenko A. G. and Dmitriev V. P. (2018b). Step-by-step synthesis of a heteroatom-doped carbon-based electrocatalyst for the oxygen reduction reaction. *Electrochemistry Communications*, **88**, 83–87.

Laurent S., Forge D., Port M., Roch A., Robic C., Vander Elst L. and Muller R. N. (2008). Magnetic iron oxide nanoparticles: synthesis, stabilization, vectorization, physicochemical characterizations, and biological applications. *Chemical Reviews*, **108**(6), 2064–2110.

Layek A., Banerjee S., Manna B. and Chowdhury A. (2016). Synthesis of rare-earth doped ZnO nanorods and their defect–dopant correlated enhanced visible-orange luminescence. *RSC Advances*, **6**(42), 35892–35900.

Li Y., Jiang R., Liu T., Lv H., Zhou L. and Zhang X. (2014). One-pot synthesis of grass-like Fe_3O_4 nanostructures by a novel microemulsion-assisted solvothermal method. *Ceramics International*, **40**(1), 1059–1063.

Li J., Xia W., Tang J., Tan H., Wang J., Kaneti Y. V., Bando Y., Wang T., He J. and Yamauchi Y. (2019). MOF nanoleaves as new sacrificial templates for the fabrication of nanoporous Co–N x/C electrocatalysts for oxygen reduction. *Nanoscale Horizons*, **4**(4), 1006–1013.

Li J., Cao F., Yin H. L., Huang Z. J., Lin Z. T., Mao N., Sun B. and Wang G. (2020). Ferroptosis: past, present and future. *Cell Death & Disease*, **11**(2), 1–13.

Lian H., Hou Z., Shang M., Geng D., Zhang Y. and Lin J. (2013). Rare earth ions doped phosphors for improving efficiencies of solar cells. *Energy*, **57**, 270–283.

Litkohi H. R., Bahari A. and Gatabi M. P. (2020). Improved oxygen reduction reaction in PEMFCs by functionalized CNTs supported Pt–M (M = Fe, Ni, Fe–Ni) bi-and tri-metallic nanoparticles as efficient electrocatalyst. *International Journal of Hydrogen Energy*, **45**(43), 23543–23556.

Liu B., Zhang H. and Ding Y. (2018). Au-Fe_3O_4 heterostructures for catalytic, analytical, and biomedical applications. *Chinese Chemical Letters*, **29**(12), 1725–1730.

Lohrke J., Frenzel T., Endrikat J., Alves F. C., Grist T. M., Law M., Lee J. M., Leiner T., Li K. C., Nikolaou K. and Prince M. R. (2016). 25 years of contrast-enhanced MRI: developments, current challenges and future perspectives. *Advances in Therapy*, **33**(1), 1–28.

Lu J. F. and Tsai C. J. (2014). Hydrothermal phase transformation of hematite to magnetite. *Nanoscale Research Letters*, **9**(1), 1–8.

Lu L., Li R., Peng T., Fan K. and Dai K. (2011). Effects of rare earth ion modifications on the photoelectrochemical properties of ZnO-based dye-sensitized solar cells. *Renewable Energy*, **36**(12), 3386–3393.

Luo Y., Habrioux A., Calvillo L., Granozzi G. and Alonso-Vante N. (2014). Yttrium oxide/gadolinium oxide-modified platinum nanoparticles as cathodes for the oxygen reduction reaction. *ChemPhysChem*, **15**(10), 2136–2144.

Luo F., Zhang Q., Qu K., Guo L., Hu H., Yang Z., Cai W. and Cheng H. (2019). Decorated PtRu electrocatalyst for concentrated direct methanol fuel cells. *ChemCatChem*, **11**(4), 1238–1243.

Lux K. W. and Cairns E. J. (2006). Lanthanide–platinum intermetallic compounds as anode electrocatalysts for direct ethanol PEM fuel cells: i. synthesis and characterization of nanopowders. *Journal of the Electrochemical Society*, **153**(6), A1132.

Mahajan K. D., Fan Q., Dorcéna J., Ruan G. and Winter J. O. (2013). Magnetic quantum dots in biotechnology–synthesis and applications. *Biotechnology Journal*, **8**(12), 1424–1434.

Massart R. (1981). Preparation of aqueous magnetic liquids in alkaline and acidic media. *IEEE Transactions on Magnetics*, **17**(2), 1247–1248.

Mazario E., Sánchez-Marcos J., Menéndez N., Herrasti P., García-Hernández M. and Muñoz-Bonilla A. (2014). One-pot electrochemical synthesis of polydopamine coated magnetite nanoparticles. *RSC Advances*, **4**(89), 48353–48361.

McKeown C. and Rhen F. M. (2018). Fe–Pt thin film for oxygen reduction reaction. *Journal of Applied Electrochemistry*, **48**(9), 1009–1017.

Miles M. H. (1975). Evaluation of electrocatalysts for water electrolysis in alkaline solutions. *Journal of Electroanalytical Chemistry and Interfacial Electrochemistry*, **60**(1), 89–96.

Mirzaei H. and Darroudi M. (2017). Zinc oxide nanoparticles: biological synthesis and biomedical applications. *Ceramics International*, **43**(1), 907–914.

Mohankumar P., Ajayan J., Mohanraj T. and Yasodharan R. (2021). Recent developments in biosensors for healthcare and biomedical applications: a review. *Measurement*, **167**, 108293.

Moshfegh A. Z. (2009). Nanoparticle catalysts. *Journal of Physics D: Applied Physics*, **42**(23), 233001.

Mourdikoudis S. and Liz-Marzán L. M. (2013). Oleylamine in nanoparticle synthesis. *Chemistry of Materials*, **25**(9), 1465–1476.

Mourdikoudis S., Kostopoulou A. and LaGrow A. P. (2021). Magnetic nanoparticle composites: synergistic effects and applications. *Advanced Science*, **8**, 2004951, 1–57, doi: 10.1002/advs.202004951

Mudshinge S. R., Deore A. B., Patil S. and Bhalgat C. M. (2011). Nanoparticles: emerging carriers for drug delivery. *Saudi Pharmaceutical Journal*, **19**(3), 129–141.

Na H. B., Song I. C. and Hyeon T. (2009). Inorganic nanoparticles for MRI contrast agents. *Advanced Materials*, **21**(21), 2133–2148.

Nasrazadani S. and Raman A. (1993). The application of infrared spectroscopy to the study of rust systems—II. Study of cation deficiency in magnetite (Fe_3O_4) produced during its transformation to maghemite (γ-Fe_2O_3) and hematite (α-Fe_2O_3). *Corrosion Science*, **34**(8), 1355–1365.

Nejati K., Dadashpour M., Gharibi T., Mellatyar H. and Akbarzadeh A. (2021). Biomedical applications of functionalized gold nanoparticles: a review. *Journal of Cluster Science*, 1–16. doi: 10.1007/s10876-020-01955-9

Neto A. O., Watanabe A. Y., Brandalise M., Tusi M. M., Rodrigues R. M. D. S., Linardi M., Spinacé E. V. and Forbicini C. A. (2009). Preparation and characterization of Pt–Rare Earth/C electrocatalysts using an alcohol reduction process for methanol electro-oxidation. *Journal of Alloys and Compounds*, **476**(1–2), 288–291.

Nishanth K. G., Sridhar P. and Pitchumani S. (2011). Enhanced oxygen reduction reaction activity through spillover effect by Pt–$Y(OH)_3$/C catalyst in direct methanol fuel cells. *Electrochemistry Communications*, **13**(12), 1465–1468.

O'regan B. and Grätzel M. (1991). A low-cost, high-efficiency solar cell based on dye-sensitized colloidal TiO_2 films. *Nature*, **353**(6346), 737–740.

Oh H. D. and Lee S. W. (2013). Facile synthesis of QD-anchored composite particles with magnetite cluster cores ($nFe_3O_4@SiO_2@QDs$). *Materials Research Bulletin*, **48**(6), 2191–2195.

Osial M., Rybicka P., Pękała M., Cichowicz G., Cyrański M. K. and Krysiński P. (2018). Easy synthesis and characterization of holmium-doped SPIONs. *Nanomaterials*, **8**(6), 430.

Parihar V., Raja M. and Paulose R. (2018). A brief review of structural, electrical and electrochemical properties of zinc oxide nanoparticles. *Reviews on Advanced Materials Science*, **53**(2), 119–130.

Parmar J., Vilela D., Villa K., Wang J. and Sánchez S. (2018). Micro-and nanomotors as active environmental microcleaners and sensors. *Journal of the American Chemical Society*, **140**(30), 9317–9331.

Peera S. G., Lee T. G. and Sahu A. K. (2019). Pt-rare earth metal alloy/metal oxide catalysts for oxygen reduction and alcohol oxidation reactions: an overview. *Sustainable Energy & Fuels*, **3**(8), 1866–1891.

Pereira M. C., Oliveira L. C. A. and Murad E. (2012). Iron oxide catalysts: Fenton and Fentonlike reactions–a review. *Clay Minerals*, **47**(3), 285–302.

Perera T. S. H., Han Y., Lu X., Wang X., Dai H. and Li S. (2015). Rare earth doped apatite nanomaterials for biological application. *Journal of Nanomaterials*, **2015**, 5. doi: 10.1155/2015/705390

Petran A., Radu T., Borodi G., Nan A., Suciu M. and Turcu R. (2018). Effects of rare earth doping on multi-core iron oxide nanoparticles properties. *Applied Surface Science*, **428**, 492–499.

Pimonova Y. A., Budnyk A. P., Yohannes W., Bugaev A. L. and Lastovina T. A. (2019a). Iron-/nitrogen-doped electrocatalytically active carbons for the oxygen reduction reaction with low amounts of cobalt. *ACS Omega*, **4**(22), 19548–19555.

Pimonova Y. A., Lastovina T. A., Budnyk A. P., Kudryavtsev E. A. and Yapryntsev M. N. (2019b). Cobalt-based ZIF-68 and ZIF-69 as the precursors of non-platinum electrocatalysts for oxygen reduction. *Mendeleev Communications*, **29**, 544–546.

Polman A., Knight M., Garnett E. C., Ehrler B. and Sinke W. C. (2016). Photovoltaic materials: Present efficiencies and future challenges. *Science*, **352**(6283). https://www.science.org/doi/10.1126/science.aad4424

Ponmani K., Nayeemunisa S. M., Kiruthika S. and Muthukumaran B. (2016). Electrochemical characterization of platinum-based anode catalysts for membraneless fuel cells. *Ionics*, **22**(3), 377–387.

Prijic S. and Sersa G. (2011). Magnetic nanoparticles as targeted delivery systems in oncology. *Radiology and Oncology*, **45**(1), 1–16. doi: 10.2478/v10019-011-0001-z

Qiu P., Zhou N., Chen H., Zhang C., Gao G. and Cui D. (2013). Recent advances in lanthanide-doped upconversion nanomaterials: synthesis, nanostructures and surface modification. *Nanoscale*, **5**(23), 11512–11525.

Rahimi N., Pax R. A. and Gray E. M. (2016). Review of functional titanium oxides. I: TiO_2 and its modifications. *Progress in Solid State Chemistry*, **44**(3), 86–105.

Rajakumar G., Mao L., Bao T., Wen W., Wang S., Gomathi T., Gnanasundaram N., Rebezov M., Shariati M. A., Chung I. M. and Thiruvengadam M. (2021). Yttrium oxide nanoparticle synthesis: an overview of methods of preparation and biomedical applications. *Applied Sciences*, **11**(5), 2172.

Rajan A. and Sahu N. K. (2020). Review on magnetic nanoparticle-mediated hyperthermia for cancer therapy. *Journal of Nanoparticle Research*, **22**(11), 1–25.

Ram P., Gören A., Ferdov S., Silva M. M., Singhal R., Costa C. M., Sharma R. K. and Lanceros-Mendez S. (2016). Improved performance of rare earth doped $LiMn_2O_4$ cathodes for lithium-ion battery applications. *New Journal of Chemistry*, **40**(7), 6244–6252.

Ramakrishnan V. and John N. S. (2019). Chemical synthesis of hybrid nanoparticles based on metal–metal oxide systems. In: Morphology Design Paradigms for Supercapacitors Boddula I.R., Ahmer M.F. and Asiri I.R., CRC Press, Boca Raton, pp. 163–202.

Reyes-Ortega F., Delgado Á. V. and Iglesias G. R. (2021). Modulation of the magnetic hyperthermia response using different superparamagnetic iron oxide nanoparticle morphologies. *Nanomaterials*, **11**(3), 627. doi: 10.3390/nano11030627

Rhodehouse M. L., Bell T., Smetana V., Mudring A. V. and Meyer G. H. (2018). From the nonexistent polar intermetallic Pt_3Pr_4 via Pt_2–x Pr_3 to Pt/Sn/Pr ternaries. *Inorganic Chemistry*, **57**(16), 9949–9961.

Rhodehouse M. L., Smetana V., Celania C., Mudring A. V. and Meyer G. H. (2020). Ternary Polar Intermetallics within the Pt/Sn/R Systems (R = La–Sm): Stannides or Platinides? *Inorganic Chemistry*, **59**(10), 7352–7359.

Rice K. P., Russek S. E., Geiss R. H., Shaw J. M., Usselman R. J., Evarts E. R., Silva T. J., Nembach H. T., Arenholz E. and Idzerda Y. U. (2015). Temperature-dependent structure of Tb-doped magnetite nanoparticles. *Applied Physics Letters*, **106**(6), 062409.

Rosalbino F., Delsante S., Borzone G. and Angelini E. (2008). Electrocatalytic behaviour of Co–Ni–R (R = Rare earth metal) crystalline alloys as electrode materials for hydrogen evolution reaction in alkaline medium. *International Journal of Hydrogen Energy*, **33** (22), 6696–6703.

Roy C., Knudsen B. P., Pedersen C. M., Velazquez-Palenzuela A., Christensen L. H., Damsgaard C. D., Stephens I. E. and Chorkendorff I. (2018). Scalable synthesis of carbon-supported platinum–lanthanide and– rare-earth alloys for oxygen reduction. *ACS Catalysis*, **8**(3), 2071–2080.

Ruhle S., Anderson A. Y., Barad H. N., Kupfer B., Bouhadana Y., Rosh-Hodesh E. and Zaban A. (2012). All-oxide photovoltaics. *The Journal of Physical Chemistry Letters*, **3**(24), 3755–3764.

Rümenapp C., Gleich B. and Haase A. (2012). Magnetic nanoparticles in magnetic resonance imaging and diagnostics. *Pharmaceutical Research*, **29**(5), 1165–1179.

Ryoo R., Kim J., Jo C., Han S. W., Kim J. C., Park H., Han J., Shin H. S. and Shin J. W. (2020). Rare-earth–platinum alloy nanoparticles in mesoporous zeolite for catalysis. *Nature*, **585**(7824), 221–224.

Sangaiya P. and Jayaprakash R. (2018). A review on iron oxide nanoparticles and their biomedical applications. *Journal of Superconductivity and Novel Magnetism*, **31**(11), 3397–3413.

Santoro T. A. B., Neto A. O., Forbicini C. D. O., Linardi M., Rodríguez J. L. and Pastor E. (2012). Ethanol Electrooxidation on Pt with Lanthanum Oxide as Cocatalyst in a DAFC. *International Journal of Electrochemistry*, **2012**, 674150. doi: 10.1155/2012/674150

Sarma S. C., Subbarao U., Khulbe Y., Jana R. and Peter S. C. (2017). Are we underrating rare earths as an electrocatalyst? The effect of their substitution in palladium nanoparticles enhances the activity towards ethanol oxidation reaction. *Journal of Materials Chemistry A*, **5**(44), 23369–23381.

Sazali N., Wan Salleh W. N., Jamaludin A. S. and Mhd Razali M. N. (2020). New perspectives on fuel cell technology: a brief review. *Membranes*, **10**(5), 99. doi: 10. 3390/membranes10050099

Schick I., Lorenz S., Gehrig D., Tenzer S., Storck W., Fischer K., Strand D., Laquai F. and Tremel W. (2014). Inorganic Janus particles for biomedical applications. *Beilstein Journal of Nanotechnology*, **5**(1), 2346–2362.

Serga V., Burve R., Maiorov M., Krumina A., Skaudžius R., Zarkov A., Kareiva A. and Popov A. I. (2020). Impact of gadolinium on the structure and magnetic properties of nanocrystalline powders of iron oxides produced by the extraction-pyrolytic method. *Materials*, **13**(18), 4147.

Shao M., Chang Q., Dodelet J. P. and Chenitz R. (2016). Recent advances in electrocatalysts for oxygen reduction reaction. *Chemical Reviews*, **116**(6), 3594–3657.

Silva R. L., Araújo J. H., Silva L. M. and Morales M. A. (2018). Surface effect in PVP coated Sm doped magnetite nanoparticles prepared by the polyol method. *Ceramics International*, **44**(11), 13050–13054.

Sinmyo R., Bykova E., Ovsyannikov S. V., McCammon C., Kupenko I., Ismailova L. and Dubrovinsky L. (2016). Discovery of Fe_7O_9: a new iron oxide with a complex monoclinic structure. *Scientific Reports*, **6**(1), 1–7.

Soldatov M. A., Göttlicher J., Kubrin S. P., Guda A. A., Lastovina T. A., Bugaev A. L., Rusalev Y. V., Soldatov A. V. and Lamberti C. (2018). Insight from x-ray absorption spectroscopy to octahedral/tetrahedral site distribution in Sm-doped iron oxide magnetic nanoparticles. *The Journal of Physical Chemistry C*, **122**(15), 8543–8552.

Srinivasan R., Yogamalar R. and Bose A. C. (2010). Structural and optical studies of yttrium oxide nanoparticles synthesized by co-precipitation method. *Materials Research Bulletin*, **45**(9), 1165–1170.

Stamenkovic V. R., Mun B. S., Mayrhofer K. J., Ross P. N. and Markovic N. M. (2006). Effect of surface composition on electronic structure, stability, and electrocatalytic properties of Pt-transition metal alloys: Pt-skin versus Pt-skeleton surfaces. *Journal of the American Chemical Society*, **128**(27), 8813–8819.

Strijkers G. J., Mulder W. J. M., van Tilborg G. A. F. and Nicolay K. (2007). MRI contrast agents: current status and future perspectives. *Anti-Cancer Agents in Medicinal Chemistry*, **7**(3), 291–305.

Su H., Price C. A. H., Jing L., Tian Q., Liu J. and Qian K. (2019). Janus particles: design, preparation, and biomedical applications. *Materials Today Bio*, **4**, 100033. doi: 10.1016/j.mtbio.2019.100033

Sun S. and Zeng H. (2002). Size-controlled synthesis of magnetite nanoparticles. *Journal of the American Chemical Society*, **124**(28), 8204–8205.

Sun P., Zhang H., Liu C., Fang J., Wang M., Chen J., Zhang J., Mao C. and Xu S. (2010). Preparation and characterization of Fe_3O_4/CdTe magnetic/fluorescent nanocomposites and their applications in immuno-labeling and fluorescent imaging of cancer cells. *Langmuir*, **26**(2), 1278–1284.

Tan M., Del Rosal B., Zhang Y., Rodríguez E. M., Hu J., Zhou Z., Fan R., Ortgies D. H., Fernández N., Chaves-Coira I. and Núñez Á. (2018). Rare-earth-doped fluoride nanoparticles with engineered long luminescence lifetime for time-gated in vivo optical imaging in the second biological window. *Nanoscale*, **10**(37), 17771–17780.

Tedeschi E., Caranci F., Giordano F., Angelini V., Cocozza S. and Brunetti A. (2017). Gadolinium retention in the body: what we know and what we can do. *La Radiologia Medica*, **122**(8), 589–600.

Thakur D., Deng S., Baldet T. and Winter J. O. (2009). pH sensitive CdS–iron oxide fluorescent–magnetic nanocomposites. *Nanotechnology*, **20**(48), 485601.

Thiel C. W., Böttger T. and Cone R. L. (2011). Rare-earth-doped materials for applications in quantum information storage and signal processing. *Journal of Luminescence*, **131**, 353–361.

Tian J. and Cao G. (2016). Design, fabrication and modification of metal oxide semiconductor for improving conversion efficiency of excitonic solar cells. *Coordination Chemistry Reviews*, **320**, 193–215.

Tsitovich P. B., Burns P. J., McKay A. M. and Morrow J. R. (2014). Redox-activated MRI contrast agents based on lanthanide and transition metal ions. *Journal of Inorganic Biochemistry*, **133**, 143–154.

Ulyankina A., Molodtsova T., Gorshenkov M., Leontyev I., Zhigunov D., Konstantinova E., Lastovina T., Tolasz J., Henych J., Licciardello N. and Cuniberti G. (2021). Photocatalytic degradation of ciprofloxacin in water at nano-ZnO prepared by pulse alternating current electrochemical synthesis. *Journal of Water Process Engineering*, **40**, 101809.

Védrine J. C. (2019). Metal oxides in heterogeneous oxidation catalysis: state of the art and challenges for a more sustainable world. *ChemSusChem*, **12**(3), 577–588.

Wagle D. V., Rondinone A. J., Woodward J. D. and Baker G. A. (2017). Polyol synthesis of magnetite nanocrystals in a thermostable ionic liquid. *Crystal Growth & Design*, **17**(4), 1558–1567.

Wan X., Liu X., Li Y., Yu R., Zheng L., Yan W., Wang H., Xu M. and Shui J. (2019). Fe–N–C electrocatalyst with dense active sites and efficient mass transport for high-performance proton exchange membrane fuel cells. *Nature Catalysis*, **2**(3), 259–268.

Wang Y. X. J. (2011). Superparamagnetic iron oxide based MRI contrast agents: current status of clinical application. *Quantitative Imaging in Medicine and Surgery*, **1**(1), 35–40.

Wang D., He J., Rosenzweig N. and Rosenzweig Z. (2004). Superparamagnetic Fe_2O_3 beads–CdSe/ZnS quantum dots core– shell nanocomposite particles for cell separation. *Nano Letters*, **4**(3), 409–413.

Wang F., Zheng Y. and Guo Y. (2010). The promoting effect of europium on PtSn/C catalyst for ethanol oxidation. *Fuel Cells*, **10**(6), 1100–1107.

Wang M., Wang C., Young K. L., Hao L., Medved M., Rajh T., Fry H. C., Zhu L., Karczmar G. S., Watson C. and Jiang J. S. (2012). Cross-linked heterogeneous nanoparticles as bifunctional probe. *Chemistry of Materials*, **24**(13), 2423–2425.

Wang Y., Zhu G., Xin S., Wang Q., Li Y., Wu Q., Wang C., Wang X., Ding X. and Geng W. (2015). Recent development in rare earth doped phosphors for white light emitting diodes. *Journal of Rare Earths*, **33**(1), 1–22.

Wang L., Kaeppler A., Fischer D. and Simmchen J. (2019). Photocatalytic TiO_2 micromotors for removal of microplastics and suspended matter. *ACS Applied Materials & Interfaces*, **11**(36), 32937–32944.

Wolff S. D. and Balaban R. S. (1989). Magnetization transfer contrast (MTC) and tissue water proton relaxation in vivo. *Magnetic Resonance in Medicine*, **10**(1), 135–144.

Wu W., Jiang C. Z. and Roy V. A. (2016). Designed synthesis and surface engineering strategies of magnetic iron oxide nanoparticles for biomedical applications. *Nanoscale*, **8**(47), 19421–19474.

Xiaobo C. (2009). Titanium dioxide nanomaterials and their energy applications. *Chinese Journal of Catalysis*, **30**(8), 839–851.

Xiao Y. D., Paudel R., Liu J., Ma C., Zhang Z. S. and Zhou S. K. (2016). MRI contrast agents: classification and application. *International Journal of Molecular Medicine*, **38**(5), 1319–1326.

Xiong L., Guo Y., Wen J., Liu H., Yang G., Qin P. and Fang G. (2018). Review on the application of SnO_2 in perovskite solar cells. *Advanced Functional Materials*, **28**(35), 1802757.

Xu Z., Shen C., Hou Y., Gao H. and Sun S. (2009). Oleylamine as both reducing agent and stabilizer in a facile synthesis of magnetite nanoparticles. *Chemistry of Materials*, **21**(9), 1778–1780.

Xu F., Geiger J. H., Baker G. L. and Bruening M. L. (2011). Polymer brush-modified magnetic nanoparticles for His-tagged protein purification. *Langmuir*, **27**(6), 3106–3112.

Xu H., Ci S., Ding Y., Wang G. and Wen Z. (2019). Recent advances in precious metal-free bifunctional catalysts for electrochemical conversion systems. *Journal of Materials Chemistry A*, **7**(14), 8006–8029.

Yao J., Huang C., Liu C. and Yang M. (2020). Upconversion luminescence nanomaterials: A versatile platform for imaging, sensing, and therapy. *Talanta*, **208**, 120157.

Yan J. and Saunders B. R. (2014). Third-generation solar cells: a review and comparison of polymer: fullerene, hybrid polymer and perovskite solar cells. *RSC Advances*, **4**(82), 43286–43314.

Yang L., Zhou Z., Liu H., Wu C., Zhang H., Huang G., Ai H. and Gao J. (2015). Europium-engineered iron oxide nanocubes with high T_1 and T_2 contrast abilities for MRI in living subjects. *Nanoscale*, **7**(15), 6843–6850.

Yi D. K., Selvan S. T., Lee S. S., Papaefthymiou G. C., Kundaliya D. and Ying J. Y. (2005). Silica-coated nanocomposites of magnetic nanoparticles and quantum dots. *Journal of the American Chemical Society*, **127**(14), 4990–4991.

Yildirimer L., Thanh N. T., Loizidou M. and Seifalian A. M. (2011). Toxicology and clinical potential of nanoparticles. *Nano Today*, **6**(6), 585–607.

Yoo S. J., Kim S. K., Jeon T. Y., Hwang S. J., Lee J. G., Lee S. C., Lee K. S., Cho Y. H., Sung Y. E. and Lim T. H. (2011). Enhanced stability and activity of Pt–Y alloy catalysts for electrocatalytic oxygen reduction. *Chemical Communications*, **47**(41), 11414–11416.

Yuhua W., Ge Z., Shuangyu X., Qian W., Yanyan L., Quansheng W., Chuang W., Xicheng W., Xin D. and Wanying G (2015). Recent development in rare earth doped phosphors for white light emitting diodes. *Journal of Rare Earths*, **33**(1), 1–12.

Yu Y., Chen G., Zhou Y. and Han Z. (2015). Recent advances in rare-earth elements modification of inorganic semiconductor-based photocatalysts for efficient solar energy conversion: a review. *Journal of Rare Earths*, **33**(5), 453–462. doi: 10.1016/S1002-0721(14)60440-3

Yu X., Marks T. J. and Facchetti A. (2016). Metal oxides for optoelectronic applications. *Nature Materials*, **15**(4), 383–396.

Yu H., Peng Y., Yang Y. and Li Z. Y. (2019). Plasmon-enhanced light–matter interactions and applications. *NPJ Computational Materials*, **5**(1), 1–14.

Zahmouli N., Hjiri M., Leonardi S. G., El Mir L., Neri G., Iannazzo D., Espro C. and Aida M. S. (2020). High performance Gd-doped γ-Fe_2O_3 based acetone sensor. *Materials Science in Semiconductor Processing*, **116**, 105154.

Zahoor A., Ghouri Z. K., Hashmi S., Raza F., Ishtiaque S., Nadeem S., Ullah I. and Nahm K. S. (2019). Electrocatalysts for lithium–air batteries: current status and challenges. *ACS Sustainable Chemistry & Engineering*, **7**(17), 14288–14320.

Zeng J., Jing L., Hou Y., Jiao M., Qiao R., Jia Q., Liu C., Fang F., Lei H. and Gao M. (2014). Anchoring group effects of surface ligands on magnetic properties of Fe_3O_4 nanoparticles: towards high performance MRI contrast agents. *Advanced Materials*, **26**(17), 2694–2698.

Zhang Z. (2020). Research progress of gold core-shell structured nanoparticles in tumor therapy. In: Journal of Physics: Conference Series, IOP Publishing, **1699**, p. 012007. https://iopscience.iop.org/issue/1742-6596/1699/1

Zhang J., Peng W., Chen Z., Chen H. and Han L. (2012). Effect of cerium doping in the TiO_2 photoanode on the electron transport of dye-sensitized solar cells. *The Journal of Physical Chemistry C*, **116**(36), 19182–19190.

Zhang Q., Wang C. F., Ling L. T. and Chen S. (2014). Fluorescent nanomaterial-derived white light-emitting diodes: what's going on. *Journal of Materials Chemistry C*, **2**(22), 4358–4373.

Zhang H., Malik V., Mallapragada S. and Akinc M. (2017). Synthesis and characterization of Gd-doped magnetite nanoparticles. *Journal of Magnetism and Magnetic Materials*, **423**, 386–394.

Zhang T., Wang M. and Yang H. (2018). A review of the energy performance and life-cycle assessment of building-integrated photovoltaic (BIPV) systems. *Energies*, **11**(11), 3157.

Zhao B., Wang J., Li H., Jia X., Dong L. and Den Engelsen D. (2016). Rare Earth-Activated Y_2O_3 Phosphors with Novel Morphology for Dye-Sensitized Solar Cells. *ChemistrySelect*, **1**(6), 1136–1139.

Zhao S., Yu X., Qian Y., Chen W. and Shen J. (2020). Multifunctional magnetic iron oxide nanoparticles: an advanced platform for cancer theranostics. *Theranostics*, **10**(14), 6278.

Zheng H., Su R., Gao Z., Qi W., Huang R., Wang L. and He Z. (2014). Magnetic–fluorescent nanocomposites as reusable fluorescence probes for sensitive detection of hydrogen peroxide and glucose. *Analytical Methods*, **6**(16), 6352–6357.

Zhong C. J., Luo J., Njoki P. N., Mott D., Wanjala B., Loukrakpam R., Lim S., Wang L., Fang B. and Xu Z. (2008). Fuel cell technology: nano-engineered multimetallic catalysts. *Energy & Environmental Science*, **1**(4), 454–466.

Zhu X., Lin H., Wang L., Tang X., Ma L., Chen Z. and Gao J. (2017). Activatable T_1 relaxivity recovery nanoconjugates for kinetic and sensitive analysis of matrix metalloprotease 2. *ACS Applied Materials & Interfaces*, **9**(26), 21688–21696.

Index

IWA Publishing's authorised EU representative for General Product Safety Regulations is Diane D'Arras, 15 rue Duret, 75116 Paris, France, e-mail: safety@iwap.co.uk.

Printed and bound by CPI Group (UK) Ltd, Croydon, CR0 4YY
12/05/2026
02108863-0002